VOLUME ONE HUNDRED AND TWENTY TWO

ADVANCES IN FOOD AND NUTRITION RESEARCH

Novel Strategies to Improve the Eating and Nutritional Properties of Food Products

VOLUME ONE HUNDRED AND TWENTY TWO

ADVANCES IN FOOD AND NUTRITION RESEARCH

Novel Strategies to Improve the Eating and Nutritional Properties of Food Products

Edited by

WANGANG ZHANG

Jiangsu Synergetic Innovation Center of Meat Processing and Quality Control; State Key Laboratory of Meat Quality Control and Cultured Meat Development, MOE, College of Food Science and Technology, Nanjing Agricultural University, Nanjing, P.R. China

ELSEVIER

ACADEMIC PRESS
An imprint of Elsevier

Academic Press is an imprint of Elsevier
50 Hampshire Street, 5th Floor, Cambridge, MA 02139, United States
125 London Wall, London, EC2Y 5AS, United Kingdom

First edition 2026

ISBN: 978-0-443-46992-3
ISSN: 1043-4526

For information on all Academic Press publications
visit our website at https://www.elsevier.com/books-and-journals

Publisher: Zoe Kruze
Acquisition Editor: Mariana Kuhl
Editorial Project Manager: Saloni Vohra
Production Project Manager: Abdulla Sait
Cover Designer: Muthukumaran Thangaraj

Typeset by MPS Limited, India

Contents

Contributors

E.D.N.S. Abeyrathne
Department of Animal Science, Uva Wellassa University, Badulla, Sri Lanka

Rubén Agregán
Fundación Centro Tecnolóxico da Carne, Ourense, Spain

D.U. Ahn
Department of Animal Science, Iowa State University, Ames, IA, United States

Sher Ali
Institute of Food Sciences and Technology, Sindh Agriculture University, Tandojam, Sindh, Pakistan

Fawze Alnadari
Functional Protein Laboratory, Research and Development Center, Jiangsu Tianmeijian Nature Bioengineering Co., Ltd., Nanjing, Jiangsu, P.R. China

T. Astruc
INRAE, UR370 Qualité des Produits Animaux (QuaPA), Saint-Genès-Champanelle, France

Anthony Pius Bassey
State Key laboratory of Meat Quality Control and Culture Meat Development, MOST: College of Food Science and Technology, Nanjing Agriculture University, Nanjing, Jiangsu, P.R. China

Evans Frimpong Boateng
Department of Food Science and Technology, University of Energy and Natural Resources, Sunyani, Bono, Ghana

Paa Kwesi Bordoh
Department of Horticulture and Crop Production, Postharvest Unit, University of Energy and Natural Resources, Sunyani, Bono, Ghana

Paulo Cezar Bastianello Campagnol
Universidade Federal de Santa Maria (UFSM), Santa Maria, Rio Grande do Sul, Brazil

Alexandre José Cichoski
Universidade Federal de Santa Maria (UFSM), Santa Maria, Rio Grande do Sul, Brazil

Rita Dankwa
Department of Food Science and Technology, University of Energy and Natural Resources, Sunyani, Bono, Ghana

Bibiana Alves dos Santos
Universidade Federal de Santa Maria (UFSM), Santa Maria, Rio Grande do Sul, Brazil

Mustafa M. Farouk
Food Technology & Processing Team, Bioeconomy Science Institute, Palmerston North, New Zealand

Priscila Rossato Fracari
Universidade Federal de Santa Maria (UFSM), Santa Maria, Rio Grande do Sul, Brazil

X. Huang
College of Food Science and Technology, Huazhong Agricultural University, Wuhan, P.R. China

Lei Jin
Jiangsu Synergetic Innovation Center of Meat Processing and Quality Control; State Key Laboratory of Meat Quality Control and Cultured Meat Development, MOE, College of Food Science and Technology, Nanjing Agricultural University, Nanjing, P.R. China

Muhammad Ammar Khan
Department of Food Science & Technology, Faculty of Agriculture & Environment, The Islamia University of Bahawalpur, Bahawalpur, Pakistan

José Manuel Lorenzo
Centro TecnolÓgico de la Carne de Galicia; Área de Tecnología de los Alimentos, Facultad de Ciencias de Ourense, Universidad de Vigo, Ourense, Spain

Óscar López-Campos
Lacombe Research and Development Centre, Agriculture and Agri-Food Canada, Lacombe, Alberta, Canada

Chao Ma
Jiangsu Synergetic Innovation Center of Meat Processing and Quality Control; State Key Laboratory of Meat Quality Control and Cultured Meat Development, MOE, College of Food Science and Technology, Nanjing Agricultural University, Nanjing, P.R. China

Asadullah Marri
Institute of Food Sciences and Technology, Sindh Agriculture University, Tandojam, Sindh, Pakistan

M. Meurillon
INRAE, UR370 Qualité des Produits Animaux (QuaPA), Saint-Genès-Champanelle, France

Tahseen Fatima Miano
Institute of Food Sciences and Technology, Sindh Agriculture University, Tandojam, Sindh, Pakistan

P.-S. Mirade
INRAE, UR370 Qualité des Produits Animaux (QuaPA), Saint-Genès-Champanelle, France

Mustapha Muhammad Nasiru
College of Food Science and Engineering, Bohai University, Jinzhou, Liaoning, P.R. China

Gema Nieto
Department of Food Technology, Nutrition and Food Science, Veterinary Faculty, Campus de Excelencia Internacional de Ámbito Regional (CEIR), Campus Mare Nostrum (CMN), University of Murcia, Murcia, Spain

Matthew G. Nosworthy
Guelph Research and Development Centre, Agriculture and Agri-Food Canada, Guelph, Canada

James Owusu-Kwarteng
Department of Food Science and Technology, University of Energy and Natural Resources, Sunyani, Bono, Ghana

S. Portanguen
INRAE, UR370 Qualité des Produits Animaux (QuaPA), Saint-Genès-Champanelle, France

Nuria Prieto
Lacombe Research and Development Centre, Agriculture and Agri-Food Canada, Lacombe, Alberta, Canada

Nasir Rajput
Department of Poultry Science, Faculty of Animal Husbandry & Veterinary Sciences, Sindh Agriculture University, Tandojam, Sindh, Pakistan

Carolina E. Realini
Food Technology & Processing Team, Bioeconomy Science Institute, Palmerston North, New Zealand

Nida Shaikh
Institute of Food Sciences and Technology, Sindh Agriculture University, Tandojam, Sindh, Pakistan

L. Théron
INRAE, UR370 Qualité des Produits Animaux (QuaPA), Saint-Genès-Champanelle, France

Márcio Vargas-Ramella
Universidade do Estado de Santa Catarina (UDESC), Laguna, Santa Catarina, Brazil

Manuel Viuda-Martos
Research Group in Innovations in Food Products (IPOA), Institute for Agri-Food and Agri-Environmental Research and Innovation, Miguel Hernández University (CIAGRO-UMH), Orihuela, Alicante, Spain

Roger Wagner
Universidade Federal de Santa Maria (UFSM), Santa Maria, Rio Grande do Sul, Brazil

Wenxuan Wang
Jiangsu Synergetic Innovation Center of Meat Processing and Quality Control; State Key Laboratory of Meat Quality Control and Cultured Meat Development, MOE, College of Food Science and Technology, Nanjing Agricultural University, Nanjing, P.R. China

Lujuan Xing
Jiangsu Synergetic Innovation Center of Meat Processing and Quality Control; State Key Laboratory of Meat Quality Control and Cultured Meat Development, MOE, College of Food Science and Technology, Nanjing Agricultural University, Nanjing, P.R. China

Qian Xu
College of Food Science and Engineering, Production & Construction Group Key Laboratory of Special Agricultural Products Further Processing in Southern Xinjiang, Tarim University, Alar, P.R. China

Xianglin Yan
College of Food Science and Engineering, Production & Construction Group Key Laboratory of Special Agricultural Products Further Processing in Southern Xinjiang, Tarim University, Alar, P.R. China

Renyu Zhang
Food Technology & Processing Team, Bioeconomy Science Institute, Palmerston North, New Zealand

Wangang Zhang
Jiangsu Synergetic Innovation Center of Meat Processing and Quality Control; State Key Laboratory of Meat Quality Control and Cultured Meat Development, MOE, College of Food Science and Technology, Nanjing Agricultural University, Nanjing, P.R. China

CHAPTER ONE

Pathway-oriented strategies for nitrosamine mitigation in food products: Plant-based inhibitors and emerging processing technologies

Priscila Rossato Fracari[a], Bibiana Alves dos Santos[a], Márcio Vargas-Ramella[b], José Manuel Lorenzo[c,d], Alexandre José Cichoski[a], Roger Wagner[a], and Paulo Cezar Bastianello Campagnol[a,*]

[a]Universidade Federal de Santa Maria (UFSM), Santa Maria, Rio Grande do Sul, Brazil
[b]Universidade do Estado de Santa Catarina (UDESC), Laguna, Santa Catarina, Brazil
[c]Centro TecnolÓgico de la Carne de Galicia, Ourense, Spain
[d]Área de Tecnología de los Alimentos, Facultad de Ciencias de Ourense, Universidad de Vigo, Ourense, Spain
*Corresponding author. e-mail address: paulo.campagnol@ufsm.br

Contents

Advances in Food and Nutrition Research, Volume 122
ISSN 1043-4526, https://doi.org/10.1016/bs.afnr.2026.03.004

Abstract

Nitrosamines are a food safety concern due to their established toxicological relevance and their occurrence in widely consumed processed foods. In cured products, nitrite has a paradoxical role, acting both as a nitrosamine precursor and as an essential additive for safety, color, flavor, and oxidative stability. As a result, strategies based solely on nitrite reduction are increasingly viewed as insufficient and potentially detrimental to food safety. This chapter examines innovative approaches for reducing nitrosamine formation in foods, shifting the focus from additive elimination toward pathway-oriented control of nitrosation reactions. The chemical basis of nitrosamine formation is first reviewed, emphasizing the matrix- and process-dependent nature of nitrosation. Current mitigation strategies and their limitations are then discussed, highlighting the need for integrated solutions. Particular attention is given to plant-based extracts as natural inhibitors of nitrosation, as well as to emerging processing technologies that can modulate the physicochemical environment in which nitrosamines form. The chapter distinguishes between technologies with indirect mitigation potential and those supported by direct experimental evidence, avoiding overgeneralization. Synergistic strategies combining formulation-level inhibitors with process-level interventions are evaluated from mechanistic, technological, sensory, and risk management perspectives. Overall, this chapter argues that sustainable nitrosamine mitigation cannot be achieved through indiscriminate nitrite reduction, but rather through rational design of integrated processing and formulation systems that decouple nitrosamine formation from nitrite's essential technological functions. By consolidating current evidence and identifying key research gaps, this work provides a structured framework to guide future innovation in safer food processing.

1. Introduction

Nitrosamines have long occupied a central position in discussions on food safety due to their well-established toxicological relevance and their occurrence in widely consumed food products. Among these, cured and processed foods have received particular attention, as nitrite and nitrate, key contributors to product safety and quality, are also involved in nitrosation pathways leading to nitrosamine formation. This dual role has placed nitrosamines at the intersection of chemical risk assessment, food processing technology, and regulatory decision-making (Schrenk et al., 2023).

In recent years, growing consumer demand for reformulated and "clean-label" products has intensified pressure to reduce or eliminate nitrite from food formulations (Kim et al., 2025). While such efforts are often well-intentioned, they frequently overlook the multifunctional technological role of nitrite, particularly its critical contribution to microbiological safety, including control of *Clostridium botulinum*, as well as its impact on color stability, flavor development, and oxidative control (Lee et al., 2018). Consequently, strategies based on indiscriminate nitrite reduction raise legitimate concerns regarding food safety and product integrity.

This tension exposes a fundamental limitation of many current mitigation narratives: the implicit assumption that lowering nitrite content is synonymous with reducing nitrosamine risk. Extensive evidence demonstrates that nitrosamine formation is not solely determined by nitrite concentration but by a complex network of chemical pathways influenced by matrix composition, redox balance, precursor availability, and processing conditions (Bayne et al., 2023; Rot et al., 2025). Approaches that focus exclusively on additive reduction therefore risk addressing the symptom rather than the underlying reaction mechanisms.

Despite decades of research on nitrosamine chemistry, comparatively little attention has been devoted to integrating mechanistic understanding with modern food processing technologies and formulation strategies. In particular, the potential of plant-based inhibitors and emerging non-thermal or mild processing technologies to reshape nitrosation pathways has been explored unevenly, often without clear differentiation between indirect mitigation, direct chemical intervention, and technological readiness. This fragmentation has limited the translation of scientific insights into robust, transferable mitigation strategies.

The objective of this chapter was to critically examine innovative approaches for reducing nitrosamine formation in foods while preserving the essential technological functions of nitrite. Rather than advocating for additive elimination, this work adopts a pathway-oriented perspective, focusing on strategies that suppress nitrosation reactions through integrated formulation and processing design. By evaluating plant-based extracts, emerging processing technologies, and their synergistic application, this chapter aimed to provide a structured, evidence-based framework for rational nitrosamine mitigation that is aligned with food safety, quality, and industrial feasibility.

2. Nitrosamines in food products

Nitrosamines are a chemically diverse group of N-nitroso compounds formed primarily through reactions between nitrosating agents and nitrogen-containing precursors (Scanlan, 1983). Their relevance in food science stems from both their widespread occurrence in commonly consumed foods and their well-established toxicological significance. Since the early identification of volatile nitrosamines in cured meats in the 1960s and 1970s, extensive research has demonstrated that food processing conditions, ingredient composition, and storage parameters played decisive roles in determining nitrosamine formation and accumulation (Rot et al., 2025).

2.1 Chemical structures and classification of nitrosamines

Nitrosamines constitute a chemically heterogeneous family of N-nitroso compounds unified by the presence of the functional N–N = O moiety, formed through the nitrosation of nitrogen-containing precursors. Although often discussed as a single class, nitrosamines exhibit substantial structural diversity, which directly influences their physicochemical behavior, metabolic fate, analytical detectability, and toxicological relevance. In food systems, understanding this structural diversity is not merely taxonomic; it is essential for interpreting formation pathways and risk (Beard & Swager, 2021).

From a structural standpoint, nitrosamines are most commonly classified into volatile nitrosamines (VNAs) and non-volatile nitrosamines (NVNAs). This classification, while operationally useful, reflects fundamental differences in molecular size, polarity, and thermal stability rather than volatility alone. VNAs are typically low–molecular–weight, relatively non-polar compounds, including N-nitrosodimethylamine (NDMA), N-nitrosodiethylamine (NDEA), N-nitrosopyrrolidine (NPYR), and N-nitrosopiperidine (NPIP). Their small size and weak intermolecular interactions facilitate volatilization during thermal processing and analytical detection, which partly explains their historical predominance in the literature.

In contrast, NVNAs encompass structurally larger and more polar compounds, often derived from amino acids or peptides, such as N-nitrosoproline *and* N-nitrososarcosine. These compounds tend to remain tightly associated with the food matrix due to hydrogen bonding and ionic interactions, resulting in lower volatility and reduced analytical recovery under conventional methods (Herrmann et al., 2015). Importantly, NVNAs should not be regarded as toxicologically irrelevant; rather, they may act as latent reservoirs of nitroso functionality, capable of undergoing thermal or enzymatic transformation into VNAs during cooking or digestion.

At the molecular level, the defining N–N = O group confers a unique electronic structure that governs nitrosamine reactivity. The N–N bond is relatively weak, and the nitroso group can undergo metabolic activation via α-hydroxylation, producing highly reactive alkylating intermediates. This structural feature underlies the potent carcinogenicity observed for many VNAs, particularly NDMA and NDEA, which readily form DNA adducts after metabolic activation. Consequently, structural simplicity often correlates with higher toxicological concern, a paradoxical but well-established phenomenon in nitrosamine chemistry.

Structural classification can also be approached based on the nature of the amine precursor. Nitrosamines derived from cyclic secondary amines, such as pyrrolidine and piperidine, form NPYR and NPIP, respectively, which are frequently detected in thermally processed meat products. These cyclic nitrosamines exhibit greater thermal stability than their acyclic analogues, making them particularly relevant in fried, grilled, or roasted foods. In contrast, nitrosamines derived from simple dialkylamines, such as NDMA, may form readily but can also be partially lost through volatilization during high-temperature processing, complicating exposure assessment.

Another relevant structural distinction concerns substitution patterns and steric effects around the nitrosated nitrogen. Sterically hindered secondary amines exhibit lower nitrosation reactivity, while small, unhindered amines favor rapid nitrosamine formation under suitable conditions. This principle partially explains why specific nitrosamines dominate in certain food categories despite the presence of multiple potential amine precursors. For example, the prevalence of NPYR in cured and fermented meats is closely linked to the availability of proline and hydroxyproline-derived amines, as well as to processing conditions that favor cyclization reactions.

From an analytical and regulatory perspective, structural classification has historically biased monitoring efforts toward VNAs, largely because of their volatility and established carcinogenic profiles. However, advances in mass spectrometry have increasingly revealed the presence of NVNAs and structurally diverse nitroso compounds that were previously overlooked.

This expanding analytical window has challenged the traditional focus on a limited subset of nitrosamines and highlighted the need for a broader, structure-informed assessment of nitrosation products in foods.

Critically, the structural diversity of nitrosamines implies that formation, stability, and toxicity cannot be generalized across the class. Each nitrosamine results from a specific interaction among precursor chemistry, matrix composition, and processing conditions. Therefore, any discussion of nitrosamine mitigation strategies must be grounded in a clear understanding of which structural classes are favored under given technological scenarios, a premise that underpins the mechanistic analysis developed in subsequent sections of this chapter.

2.2 Mechanisms of nitrosamine formation

Nitrosamine formation in food systems is governed by a network of nitrosation reactions that are often described in simplified chemical terms but, in practice, are strongly modulated by the physicochemical complexity of real food matrices. While the classical model of nitrosation, derived largely from aqueous and gastric-simulation systems, provides a useful conceptual framework, it frequently overestimates nitrosamine formation when extrapolated to processed foods, particularly meat products (Zhou et al., 2025).

At the core of nitrosamine formation lies nitrite chemistry. Under acidic conditions, nitrite is protonated to form nitrous acid (HNO_2), an unstable intermediate that decomposes into reactive nitrosating species, primarily dinitrogen trioxide (N_2O_3) and, to a lesser extent, nitrosonium ions (NO^+). These species can react with unprotonated secondary amines via electrophilic substitution, yielding N-nitrosamines. In model systems, the rate of this reaction is maximized at moderately acidic pH values (approximately pH 3–4), where both nitrosating agents and nucleophilic amines coexist in reactive forms (Wainright, 1986).

However, food matrices introduce multiple competing pathways that substantially alter this chemistry. In cured meats, for example, nitrite-derived reactive nitrogen species are not exclusively directed toward nitrosation reactions. A significant fraction is diverted toward nitric oxide (NO) formation, particularly in the presence of reducing agents such as ascorbate, erythorbate, endogenous thiols, and phenolic compounds. This shift toward NO generation is technologically desirable for cured color development but simultaneously reduces the availability of nitrosating species capable of forming nitrosamines (Bonifacie et al., 2021).

The competition between nitrosation and NO-forming pathways represents a critical mechanistic bottleneck that is often underappreciated in simplified models. In real food systems, especially those rich in reductants and metal-binding compounds, nitrosation efficiency may be substantially attenuated despite favorable pH and nitrite concentrations. This phenomenon helps explain the frequently weak or inconsistent correlation observed between residual nitrite levels and measured nitrosamine concentrations in industrial surveys.

Another important factor differentiating real food matrices from simplified model systems is the role of metal ions, particularly iron. In meat products, heme iron is abundant and actively participates in nitrite-derived chemistry. When the local redox environment is predominantly reducing, due to the presence of compounds such as ascorbate, thiols, and other endogenous or added antioxidants, iron-catalyzed reactions favor the conversion of nitrite into nitric oxide. This pathway diverts reactive nitrogen away from nitrosation, thereby suppressing nitrosamine formation. In contrast, under more oxidizing local conditions, characterized by limited reducing capacity, higher oxygen availability, and ongoing lipid oxidation, iron-mediated chemistry promotes the generation of nitrosating species, increasing the probability of nitrosamine formation at specific micro-sites within the matrix. Because redox conditions vary spatially within real foods, iron can simultaneously promote localized nitrosation while limiting overall nitrosamine accumulation by forming nitric oxide, scavenging radicals, and complexing with metal–ligand complexes at the bulk level (de La Pomélie et al., 2018). As a result, nitrosamine formation becomes spatially heterogeneous and cannot be reliably predicted using global parameters such as average pH or residual nitrite concentration alone.

Thermal processing introduces an additional layer of complexity. At elevated temperatures, several mechanisms may operate simultaneously: (i) accelerated decomposition of nitrosating species, (ii) increased availability of secondary amines through protein degradation and lipid–amine interactions, and (iii) volatilization or thermal degradation of already formed nitrosamines. Consequently, high-temperature treatments such as frying or grilling may both promote and mask nitrosamine formation, depending on the balance between formation and loss processes (Rot et al., 2025; Xie et al., 2023). This duality complicates risk assessment and partially explains the wide variability reported for nitrosamine concentrations in thermally processed foods.

Lipid oxidation further modulates nitrosation pathways. Reactive carbonyl compounds generated during lipid peroxidation can interact synergistically with nitrite-derived species, indirectly facilitating nitrosamine formation. Conversely, NO generated from nitrite may act as a chain-breaking radical scavenger, inhibiting lipid oxidation and thereby limiting secondary pathways that contribute to (Altomare et al., 2021). These opposing effects highlight the dynamic interplay between oxidation and nitrosation, which static chemical models cannot adequately capture.

The nature and the availability of amine precursors are central determinants of nitrosamine formation in foods, with secondary amines serving as the critical, rate-limiting substrates for nitrosation reactions. In food systems, these secondary amines arise predominantly from amino acid degradation, biogenic amine formation during fermentation, and intramolecular cyclization reactions involving specific amino acids, most notably proline and hydroxyproline, which give rise to pyrrolidine-type structures, and lysine-derived pathways that contribute to piperidine formation (Drabik-Markiewicz et al., 2009). These reactions explain the recurrent dominance of cyclic nitrosamines such as NPYR and NPIP in thermally processed and fermented meat products. Importantly, nitrosamine formation is highly selective with respect to amine structure. Secondary amines exhibit optimal nucleophilicity and stability for reaction with nitrosating species, whereas primary amines generally do not accumulate as nitrosamines due to rapid decomposition into unstable diazonium intermediates. Tertiary amines, in turn, are largely unreactive toward direct nitrosation under typical food processing conditions. As a result, the presence, localization, and chemical accessibility of specific secondary amines, rather than total nitrogen or protein content, largely determine nitrosamine formation potential in real food matrices. Furthermore, the reactivity of secondary amines is modulated by their protonation state, steric environment, and spatial distribution within the food matrix. Amines confined within lipid-rich or protein-dense microdomains may exhibit enhanced exposure to nitrosating species during thermal processing, amplifying localized nitrosamine formation. This precursor-specific and microenvironment-dependent behavior further explains why nitrosamine occurrence cannot be reliably predicted from bulk compositional parameters alone.

Taken together, these observations underscore that nitrosamine formation in foods cannot be described solely by nitrite concentration, pH, and temperature. Instead, it emerges from a matrix-dependent balance between nitrosating species generation, competitive NO formation, redox conditions, metal-catalyzed reactions, precursor availability, and processing severity. The frequent reliance on simplified nitrosation models has therefore led to overly

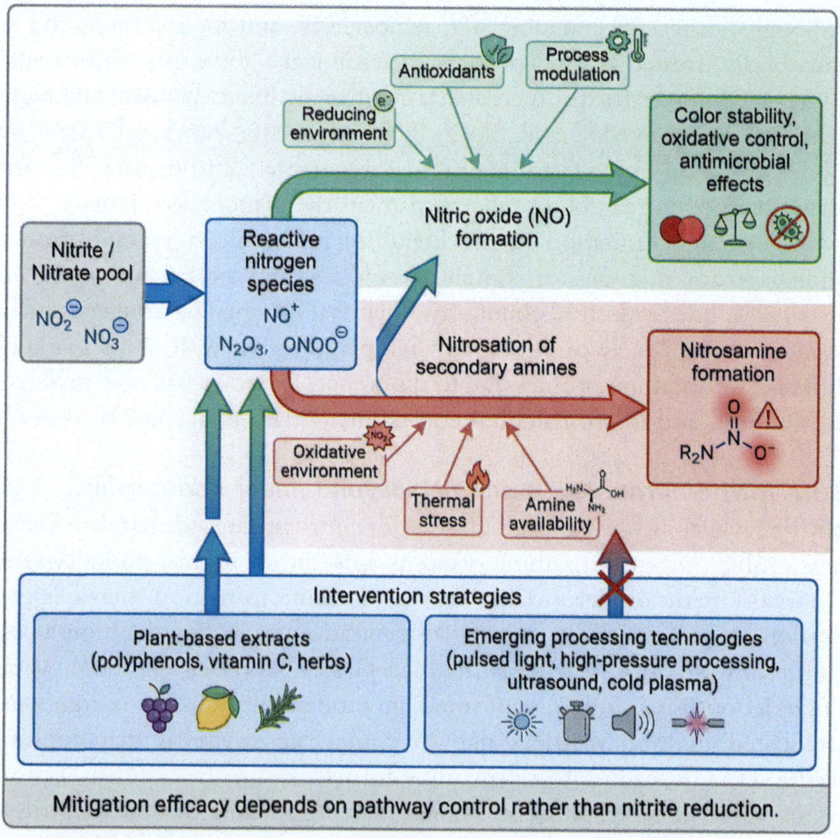

Fig. 1 Pathway-oriented framework of nitrosamine formation and mitigation in foods.

generalized assumptions about nitrosamine risk in foods. These competing pathways and their modulation by matrix- and process-dependent factors are schematically summarized in Fig. 1.

A mechanistic understanding that explicitly accounts for these competing pathways is essential for the rational design of mitigation strategies. Such an understanding provides the conceptual foundation for integrated approaches combining formulation adjustments, antioxidant systems, and emerging processing technologies, topics that will be explored in subsequent sections of this chapter.

2.3 Factors influencing nitrosamine formation

Nitrosamine formation in food systems arises from the interplay of formulation, processing, and storage parameters rather than from isolated variables.

Although nitrite concentration, pH, temperature, and food composition are traditionally treated as independent determinants, extensive experimental evidence demonstrates that their effects are strongly interdependent and highly matrix-specific (De Mey et al., 2017; Rot et al., 2025). As a result, predictive models based on single-factor approaches frequently fail to explain the variability observed in real food products, particularly in processed meats.

A recurring limitation in the literature is the direct extrapolation of findings from aqueous or simplified model systems to complex food matrices. Under industrial conditions, identical nominal parameters, such as residual nitrite levels or processing temperatures, may lead to markedly different nitrosamine profiles due to differences in redox balance, precursor distribution, and microstructural constraints (Gilani et al., 2024).

2.3.1 Nitrite/nitrate levels and pH: Beyond linear relationships

Nitrite availability is a prerequisite for nitrosamine formation, yet its relationship with nitrosamine yield is not linear in real food systems. Classical kinetic models predict that nitrosamine formation increases proportionally to the square of nitrite concentration, reflecting the bimolecular nature of nitrosation reactions (del Pilar García Santos et al., 2003). While this relationship is often confirmed in model solutions, it is frequently attenuated in food matrices due to competing reactions that consume nitrite without leading to nitrosamine formation.

In cured meat products, a substantial proportion of added nitrite is diverted toward NO formation, metal chelation, or oxidative degradation, thereby reducing the fraction available for nitrosation (Honikel, 2008). Consequently, increases in nitrite concentration within regulatory limits do not necessarily result in proportional increases in nitrosamine levels, a phenomenon repeatedly observed in experimental studies (Drabik-Markiewicz et al., 2011).

pH exerts a similarly complex influence. Acidic conditions favor the formation of nitrosating species such as nitrous acid and dinitrogen trioxide, theoretically enhancing nitrosamine formation. However, in fermented and cured foods, acidification often coincides with increased NO generation and enhanced reductive capacity, which can divert reactive nitrogen species away from nitrosation pathways (Shakil et al., 2022). Moreover, extreme pH values alter the protonation state and nucleophilicity of secondary amines, further complicating the relationship between pH and nitrosamine formation (Rot et al., 2025).

2.3.2 Processing and storage conditions: Formation, transformation, and loss

Processing conditions play a decisive role in determining nitrosamine profiles, not only by promoting their formation but also by influencing their transformation and loss. Thermal treatments such as frying, grilling, and roasting have been consistently associated with increased concentrations of volatile nitrosamines, particularly NDMA and NPYR (Yurchenko & Mölder, 2007). These effects are attributed to accelerated nitrosation kinetics and increased availability of secondary amines resulting from protein and lipid degradation.

At the same time, high temperatures favor the volatilization and thermal degradation of low–molecular-weight nitrosamines, potentially leading to an underestimation of their formation potential when only post-processing concentrations are measured (Yurchenko & Mölder, 2007). This dual effect explains the substantial variability observed across studies that employ different cooking methods, temperatures, and analytical protocols.

Storage conditions introduce additional temporal complexity. Residual nitrite, oxygen availability, and storage temperature collectively influence slow nitrosation reactions during shelf life. Several studies have demonstrated that nitrosamine levels may increase during storage, even in the absence of further thermal treatment, particularly in products containing residual nitrite and biogenic amine precursors (Honikel, 2008; Rot et al., 2025).

2.3.3 Food composition and additives: Matrix-controlled reactivity

Food composition fundamentally governs nitrosamine formation by controlling the availability, localization, and reactivity of amine precursors and nitrosating agents. Protein-rich matrices provide amino acids and peptides that can generate secondary amines through enzymatic degradation, fermentation, or thermal reactions. Amino acids such as proline, hydroxyproline, and lysine have been identified as key precursors of cyclic nitrosamines, including NPYR and NPIP (Drabik-Markiewicz et al., 2009).

Lipid content further modulates nitrosation pathways by influencing oxidative chemistry. Lipid oxidation generates reactive carbonyl compounds that may interact synergistically with nitrite-derived species, indirectly promoting nitrosamine formation (Steinberg, 2019). Conversely, NO generated from nitrite can inhibit lipid oxidation by acting as a radical scavenger, thereby suppressing secondary pathways associated with nitrosation (Hogg & Kalyanaraman, 1999).

Additives exert both inhibitory and promotive effects depending on their chemical nature and interaction with the food matrix. Reducing agents such as ascorbic acid and erythorbate are well-established inhibitors of nitrosamine formation because they convert nitrosating species into nitric oxide (Shakil et al., 2022). In contrast, smoke constituents, certain spices, and metal ions may enhance nitrosation under specific conditions by increasing precursor availability or catalytic activity (Sen et al., 1974).

2.4 Occurrence of nitrosamines in food products: A matrix- and process-driven perspective

The occurrence of nitrosamines in foods is highly heterogeneous and cannot be rationalized solely by the presence or amount of nitrite or nitrate. Instead, reported concentrations reflect the convergence of matrix composition, dominant processing steps, and post-processing handling, which together determine the balance between nitrosation reactions, competing chemical pathways, and compound loss. Consequently, comparative analyses across food categories consistently reveal patterns that are more coherently explained by mechanistic and technological factors than by additive use alone (Rot et al., 2025; Scanlan, 1983).

Processed meat products consistently exhibit the highest frequency and structural diversity of detected nitrosamines, particularly volatile compounds such as *N*-nitrosodimethylamine (NDMA), *N*-nitrosopyrrolidine (NPYR), and *N*-nitrosopiperidine (NPIP). This prevalence is not simply a function of nitrite addition, but rather the outcome of multiple converging conditions, including abundant secondary amine precursors derived from proteins and biogenic amines, the presence of heme iron, and processing regimes that impose thermal and oxidative stress (De Mey et al., 2017; Honikel, 2008).

Within this category, substantial product-to-product variability is observed. Bacon and fried or grilled sausages frequently display higher nitrosamine concentrations than cooked hams or emulsified products, a difference causally linked to intense surface heating and lipid-rich matrices that favor both the formation of cyclic amine precursors and the thermal stability of NPYR and NPIP (Yurchenko & Mölder, 2007). In contrast, fermented sausages often exhibit variable, and sometimes low, nitrosamine levels despite prolonged curing periods, reflecting the counterbalancing effects of acidification, enhanced NO formation, and antioxidant systems associated with fermentation (De Mey et al., 2017). Importantly, large-scale surveys of commercial meat products frequently report weak or inconsistent correlations between residual nitrite content and measured nitrosamine concentrations, reinforcing the concept that processing severity and matrix chemistry dominate nitrosamine occurrence under real production conditions (Rot et al., 2025).

Fish and fish products represent a distinct but mechanistically coherent case. Nitrosamines are most commonly detected in dried, salted, or smoked fish, where formation is driven less by intentional nitrite addition and more by the high endogenous availability of reactive amines, particularly dimethylamine generated from the degradation of trimethylamine oxide. When combined with nitrite contamination, microbial nitrate reduction, or exposure to nitrogen oxides during smoking, these amines provide an efficient substrate for NDMA formation (Yurchenko & Mölder, 2006; Zou et al., 1994). Thermal drying and smoking further promote the formation of nitrosamines by concentrating precursors and introducing reactive nitrogen species. At the same time, the relative absence of heme iron and the lower lipid content in many fish matrices shift the nitrosamine profile toward NDMA dominance rather than cyclic nitrosamines. This contrasts with meat products, underscoring the role of precursor specificity and matrix composition in determining nitrosamine distribution (Sen et al., 1985).

Beer and malt-based beverages illustrate a fundamentally different pathway for nitrosamine formation, in which it is largely decoupled from final product formulation and instead associated with upstream ingredient processing. Historically, NDMA formation in beer has been linked to malt kilning and drying, during which nitrogen oxides generated by combustion gases react with amine precursors present in barley malt. The finished beer matrix itself is generally unfavorable for nitrosamine formation due to low nitrite availability, dilution effects, and the presence of polyphenolic compounds with nitrosation-inhibitory properties. The substantial reduction in NDMA levels in modern beers following targeted technological modifications in malting and kilning provides a clear demonstration that process control, rather than ingredient substitution, can effectively mitigate nitrosamine formation (Lachenmeier & Fügel, 2007).

In contrast to animal-derived products, vegetables and other plant-based foods typically exhibit low nitrosamine concentrations despite sometimes high nitrate levels, particularly in leafy greens. The limited availability of secondary amines explains this apparent paradox, the predominance of nitrate over nitrite, and the abundance of endogenous antioxidants such as ascorbic acid and polyphenols, which suppress nitrosation reactions (Karwowska & Kononiuk, 2020). Nonetheless, nitrosamine formation in vegetables is not impossible and may occur under specific conditions, including improper storage, microbial spoilage, or high-temperature cooking, which promote nitrate reduction and localized nitrosation (Niklas et al., 2023). These cases highlight that plant-based matrices are conditionally, rather than inherently, protected against nitrosamine formation.

Dairy products generally display the lowest nitrosamine occurrence among major food categories, reflecting their low nitrite content and limited pool of reactive secondary amine precursors. Trace levels have, however, been reported in certain cheeses and fermented dairy products, where microbial nitrate reduction and protein degradation during ripening can create localized environments conducive to nitrosation (Gray et al., 1979; Spirić et al., 2025). Such occurrences are sporadic and highly product-specific, underscoring the dominant influence of microbial ecology and process control on raw material composition.

Across food categories, a consistent pattern emerges in which nitrosamine occurrence is governed primarily by processing intensity and precursor availability rather than by the mere presence of nitrite or nitrate. High-temperature treatments, lipid oxidation, and amine-rich matrices favor nitrosamine formation, whereas antioxidant-rich systems and processing conditions that promote NO formation tend to suppress it. This matrix- and process-driven perspective explains the wide variability reported in surveillance studies and underscores the limitations of risk assessments based solely on additive levels.

Understanding these causal relationships is essential for the rational development of mitigation strategies. Rather than indiscriminate reductions in nitrite use, effective control of nitrosamine formation requires matrix-specific interventions that address the dominant formation pathways in each food category, an approach that underpins the integrated strategies discussed in subsequent sections of this chapter.

2.5 Health risks associated with dietary exposure to nitrosamines

Nitrosamines are among the most thoroughly investigated food-related contaminants with respect to their potential adverse effects on human health, particularly carcinogenicity. Experimental evidence accumulated over several decades has unequivocally demonstrated that several volatile nitrosamines, notably N-nitrosodimethylamine (NDMA) and N-nitrosodiethylamine (NDEA), were potent genotoxic carcinogens in multiple animal species (Eisenbrand et al., 2024). Their toxicological relevance is not derived from the parent compounds themselves but from their metabolic activation, primarily through cytochrome P450–mediated α-hydroxylation, which generates highly reactive alkylating intermediates capable of forming DNA adducts and inducing mutagenesis (Li & Hecht, 2022). This well-characterized mode of action provides a strong mechanistic basis supporting concerns related to dietary exposure.

Based on this experimental evidence, several nitrosamines have been classified by the International Agency for Research on Cancer as probably or possibly carcinogenic to humans (International Agency for Research on Cancer IARC, 2010). Importantly, this classification reflects hazard identification rather than quantitative dietary risk. Most toxicological studies employ exposure levels that exceed those typically encountered through food consumption, which introduces uncertainty when extrapolating dose–response relationships to realistic dietary scenarios. Nevertheless, given the genotoxic nature of nitrosamines and the absence of a clearly defined threshold for such compounds, a precautionary approach to exposure from food sources has been widely adopted.

Further complexity arises from limitations in exposure assessment. Dietary intake estimates often rely on surrogate markers such as processed meat consumption or nitrite intake rather than direct measurement of nitrosamine exposure. In addition, analytical monitoring typically targets a limited subset of volatile nitrosamines, potentially underestimating total exposure to NOCs. Endogenous nitrosation, occurring in the acidic environment of the stomach from dietary nitrite and amine precursors, may represent a substantial and highly variable source of internal exposure that is not captured by food composition data alone (Mirvish, 1995). As a result, the relationship between measured dietary nitrosamine intake and actual biological exposure is inherently uncertain.

In recent years, attention has also been drawn to potential non-cancer health effects associated with chronic nitrosamine exposure, including metabolic dysfunction, oxidative stress, and cardiovascular alterations. Although experimental and epidemiological evidence supporting these outcomes is emerging, it remains considerably less robust than that linking nitrosamines to carcinogenic risk and should therefore be interpreted cautiously (Sokal-Dembowska et al., 2025). Nonetheless, these findings suggest that the health implications of nitrosamine exposure may extend beyond cancer and reinforce the importance of continued mitigation efforts.

Taken together, the available evidence supports a balanced interpretation of nitrosamine-related health risks. The carcinogenic potential of specific nitrosamines is firmly established at the hazard level, whereas exposure variability, matrix effects, endogenous formation, and methodological limitations in epidemiological studies modulate quantitative dietary risk. This uncertainty does not diminish the relevance of risk management but rather highlights the need for mitigation strategies grounded in mechanistic understanding. Instead of indiscriminate reductions in nitrite use, which may compromise microbiological safety and product quality, effective risk reduction should focus on approaches that shift the nitrosation balance away from harmful pathways. This integrated perspective provides the conceptual foundation for the mitigation strategies discussed in the subsequent sections of this chapter.

3. Current approaches for nitrosamine mitigation in foods

Building on the mechanistic basis and occurrence patterns discussed in Section 2, mitigation strategies for nitrosamines must be evaluated within a constrained technological space. In many products, particularly cured meats, nitrite and nitrate are not discretionary ingredients but functional inputs that contribute to microbial safety, color stability, flavor development, and oxidative control. This dual role creates a risk-management challenge: reducing nitrosation potential without compromising product safety and sensory quality. In practice, current control approaches combine regulatory limits, formulation measures (e.g., reductants and antioxidants), and process design choices to reduce conditions that favor nitrosamine formation. Regulatory frameworks provide the boundary conditions for these interventions by defining permissible additive levels, monitoring targets, and, in some cases, action thresholds for specific nitrosamines. Accordingly, an understanding of regulatory logic is essential before assessing technological solutions.

3.1 Regulatory frameworks and permissible limits

An important asymmetry characterizes the regulatory control of nitrosamines in foods. While public health concerns center on a set of highly potent compounds (e.g., NDMA and related volatile nitrosamines), most regulatory instruments do not directly limit nitrosamines across all foods. Instead, risk management is largely implemented indirectly by controlling the precursors and enabling conditions, principally nitrite and nitrate use, while requiring good manufacturing practices that reduce nitrosation potential. This structure reflects both analytical realities (historically limited routine monitoring of a broad spectrum of N-nitroso compounds) and the technological dependence of certain food categories on nitrite/nitrate for safety and quality.

Permissible limits for nitrite and nitrate vary across jurisdictions and food categories, and are periodically revised to reflect evolving toxicological evidence, technological needs, and consumption patterns. In the European Union, regulatory control has traditionally emphasized permissible levels of food additives, including nitrite (E249–E250) and nitrate (E251–E252), with product-specific maximum limits and conditions of use (European Commission, 2023). From a practical standpoint, this approach targets the controllable input that most strongly determines nitrosation capacity in processed foods, while acknowledging that nitrite reduction is bound by microbiological safety requirements, particularly in cured meat systems. A notable limitation of this precursor-focused strategy is that it does not guarantee uniform reductions in nitrosamine occurrence, because formation is strongly matrix- and process-dependent, and may be influenced by factors that are not directly constrained by additive limits (e.g., thermal load, precursor amines, and redox environment) (Drabik-Markiewicz et al., 2011; Rot et al., 2025).

In the United States, the regulatory logic similarly relies on controlling nitrite usage in cured meats through maximum allowable levels and compliance mechanisms embedded in meat inspection and processing standards, complemented by industry practices that incorporate nitrosation inhibitors (notably ascorbate/erythorbate) as part of standard curing systems. This model recognizes that direct nitrosamine limits are difficult to operationalize universally across diverse product categories and processing conditions. Consequently, the regulatory signal is often expressed as "process and formulation expectations" rather than as broadly enforced numerical limits for nitrosamines in all foods.

Internationally, Codex Alimentarius standards have also focused on the conditions of use for nitrite and nitrate in specific food categories, aiming to harmonize safety and technological functionality while minimizing exposure to undesirable reaction products. However, harmonization remains incomplete across jurisdictions, and differences persist in allowable additive levels, product definitions, and enforcement intensity. These divergences complicate cross-country comparisons of nitrosamine occurrence data and underscore the need for mitigation strategies that are robust to regulatory variability.

A second characteristic of current frameworks is the targeted treatment of nitrosamines in specific contexts where exposure has been historically significant or analytically tractable. Beer provides a clear example: although modern industrial practices have greatly reduced NDMA formation during malting through process redesign, historical regulatory and industry attention to NDMA in malt-based beverages illustrates how regulation can catalyze technological change even when direct additive controls are not the primary lever (Lachenmeier & Fügel, 2007). This case highlights an important principle for nitrosamine mitigation: controlling nitrosation is most effectively achieved by intervening at the step where reactive nitrogen species and amine precursors converge, rather than applying generic limits at the final product level.

Despite these advances, current regulatory approaches exhibit structural gaps that are increasingly relevant for innovation. First, monitoring and compliance tend to focus on a limited set of volatile nitrosamines. In contrast, broader N-nitroso compound exposure can also arise from non-volatile species and endogenous nitrosation processes that are not routinely captured in regulatory surveillance (Herrmann et al., 2015; Ohshima & Bartsch, 1984). Second, additive limits address precursor inputs but do not explicitly incorporate modern process intensification or emerging non-thermal technologies, which may alter redox balance, precursor availability, or reaction kinetics in ways that are not easily anticipated by legacy regulations (Andrade et al., 2025).

Overall, the regulatory landscape constrains nitrosamine mitigation through precursor control and good manufacturing practice, while leaving substantial responsibility to technological design at the formulation and processing levels. This reality underscores the need to evaluate conventional mitigation measures critically and their limitations before advancing to integrated strategies targeting multiple pathways of nitrosation in complex food matrices.

3.2 Conventional mitigation strategies and their limitations

Conventional strategies for mitigating nitrosamine formation in foods have historically focused on controlling individual drivers of nitrosation rather than addressing the system-level interactions that govern their formation in complex matrices. These approaches, which remain widely implemented in industrial practice, include reductions in nitrite and nitrate usage, the incorporation of curing accelerators and reductants, and modifications of thermal processing conditions. While each strategy is supported by sound chemical or technological rationale, its effectiveness is inherently constrained by safety requirements, product functionality, and the multifactorial nature of nitrosation reactions.

Reducing nitrite and nitrate levels is the most direct and intuitive mitigation approach. Lowering the availability of nitrosating agents can, in principle, reduce the formation of nitrosamines. In practice, however, nitrite and nitrate play indispensable roles in many food systems, particularly cured meats, where they contribute to microbial safety, color development, flavor formation, and oxidative stability. Excessive reduction of nitrite compromises control of *Clostridium botulinum* and other pathogens, alters sensory attributes, and may shorten shelf life (Honikel, 2008). Moreover, studies have demonstrated that lowering nitrite concentration within regulatory limits does not consistently result in proportional reductions in nitrosamine levels, underscoring that nitrosation efficiency is not determined solely by nitrite concentration but also by its interactions with matrix composition, redox balance, and processing conditions (Drabik-Markiewicz et al., 2011).

The use of curing accelerators, most notably ascorbic acid and erythorbate, constitutes a second cornerstone of conventional mitigation. These compounds act primarily by reducing nitrosating species to nitric oxide, thereby diverting reactive nitrogen away from nitrosamine-forming pathways. This mechanism has proven highly effective in promoting cured color development and reducing nitrosamine formation under controlled conditions (Fiddler et al., 1973). It is now embedded as standard practice in many curing systems. However, the scope of this strategy is chemically narrow. Reductants do not neutralize all nitrosating intermediates, nor do they address secondary formation pathways associated with lipid oxidation, amine availability, or high-temperature processing. Their efficacy is also strongly matrix-dependent and may be diminished under severe thermal treatments or in formulations with limited reducing capacity.

Modification of thermal processing parameters represents a further technological lever for mitigation, particularly for products prone to nitrosamine formation during high-temperature cooking. Adjustments in temperature, time, and cooking method can reduce nitrosamine accumulation by limiting precursor degradation and nitrosation kinetics (Chih et al., 2025). Nonetheless, thermal mitigation is inherently constrained by the need to ensure microbiological safety, product palatability, and consumer acceptance. In some cases, lowering processing severity may reduce measured nitrosamine concentrations while simultaneously increasing microbial risk or negatively affecting texture and flavor. Additionally, as discussed in Section 2, high-temperature processes can both promote nitrosamine formation and facilitate volatilization or degradation of formed compounds, complicating the interpretation of mitigation efficacy.

In addition to formulation- and process-based strategies, the use of selected bacterial strains as starter or adjunct cultures in fermented food systems, particularly in meat products, has been investigated as an indirect biological approach for mitigating nitrosamine formation (Sun et al., 2017; Sallan et al., 2020; Xie et al., 2023). Rather than acting primarily through direct in situ degradation of nitrosamines, these microorganisms influence key steps of the nitrosation pathway by modulating nitrite metabolism, redox conditions, and the availability of amine precursors during fermentation. Lactic acid bacteria, including *Lactobacillus sakei*, *L. plantarum*, and *L. curvatus*, together with nitrate-reducing coagulase-negative staphylococci such as *Staphylococcus xylosus*, have been shown to accelerate nitrite depletion, alter nitric oxide–related reactions, and limit the accumulation of biogenic and secondary amines under controlled fermentation conditions (Sallan et al., 2020; Sun et al., 2017). These microbiologically driven shifts in nitrogen chemistry are associated with reduced formation of volatile nitrosamines in fermented meat products, although the magnitude of the effect is strongly dependent on the food matrix and processing parameters (Xie et al., 2023). While certain bacterial strains have demonstrated the ability to bind or degrade nitrosamines in vitro or in simplified model systems (Shao et al., 2021), consistent evidence for direct nitrosamine removal in complex fermented meat matrices remains limited (Laranjo et al., 2017). Overall, the contribution of bacterial cultures to nitrosamine control should therefore be viewed as pathway-modulating and matrix-dependent, providing a complementary layer of mitigation that is most effective when integrated with formulation strategies and process-level controls rather than applied as a standalone intervention.

Collectively, these conventional strategies share a common limitation: they act on isolated components of the nitrosation process rather than on the integrated chemical environment in which nitrosamines form. By targeting single variables, such as nitrite concentration, reductant presence, or processing temperature, they fail to fully account for competing reaction pathways, matrix heterogeneity, and dynamic redox interactions that shape nitrosamine formation in real foods. As a result, their effectiveness is often context-specific and insufficient to achieve consistent reductions across diverse products and processing scenarios.

This fragmentation of conventional mitigation approaches highlights a fundamental mismatch between the complexity of nitrosation chemistry and the simplicity of the control measures traditionally applied. Addressing this gap requires strategies capable of acting across multiple reaction pathways and processing stages, an issue that motivates the integrated and innovative approaches discussed in the subsequent sections of this chapter.

3.3 Why conventional mitigation strategies are insufficient

Despite their widespread adoption and demonstrated benefits under specific conditions, conventional mitigation strategies are inherently insufficient to consistently and robustly control nitrosamine formation across diverse food matrices and processing scenarios. This limitation does not arise from a lack of chemical rationale or technological soundness, but rather from a fundamental mismatch between the system-level nature of nitrosation and the localized scope of traditional control measures.

Nitrosation in foods is governed by interacting networks of reactions involving nitrite chemistry, amine availability, redox balance, metal catalysis, lipid oxidation, and processing intensity. Conventional strategies typically intervene at a single point within this network, such as reducing precursor availability, introducing reductants, or moderating thermal load, without addressing compensatory pathways that may remain active or even become more dominant under modified conditions. As a result, mitigation efficacy is often matrix-dependent, transient, or limited to specific product categories.

Another structural limitation lies in the technological constraints imposed by food safety and quality requirements. In products where nitrite and nitrate play indispensable roles, aggressive precursor reduction is neither feasible nor desirable. Similarly, reliance on curing accelerators or antioxidants is constrained by formulation compatibility, processing severity, and diminishing returns under high thermal or oxidative stress. Process-based interventions, while valuable, are likewise bounded by microbiological safety, sensory acceptance, and industrial feasibility. These constraints collectively restrict the degree to which conventional strategies can be intensified without introducing new risks or trade-offs.

Importantly, conventional approaches also tend to treat mitigation as a static intervention, applied at a single stage of production. In reality, nitrosamine formation is a dynamic phenomenon that may occur during formulation, processing, storage, cooking, and even digestion. Strategies that act only at one stage are therefore unlikely to provide comprehensive control, particularly when downstream conditions favor renewed nitrosation or transformation of latent precursors.

Taken together, these considerations indicate that further reductions in nitrosamine formation cannot be achieved solely through incremental refinements of existing measures. Instead, meaningful progress requires mitigation strategies that operate across multiple chemical pathways and processing stages, modulate the reaction environment rather than individual inputs, and remain compatible with food safety and quality objectives. This recognition provides the conceptual basis for exploring plant-based inhibitors, emerging processing technologies, and the synergistic approaches discussed in the following sections of this chapter.

4. Plant-based extracts as natural inhibitors of nitrosamine formation

The limitations of conventional mitigation strategies discussed in Section 3 have prompted growing interest in plant-based extracts as complementary tools for controlling nitrosamine formation in foods. Unlike approaches that focus primarily on reducing precursor inputs or modifying single processing parameters, plant-derived compounds can modulate the chemical environment in which nitrosation occurs by shifting reactive nitrogen chemistry and redox balance. Their relevance extends beyond consumer-driven demand for "natural" ingredients, resting instead on established principles of nitrite chemistry, NO formation, and competitive reactions that consume nitrosating species. Understanding how plant-based compounds interfere with nitrosation pathways is therefore essential before evaluating specific plant sources or technological applications.

4.1 Conceptual framework: Why plant-based compounds can modulate nitrosation

The capacity of plant-based extracts to inhibit nitrosamine formation is fundamentally rooted in their ability to alter the reaction environment governing nitrosation, rather than to eliminate individual reactants such as nitrite or amine precursors. Nitrosation in food systems is a redox-sensitive, multi-pathway process in which the balance between nitrosating species generation, NO formation, oxidation reactions, and metal-catalyzed transformations determines the net yield of nitrosamines (Xie et al., 2023). Plant-derived compounds are uniquely positioned to influence this balance because they often contain multiple functional groups that can act simultaneously along different chemical pathways.

A central aspect of this modulation lies in redox control. Many plant metabolites can behave as electron donors and intercept reactive nitrosating species, thereby reducing their availability to react with secondary amines. Importantly, mitigation in real food systems frequently occurs not by "removing nitrite," but by diverting reactive nitrogen chemistry toward NO formation, a pathway that is technologically relevant in curing and is enhanced by reducing agents (Honikel, 2008; Jo et al., 2025). Within this framework, plant-based inhibitors act less as absolute nitrosation "blockers" and more as pathway modulators that shift the nitrite-derived reaction network away from nitrosamine-forming routes and toward alternative sinks.

Metal chelation represents a second, often underappreciated mechanism. Transition metals, particularly iron, can catalyze interconversions among reactive nitrogen species and promote localized reactions that influence nitrosation outcomes. Many polyphenol-rich extracts exhibit metal-binding capacity, which may reduce the catalytic availability of free iron and dampen redox cycling in heme-rich matrices. This is relevant because iron-driven oxidative processes have been linked to conditions that favor nitrosamine formation, and antioxidants can simultaneously attenuate lipid oxidation and nitrosamine-related pathways in meat systems (Herrmann et al., 2015; Li et al., 2013).

Plant-based compounds also indirectly influence nitrosation by interacting with lipid oxidation. Lipid peroxidation generates reactive carbonyl species that can participate in secondary pathways affecting protein and amine chemistry, thereby modulating the pool and reactivity of nitrosamine precursors. Conversely, conditions that promote NO formation can contribute to an oxidative–nitrosative balance that influences both oxidation and nitrosation in processed foods (Domínguez et al., 2019).

Crucially, the effectiveness of plant-derived inhibitors cannot be predicted solely based on generic antioxidant capacity. Chemical antioxidant assays or isolated nitrite-scavenging metrics do not necessarily translate into meaningful inhibition of nitrosation in complex matrices, where kinetics, compartmentalization, and competing pathways dominate. Evidence from inhibitor screening literature supports this principle by showing that reactivity proxies (e.g., "scavenging activity") may not correlate with actual inhibition performance when multiple pathways operate simultaneously (Homšak et al., 2022).

From a conceptual standpoint, plant-based extracts therefore offer a mode of action distinct from that of conventional mitigation tools. Rather than targeting a single precursor input or one processing parameter, they can modify the chemical landscape in which nitrosation occurs. This systems-level mode of action explains both their promise and their limitations, and provides the framework for evaluating specific classes of plant-derived compounds and their performance in real food systems in the following subsections.

4.2 Chemical classes of plant-derived inhibitors and mechanisms of action

4.2.1 Polyphenolic compounds

Polyphenols constitute the most extensively studied class of plant-derived compounds for nitrosamine inhibition, largely due to their structural diversity, redox activity, and prevalence in food-grade extracts. This group encompasses flavonoids, phenolic acids, tannins, and related oligomeric structures, which differ markedly in molecular size, polarity, and reactivity. Their inhibitory effects on nitrosation arise from multiple, partially overlapping mechanisms rather than from a single dominant reaction pathway.

One primary mechanism involves the interception of nitrosating species through electron donation or nucleophilic substitution. Several polyphenols can react with dinitrogen trioxide or nitrosonium ions, thereby reducing the pool of reactive nitrogen species available for secondary amine nitrosation (Pessego et al., 2011). In contrast to ascorbate-driven pathways, this interception does not necessarily require rapid conversion to nitric oxide, allowing polyphenols to function under a wider range of pH and redox conditions.

Metal chelation represents an additional and often critical mode of action. Many polyphenols possess ortho-dihydroxyl or galloyl groups with strong affinity for transition metals, particularly iron. By limiting metal-catalyzed redox cycling, polyphenols can attenuate localized environments that favor both lipid oxidation and nitrosation, a mechanism especially relevant in heme-rich meat systems (Wu et al., 2024). This chelation effect contributes to the frequently observed decoupling between in vitro antioxidant capacity and the actual inhibition of nitrosamine formation in complex matrices.

Polyphenols also exert indirect effects by modulating lipid oxidation. By suppressing lipid peroxidation, they reduce the generation of reactive carbonyl compounds that can interact with amines or influence nitrosation pathways. However, evidence from food systems indicates that polyphenol efficacy is highly context-dependent. Structural factors, thermal stability, and interactions with proteins and lipids strongly influence their performance, and high antioxidant activity does not guarantee effective nitrosamine inhibition under severe processing conditions (Deng et al., 2022; Jin & Jin, 2025).

4.2.2 Ascorbic acid and vitamin-rich plant sources

Ascorbic acid represents a chemically distinct class of nitrosation inhibitors, with a well-defined mechanism that is both technologically entrenched and widely used. Unlike many polyphenols, ascorbate acts primarily by reducing nitrosating species to nitric oxide, thereby diverting nitrite-derived chemistry away from nitrosamine formation and toward pathways that are compatible with cured product functionality (Tannenbaum et al., 1991). This reaction is rapid and efficient under typical curing conditions, which explains the widespread regulatory endorsement and industrial adoption of ascorbate and erythorbate.

From a mechanistic standpoint, it is critical to recognize that ascorbic acid does not neutralize nitrite but rather redirects its reactivity. This distinction is particularly relevant when considering plant-based vitamin C sources, such as acerola or citrus extracts, which may provide similar reductive capacity while introducing additional matrix effects associated with co-occurring phytochemicals. In such systems, the combined presence of ascorbate and polyphenols can create synergistic or antagonistic interactions depending on concentration, processing severity, and redox environment (Li et al. 2013).

Despite its effectiveness, ascorbate-mediated inhibition is not universal. Its activity is diminished under high thermal loads, oxidative stress, or prolonged storage, where depletion of reductive capacity may occur (Yin et al., 2022). Moreover, reliance on ascorbate alone does not address alternative nitrosation pathways linked to lipid oxidation or metal-catalyzed reactions, highlighting the need to evaluate vitamin-rich plant extracts within a broader mechanistic framework rather than as direct substitutes for conventional curing accelerators.

4.2.3 Sulfur-containing compounds and other minor phytochemicals

Sulfur-containing compounds derived from plants, particularly those present in *Allium* species and *Brassica* vegetables, represent a less extensively studied but mechanistically intriguing class of nitrosation modulators. These compounds, including organosulfur molecules and thiol-containing metabolites, can influence nitrosation through redox interactions, metal binding, and competitive reactions with reactive nitrogen species (Castro et al., 2023; Choi et al., 2007).

Thiol groups exhibit a strong affinity for nitrosating agents and can form S-nitrosothiols, effectively acting as alternative sinks for nitroso functionality (Broniowska & Hogg, 2012). While this mechanism does not eliminate reactive nitrogen species, it alters their chemical fate and may reduce the probability of nitrosamine formation from secondary amines. In addition, sulfur-containing compounds can interact with metal ions, further modulating redox cycling in complex food matrices (Broniowska & Hogg, 2012).

Beyond sulfur compounds, a range of minor phytochemicals, including carotenoids, chlorophyll derivatives, and specific alkaloids, has been reported to influence nitrosation indirectly (PÉrez-GÁlvez et al., 2020). However, their contribution is often secondary and highly matrix-specific, and robust evidence from real food systems remains limited. These compounds are therefore best viewed as modulators of the overall reaction environment rather than as primary nitrosation inhibitors.

4.2.4 Integrative perspective

Across chemical classes, a consistent theme emerges: effective plant-derived inhibitors do not operate through a single reaction mechanism, and isolated chemical properties cannot predict their performance. Instead, inhibition arises from the capacity of these compounds to engage multiple pathways simultaneously, including redox modulation, metal chelation, and suppression of secondary promoters such as lipid oxidation. This multi-target mode of action distinguishes plant-based extracts from conventional mitigation tools, but also introduces variability that complicates their application in industrial settings.

Understanding these class-specific mechanisms provides a necessary foundation for evaluating experimental evidence from real food systems, where efficacy is shaped not only by chemical potential but also by processing intensity, matrix interactions, and sensory constraints. These considerations are addressed in the following section.

4.3 Evidence from food systems: Efficacy and limitations

Evidence supporting the use of plant-based extracts as inhibitors of nitrosamine formation has expanded substantially over the past two decades; however, interpreting these results requires careful distinction between results obtained in simplified model systems and those observed in real food matrices. In controlled chemical models, many plant-derived compounds demonstrate pronounced inhibitory effects on nitrosation, often attributed to efficient scavenging of nitrosating species or strong antioxidant activity. While these findings provide valuable mechanistic insight, their predictive value for complex foods is limited, as they do not capture matrix interactions, processing severity, or competing reaction pathways that dominate under industrial conditions.

In cured meat systems, where nitrosamine mitigation is most technologically relevant, plant-based extracts have shown variable but generally promising effects. Polyphenol-rich extracts derived from fruits, herbs, and plant by-products have been reported to reduce the formation of volatile nitrosamines (Deng et al., 2022; Yurchenko & MÖlder, 2007; Zhang et al., 2017). These reductions are frequently accompanied by decreased lipid oxidation, supporting the concept that suppression of oxidative pathways indirectly contributes to lower nitrosation potential. However, the magnitude and consistency of inhibition are highly dependent on formulation, extract concentration, and processing conditions. In some studies, strong inhibitory effects observed in raw or mildly processed products were attenuated or lost after high-temperature cooking, indicating that thermal degradation of bioactive compounds and intensified reaction kinetics can override extract-mediated control (Li et al. 2013; Wang et al. 2015).

Fermented meat products present a more complex and sometimes contradictory picture. The intrinsic acidification, NO formation, and antioxidant activity associated with fermentation can already suppress nitrosation to a degree, making additional inhibition by plant extracts more difficult to detect. In such systems, plant-based inhibitors may contribute marginal benefits or exhibit effects only under specific conditions, such as elevated residual nitrite or extended storage. This context-dependence underscores the risk of overgeneralizing results across product categories and highlights the importance of evaluating extracts within the specific technological framework of each food (De Mey et al., 2014; Sallan et al., 2023).

Evidence from beverage systems, particularly beer, reinforces the primacy of process control over formulation-based mitigation. While polyphenols naturally present in beer contribute to a low nitrosation potential in the final product, historical reductions in NDMA levels were achieved primarily through technological modifications in malting and kilning rather than through additive strategies (Lachenmeier & FÜgel, 2007). This case illustrates a broader principle applicable to plant-based inhibitors: when nitrosamine formation occurs upstream during ingredient processing, downstream addition of inhibitors may have limited impact unless integrated with process-level interventions.

Plant-derived foods themselves offer additional insight into the conditional nature of nitrosamine inhibition. Despite high nitrate levels in some vegetables, nitrosamine occurrence is generally low due to limited amine availability and high endogenous antioxidant content. Experimental studies indicate that under atypical conditions, such as microbial spoilage or intense thermal treatment, plant antioxidants can modulate nitrosation. However, their effectiveness is again contingent on the balance between precursor generation and inhibitory capacity (Clayton-Cuch et al., 2025).

Across food systems, a consistent limitation emerges: inhibition observed in model systems or under mild processing cannot be directly extrapolated to industrial scenarios characterized by high temperatures, oxidative stress, and complex matrix interactions. In several cases, plant-based extracts reduce nitrosamine formation only within a narrow operational window, beyond which their activity is diminished or rendered ineffective. Moreover, variability in extract composition, stability, and interactions with food components introduces additional uncertainty, complicating reproducibility and scale-up.

Taken together, evidence from real food systems supports the conclusion that plant-based extracts can meaningfully modulate nitrosamine formation, but only when their chemical mechanisms align with matrix characteristics and processing conditions. Their role is therefore best understood as that of context-dependent modulators rather than universal inhibitors. This recognition is essential for a realistic appraisal of their technological potential and directly motivates the need to evaluate constraints related to sensory quality, stability, and industrial feasibility, which are addressed in the following section.

5. Technological and sensory constraints of plant-based inhibitors

Despite their mechanistic potential, the application of plant-based extracts as nitrosamine inhibitors in real food systems is constrained by technological and sensory factors that critically limit their industrial feasibility. Evidence from experimental studies consistently shows that inhibitory effects observed under controlled conditions are strongly modulated by processing severity, matrix interactions, and consumer acceptance thresholds. One of the most prominent constraints is sensory impact: plant-derived antioxidants and polyphenol-rich extracts can impart measurable changes in flavor, aroma, and color when incorporated into food systems, particularly at concentrations required to achieve significant chemical effects. Reviews and experimental investigations indicate that such additions have been associated with altered sensory characteristics or reduced acceptability due to intense aromas or off-flavors, which helps explain why natural antioxidants have not yet been widely adopted in commercial food production despite promising antioxidant and preservative properties (Kalogianni et al., 2020; Lee et al., 2020).

Thermal stability represents a second major limitation. Several studies show that phenolic compounds undergo partial degradation during cooking, leading to a marked reduction in their inhibitory activity after heat treatment (Izham et al., 2022; Zapata et al., 2022; Zhang et al., 2023). For example, reductions in volatile nitrosamines observed in raw or mildly processed products have been reported to be attenuated or lost after high-temperature cooking, as accelerated nitrosation kinetics and precursor release outweigh the remaining antioxidant capacity. In meat systems, processing severity (e.g., frying or grilling) has been identified as a major factor influencing nitrosamine formation (Lu et al., 2022; Nabizadeh et al., 2024). Evidence suggests that the effectiveness of added plant-derived antioxidants can be diminished under intense heat due to compound degradation and intensified reaction conditions rather than extract type alone (Guo et al., 2025; Xie et al., 2023).

Matrix interactions further complicate inhibitor performance. Polyphenols readily interact with proteins and lipids, reducing their chemical availability for redox modulation or metal chelation. Studies have documented extensive interactions between polyphenolic compounds and proteins through both non-covalent and covalent bonds, which can alter protein structure and phenol functionality, thereby affecting their antioxidant and bioactivity in complex food matrices (Jakobek, 2015; Zhang et al. 2014). These interactions with proteins and other macronutrients can decrease the effective availability of polyphenols to participate in redox reactions or metal chelation, ultimately influencing their capacity to inhibit undesirable reactions such as nitrosamine formation in complex food systems.

Variability and lack of standardization constitute additional barriers to industrial adoption. Plant-based extracts exhibit substantial batch-to-batch variation due to differences in botanical origin, extraction method, and storage conditions, leading to significant disparities in phenolic composition and associated bioactivities. Studies have demonstrated that phenolic content and antioxidant capacity can vary widely between extracts obtained from different sources, extraction techniques, or growing conditions, making nominal extract concentration a poor predictor of inhibitory performance in complex food systems (PaiÉ-Ribeiro et al., 2024; Taheri et al., 2023). Such variability complicates quality control and challenges regulatory acceptance, particularly in products subject to strict compositional specifications.

Finally, interactions between plant-based inhibitors and established curing systems may produce unintended technological consequences. In cured meat technology, nitrite plays a central role not only in safety but also in color development via its reduction to NO and subsequent formation of nitrosyl hemochrome, the pigment responsible for the characteristic pink/red color of cured products. Natural extracts rich in bioactive compounds can interfere with this chemistry by altering nitrite dynamics or interacting with myoglobin and other components of the curing matrix (Tahmouzi et al., 2025). As a result, adjustments to curing formulations or processing parameters may be required to maintain desirable color and sensory characteristics if plant-derived inhibitors are incorporated. From an industrial standpoint, mitigation strategies that necessitate extensive reformulation or compromise product identity are unlikely to be adopted, regardless of their chemical effectiveness.

Taken together, evidence from food systems clearly indicates that plant-based inhibitors cannot be regarded as standalone solutions for nitrosamine mitigation. Their efficacy is constrained by sensory acceptance, thermal stability, matrix interactions, and reproducibility, underscoring the need for complementary approaches that can modulate nitrosation pathways at the process level. These limitations provide the rationale for integrating plant-based strategies with emerging processing technologies, as discussed in the following section.

6. Emerging processing technologies for nitrosamine reduction

The technological and sensory constraints discussed in Section 4 underscore a fundamental limitation of formulation-based strategies for nitrosamine mitigation. Even when plant-derived inhibitors are chemically effective, their performance in real food systems is frequently constrained by processing severity, matrix interactions, and consumer acceptance. These limitations have driven increasing interest in emerging processing technologies capable of modulating nitrosation pathways at the process level rather than at the ingredient level. Such technologies enable direct intervention in the physicochemical environment where nitrosation occurs, reducing reliance on additive substitution and expanding the operational window for effective mitigation.

6.1 Conceptual rationale: Process-level control of nitrosation

The rationale for applying emerging processing technologies to nitrosamine mitigation is a shift from precursor-oriented control to environment-oriented control of nitrosation reactions. As established in Section 2, nitrosamine formation in foods is governed by a dynamic balance between reactive nitrogen species generation, redox conditions, precursor availability, and competing reaction pathways. Conventional mitigation strategies primarily target individual inputs, such as nitrite concentration or antioxidant addition, without fundamentally altering the reaction environment in which nitrosation proceeds.

Process-based technologies offer a distinct mode of intervention by modifying key physicochemical parameters that regulate nitrosation kinetics and pathway selection. Rather than suppressing nitrosation solely through stoichiometric scavenging, these technologies can influence factors such as energy input, redox potential, mass transfer, and the formation or decay of reactive nitrogen and oxygen species. In doing so, they act upstream of individual reaction steps and can simultaneously affect multiple nitrosation-related pathways.

A central advantage of process-level control lies in its relative independence from matrix-specific sensory constraints. Technologies that reduce thermal load, alter reaction time scales, or modulate reactive species chemistry can suppress nitrosamine formation without introducing foreign compounds or altering product flavor profiles. This distinction is particularly relevant for food systems in which regulatory definitions or consumer expectations tightly constrain formulation changes. In such contexts, process interventions may offer mitigation opportunities that are not accessible through ingredient modification alone.

Emerging non-thermal and mild technologies are especially attractive in this regard. By decoupling microbial safety and quality preservation from high-temperature processing, these approaches can reduce the thermal drivers of nitrosamine formation while maintaining product integrity. In addition, some technologies generate controlled populations of reactive species that can interact with nitrite-derived intermediates, potentially redirecting nitrogen chemistry away from nitrosamine-forming routes. Importantly, these effects can be either beneficial or detrimental depending on process parameters, highlighting the need for a mechanistic understanding rather than empirical application.

From a conceptual standpoint, process-based mitigation strategies are best viewed not as replacements for formulation-level interventions, but as complementary tools capable of reshaping the chemical landscape in which nitrosation occurs. Their effectiveness depends on the extent to which they modify critical reaction determinants, such as redox balance, precursor accessibility, and reaction kinetics, without introducing new pathways for nitrosamine formation. This systems-level perspective provides the foundation for evaluating specific emerging technologies and their potential role in integrated mitigation strategies.

6.2 Classification of emerging non-thermal and mild technologies

Emerging processing technologies relevant to nitrosamine mitigation can be meaningfully classified by their mode of action on the physicochemical environment, rather than solely by equipment type. This distinction is important because technologies that appear disparate from an engineering perspective may exert similar effects on nitrosation pathways. In contrast, superficially similar processes may influence nitrosamine formation through fundamentally different mechanisms. A mechanism-oriented classification has therefore been increasingly adopted in food processing literature to facilitate comparison of emerging technologies beyond their nominal operational principles (Knorr et al., 2011).

A first class comprises technologies that primarily modulate energy input and thermal history without relying on sustained high temperatures. These approaches aim to achieve microbial safety and quality preservation while reducing the thermal drivers known to accelerate nitrosamine formation, such as protein degradation, amine release, and enhanced nitrosation kinetics. By shortening exposure time or redistributing energy in space or time, these technologies can decouple safety objectives from prolonged thermal stress, thereby limiting conditions that favor nitrosamine generation (Chacha et al., 2021).

A second class includes technologies that alter the chemistry of reactive species, particularly reactive nitrogen and oxygen species, within the food matrix. Some emerging processes generate controlled populations of radicals or excited species that can interact with nitrite-derived intermediates, influence redox balance, or redirect nitrogen chemistry toward alternative reaction pathways (Laroque et al., 2022). Depending on processing parameters, such interactions may suppress or, if improperly controlled, enhance nitrosation reactions.

A third class encompasses technologies that primarily affect mass transfer, diffusion, and structural accessibility of precursors. By altering cell permeability, protein conformation, or matrix microstructure, these processes influence the spatial distribution and reactivity of nitrosating agents and amine precursors. Changes in mass transfer dynamics can limit the frequency of reactive encounters or shift reaction loci away from microenvironments that favor nitrosation. Such indirect effects have been increasingly recognized as relevant determinants of chemical reactivity in complex food systems subjected to non-thermal processing (Knorr et al., 2011; Nowosad et al., 2021).

Importantly, many emerging technologies operate across more than one of these categories. A single process may simultaneously reduce effective thermal load, modify redox conditions, and alter mass transfer properties. This overlap underscores the limitation of rigid categorical boundaries and reinforces the importance of identifying dominant mechanisms under specific operating conditions, rather than relying solely on nominal technology classification.

From a nitrosamine mitigation perspective, this classification clarifies why emerging technologies should not be evaluated merely as substitutes for conventional heat treatments or additives. Their primary value lies in their ability to reshape the reaction environment governing nitrosation, often in ways inaccessible through formulation-based strategies alone. This framework provides a structured basis for examining individual technologies beyond their nominal engineering principles, focusing instead on how they reshape the nitrosation environment in foods.

6.3 Emerging processing technologies: Evidence, mechanisms, and maturity levels

Emerging processing technologies have been increasingly explored as complementary tools for nitrosamine mitigation, particularly in response to the recognized limitations of formulation-based strategies discussed in Section 3. Unlike conventional approaches that primarily target individual precursors, these technologies intervene at the processing environment level, altering reaction kinetics, redox conditions, and precursor accessibility. However, it is critical to emphasize that the level of scientific evidence supporting nitrosamine reduction varies substantially among technologies, and "emerging" should not be interpreted as "equivalent" in terms of mechanistic validation or technological readiness.

The following subsections critically evaluate emerging technologies based on their interactions with nitrosation pathways and the strength of available evidence, distinguishing between indirect mitigation strategies and technologies with direct experimental support. To support this distinction, a comparative, mechanism-oriented overview of emerging processing technologies, highlighting their primary intervention targets and relative evidence maturity, is presented in Fig. 2.

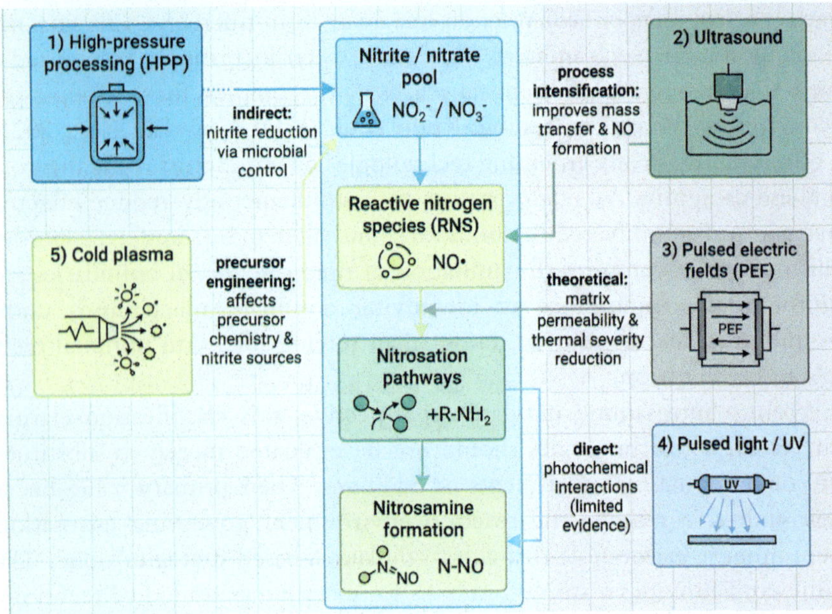

Fig. 2 Comparative mechanistic map of emerging processing technologies for nitrosamine mitigation.

6.3.1 Technologies with indirect but established mitigation potential

Several emerging technologies have demonstrated a robust capacity to reduce reliance on nitrite or to modify the processing environment in ways that are theoretically favorable for nitrosamine control, despite the absence of direct evidence showing nitrosamine degradation or suppression in foods.

High-pressure processing (HPP) is the most technologically mature example of an environment-oriented intervention with relevance to nitrosamine mitigation in meat systems. HPP is a non-thermal preservation technology in which foods are subjected to uniform isostatic pressures (commonly 300–600 MPa) for short holding times, achieving substantial microbial inactivation and extended shelf life without the application of sustained high temperatures. For example, studies in cooked ham have shown that reducing nitrite levels to as low as ~40 mg NO_2^-/kg in conjunction with HPP (600 MPa, ~3 min) did not significantly impact colour stability, odour, or texture compared with higher nitrite formulations, demonstrating that HPP can help decouple nitrite usage from product quality outcomes (Molina et al., 2025). Rather than directly altering nitrosation chemistry or generating reactive nitrogen intermediates, HPP's contribution to nitrosamine mitigation is indirect: it operates through precursor minimization and process design, expanding the formulation space for low-nitrite products without compromising microbiological safety or sensory integrity, rather than actively suppressing nitrosamine formation by chemical transformation.

Ultrasound-assisted processing occupies a similar conceptual position. High-intensity ultrasound enhances mass transfer, accelerates curing reactions, and improves the penetration of curing agents and antioxidants into meat matrices. Experimental evidence indicates that ultrasound can facilitate the development of cured color at lower nitrite concentrations and even promote nitrosylmyoglobin formation in the absence of added nitrite, likely by activating endogenous NO–related pathways (Leães et al., 2023). While these effects suggest the potential to mitigate nitrosamines through reduced nitrite availability and improved redox control, no studies to date have directly quantified the reduction in nitrosamines attributable to ultrasound treatment. Consequently, ultrasound should be regarded as a process intensification tool that supports alternative curing strategies rather than a standalone nitrosamine control technology.

Pulsed electric fields (PEF) represent a largely prospective approach in this context. Although PEF is well established for microbial inactivation in liquid foods and has shown promise in solid and semi-solid matrices, its application to cured meat systems remains limited and exploratory. Research on PEF in meat has predominantly focused on effects such as tenderization, mass transfer enhancement, and microbial safety, with studies documenting structural modifications and quality impacts under various field strengths and pulse conditions (Ghoshal, 2023). However, no published studies have directly examined the impact of PEF on nitrosamine formation or degradation in cured meats. Nevertheless, PEF's capacity to ensure microbial safety under mild thermal conditions and to facilitate mass transfer and brine penetration suggests that, in principle, it could support nitrite reduction strategies analogous to those enabled by high-pressure processing, potentially expanding the formulation space for lower-nitrite products (Andrade et al., 2025). At present, however, its relevance to nitrosamine mitigation remains conceptual rather than demonstrated, pending targeted research that directly measures nitrosamine outcomes following PEF treatment in nitrite-controlled curing systems.

Collectively, these technologies share a common characteristic: they do not directly interfere with nitrosation chemistry, but instead reshape the technological context in which nitrosation occurs. Their value lies in enabling safer formulations with reduced nitrosation potential, not in actively suppressing nitrosamine formation once precursors are present.

6.3.2 *Photonic technologies with direct experimental evidence*

Photonic technologies, particularly pulsed light (PL) and ultraviolet (UV) radiation, occupy a distinct position among emerging approaches due to their potential to interact directly with nitrosamines and nitrosation intermediates. UV radiation has long been known to induce photolytic degradation of N-nitrosamines in aqueous systems, with studies demonstrating rapid and efficient decomposition of N-nitrosodimethylamine under controlled irradiation (Afzal et al. 2016; Lee et al. 2005). However, extrapolating these findings to real food matrices is nontrivial due to limitations in light penetration, matrix opacity, and competing photochemical reactions.

In this context, pulsed light represents the only photonic technology for which direct evidence of nitrosamine reduction in a real food system has been reported. The study by Fracari et al. (2025) demonstrated that PL treatment, when applied to sliced mortadella formulated with reduced nitrite and combined with a polyphenol-rich plant extract, resulted in a significant reduction of volatile nitrosamines following storage. This work constitutes, to date, the only experimental demonstration of PL-mediated nitrosamine mitigation in a complex food matrix.

Mechanistically, the observed effects were attributed to a combination of factors, including partial photodegradation of nitrosamines or nitrosation intermediates, redirection of nitrite-derived reactive species toward NO formation, and synergistic interactions with antioxidant compounds. Importantly, the authors emphasized that PL should not be viewed as a universal nitrosamine degradation technology. Its efficacy was strongly dependent on formulation context, the presence of antioxidants, and processing parameters.

Therefore, while photonic technologies, particularly PL, represent a genuine frontier in nitrosamine mitigation research, the evidence base remains limited. Their current relevance lies in demonstrating proof of concept, rather than providing a broadly applicable industrial solution. Further studies across different food matrices and processing scenarios are required before generalization is warranted.

6.3.3 *Plasma-based approaches: Precursor engineering rather than degradation*

Cold plasma technologies introduce yet another distinct mitigation paradigm. Unlike photonic or pressure-based processes, atmospheric cold plasma primarily affects food systems by generating reactive nitrogen and oxygen species that can modify precursor chemistry rather than directly degrading nitrosamines.

To date, direct evidence linking cold plasma to nitrosamine reduction in foods is limited to a single study, reported by (Jo et al., 2025). In this work, plasma-treated vegetable powder was used as a natural nitrite source in cured meat products. When compared with conventional sodium nitrite at equivalent nitrite levels, the plasma-derived curing system resulted in significantly lower formation of N-nitrosopyrrolidine following thermal treatment.

Crucially, the plasma was not applied directly to the meat matrix. Instead, cold plasma functioned as a precursor-engineering tool, altering the chemical nature of the nitrite source and its interaction with endogenous antioxidants during processing. The observed reduction in nitrosamine formation was attributed to a shift in reactive nitrogen chemistry toward NO generation, consistent with mechanistic principles discussed in Section 2.

At the same time, the study highlighted important limitations. Plasma treatment reduced the antioxidant capacity of the vegetable substrate, underscoring the oxidative nature of plasma-generated species and the need to balance nitrite generation and antioxidant preservation carefully. Moreover, the scalability and energy efficiency of plasma-based nitrite production remain open questions.

Accordingly, cold plasma should be regarded as a highly promising but immature strategy, whose relevance to nitrosamine mitigation is indirect, formulation-adjacent, and strongly substrate-dependent.

6.3.4 Comparative synthesis: why "emerging" does not mean "equivalent"

A comparative evaluation of emerging processing technologies reveals that their contributions to nitrosamine mitigation differ not only in magnitude but also in mechanistic nature and evidentiary support. HPP, ultrasound, and PEF primarily enable safer formulations by reducing the need for nitrite or altering processing severity, without directly intervening in nitrosation chemistry (Balasubramaniam et al., 2015; Chemat et al., 2011; Toepfl et al., 2006). In contrast, photonic technologies and plasma-based approaches offer the potential to interact more directly with nitrosation pathways, either through photochemical reactions or by modifying reactive nitrogen species (GÓmez-LÓpez et al., 2007; Misra et al., 2011).

However, with the exception of pulsed light and cold plasma, each supported by a single matrix-specific study, direct experimental evidence remains sparse. This disparity underscores the importance of avoiding technological overgeneralization and highlights the need for integrated strategies that combine formulation control with carefully selected process interventions.

Within this framework, emerging technologies should be viewed not as standalone solutions, but as context-dependent tools whose effectiveness depends on matrix composition, processing intensity, and the presence of complementary mitigation measures. This perspective provides a rational foundation for the synergistic approaches discussed in the following section.

7. Synergistic strategies combining plant-based extracts and emerging technologies

The analysis presented in Sections 4 and 5 reveals a recurring pattern in current nitrosamine mitigation research: strategies based solely on formulation or processing modifications rarely yield robust, transferable solutions across food matrices. Plant-based extracts provide effective chemical inhibition of nitrosation but are constrained by sensory impact, stability, and regulatory limits. Emerging processing technologies, in turn, can reshape the physicochemical environment for nitrosation but often lack the chemical selectivity needed to suppress nitrosamine formation consistently.

These complementary strengths and limitations naturally converge toward integrated mitigation approaches that combine formulation- and process-level interventions. Rather than viewing plant-derived inhibitors and emerging technologies as competing alternatives, a synergistic framework recognizes their potential to act at distinct control points within the nitrosation network. This section examines the conceptual basis and practical implications of such combined strategies, emphasizing mechanistic complementarity, technological feasibility, and relevance for industrial food processing. The pathway-oriented logic underpinning these synergistic strategies is schematically summarized in Fig. 3.

7.1 Conceptual basis for synergistic mitigation strategies

The preceding sections of this chapter highlight a central limitation in current approaches to nitrosamine mitigation: neither formulation-based nor process-based strategies are sufficient when applied in isolation. While plant-derived extracts can effectively inhibit nitrosation through antioxidant activity, metal chelation, and redox modulation, their performance in real food systems is constrained by technological feasibility, sensory acceptability,

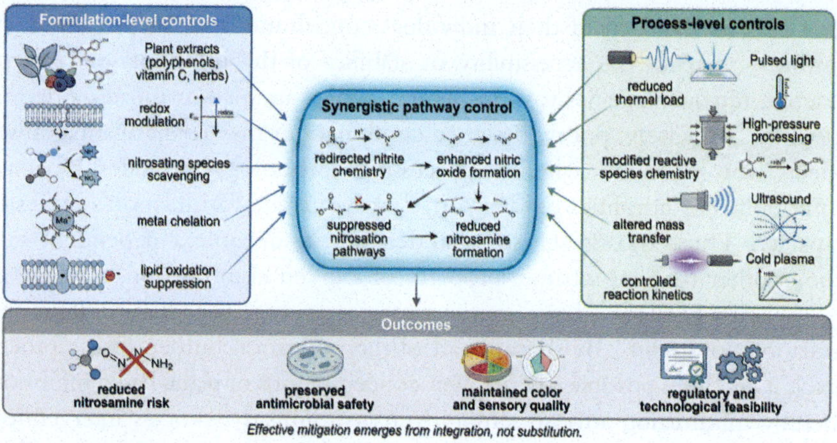

Fig. 3 Synergistic framework integrating formulation- and process-level controls of nitrosation pathways.

and regulatory boundaries. Conversely, emerging processing technologies can alter the physicochemical environment for nitrosation but often lack the chemical specificity needed to suppress nitrosamine formation across diverse matrices and processing conditions.

This intrinsic asymmetry provides the conceptual rationale for synergistic mitigation strategies. A single reaction step does not govern nitrosamine formation; rather, it is governed by a network of competing pathways involving nitrite chemistry, reactive nitrogen species, precursor availability, lipid oxidation, and thermal history. Interventions targeting only one node of this network inevitably leave alternative pathways active, limiting overall effectiveness. Synergistic strategies seek to overcome this limitation by combining complementary mechanisms that operate at different levels of the nitrosation process.

From a mechanistic perspective, synergy arises when formulation-level inhibitors and process-level interventions act on non-overlapping control points. Plant-based extracts primarily modulate chemical reactivity by scavenging nitrosating agents, chelating transition metals, and suppressing lipid oxidation, thereby reducing the potential for nitrosamine formation. Emerging processing technologies, in contrast, influence reaction kinetics and pathway selection by modifying energy input, mass transfer, redox balance, or the generation and decay of reactive species. When properly aligned, these effects can reinforce one another rather than compete.

Importantly, synergistic mitigation is not equivalent to additive inhibition. The combined effect of a natural inhibitor and a processing technology may exceed the sum of their individual contributions if the process intervention enhances the accessibility or stability of the inhibitor, or redirects reactive nitrogen species toward benign endpoints such as nitric oxide formation. Conversely, poorly designed combinations may result in antagonistic effects, for example, when a processing technology degrades bioactive compounds or amplifies oxidative stress beyond the formulation's buffering capacity. Thus, synergy must be understood as a context-dependent outcome rather than an intrinsic property of a given combination.

A key advantage of synergistic strategies is their potential to relax formulation constraints. By shifting part of the mitigation burden to the process level, it becomes possible to use lower concentrations of plant-based inhibitors, thereby minimizing adverse sensory impacts. Similarly, process interventions can be optimized for moderate intensity when supported by formulation-level controls, reducing the risk of quality degradation or unintended chemical reactions. This redistribution of control across multiple intervention points expands the practical design space for nitrosamine mitigation.

From a technological standpoint, synergistic approaches are particularly relevant for products subject to strict compositional standards or strong consumer expectations regarding flavor, color, and texture. In such cases, neither aggressive nitrite reduction nor high-dose antioxidant addition is viable on its own. Integrated strategies offer a pathway to incremental but robust risk reduction without compromising product identity or safety.

In this chapter, synergy is therefore framed not as a novel concept, but as a necessary evolution of nitrosamine mitigation thinking. Rather than asking whether a given ingredient or technology can reduce nitrosamines, the relevant question becomes how different interventions can be combined to reshape the nitrosation landscape in a controlled and predictable manner. The following sections examine documented examples of such combinations, with emphasis on mechanistic complementarity, technological feasibility, and translational relevance for industrial food processing.

7.2 Documented synergistic approaches

7.2.1 Pulsed light combined with antioxidant-rich plant extracts

The combination of pulsed light (PL) processing with antioxidant-rich plant extracts represents the most clearly documented example of synergistic nitrosamine mitigation currently available in food systems. This approach exemplifies how formulation-level chemical control and process-level environmental modulation can converge to suppress nitrosation more effectively than either strategy applied independently.

From a mechanistic standpoint, this synergy arises from the complementary modes of action of the two interventions. As discussed in Section 4, plant-derived extracts rich in polyphenols, vitamins, and other redox-active compounds primarily inhibit nitrosamine formation by scavenging nitrosating species, chelating transition metals, and attenuating lipid oxidation. These effects reduce the intrinsic nitrosation potential of the matrix but are often insufficient to fully suppress nitrosamine formation under severe processing or storage conditions. Pulsed light, in contrast, modifies the reaction environment through photochemical interactions, delivering high-energy photons that can influence reactive nitrogen chemistry without imposing substantial thermal load.

The study by (Fracari et al., 2025) provides direct experimental evidence supporting this synergistic framework. In a cured meat model system, PL treatment of sliced mortadella formulated with reduced nitrite levels resulted in a significant reduction in volatile nitrosamines when antioxidant-rich plant extracts were incorporated into the formulation. Notably, neither PL treatment alone nor extract addition alone achieved the same degree of suppression, underscoring the importance of their combined application. This finding aligns with the conceptual model presented in Section 6.1, in which interventions acting on distinct control points within the nitrosation network reinforce one another.

Several mechanistic pathways may underlie this synergy. First, PL may promote partial photodegradation of pre-formed nitrosamines or unstable nitrosation intermediates at the product surface, reducing the pool of compounds available for accumulation during storage. Second, photonic energy may redirect nitrite-derived reactive species toward nitric oxide formation rather than toward nitrosating agents, an effect that is amplified in the presence of polyphenols. Third, antioxidant extracts can buffer oxidative stress potentially induced by PL exposure, preventing secondary pathways, such as lipid oxidation, from indirectly promoting nitrosamine formation.

Importantly, the efficacy of this combined strategy is highly context-dependent. The effects observed by (Fracari et al., 2025) depended on formulation composition, extract type, and PL treatment parameters, highlighting that synergy was not an inherent property of PL or antioxidants per se but an emergent outcome of their interaction within a specific matrix. Excessive photonic exposure, for example, could degrade bioactive compounds or exacerbate oxidative reactions if antioxidant capacity is insufficient, leading to antagonistic rather than synergistic effects.

From a technological perspective, the PL–extract combination offers several advantages. Because PL is primarily a surface treatment, it complements antioxidant systems distributed throughout the matrix, enabling mitigation at multiple spatial scales. Moreover, by enhancing mitigation efficiency, this strategy permits the use of lower extract concentrations, thereby minimizing sensory impacts that often limit the practical application of plant-based inhibitors. This feature is particularly relevant for sliced and ready-to-eat products, where surface-related nitrosation processes and post-processing contamination risks coexist.

At the same time, limitations must be acknowledged. The surface-limited penetration of PL constrains its effectiveness in thick or heterogeneous products, and industrial implementation requires careful control of exposure uniformity and energy dosage. Additionally, the scarcity of studies beyond the single food matrix investigated to date precludes broad generalization. Consequently, while the PL–extract combination represents the most advanced example of synergistic nitrosamine mitigation currently documented, it should be regarded as a proof of principle rather than a universally applicable solution.

In summary, integrating PL processing with antioxidant-rich plant extracts demonstrates how synergistic strategies can overcome the limitations of isolated interventions. By simultaneously modulating nitrosation chemistry and the processing environment, this approach achieves a level of mitigation unattainable through formulation or processing alone. Its success provides a compelling template for the rational design of other integrated strategies, which are examined in the following sections.

7.2.2 Ultrasound-assisted processing combined with plant-derived curing systems

Ultrasound-assisted processing combined with plant-derived curing systems represents a mechanistically plausible but still underexplored synergistic strategy for nitrosamine mitigation. Unlike the pulsed-light–based approach discussed in Section 6.2.1, this combination lacks direct experimental evidence demonstrating reduced nitrosamine concentrations. Instead, its relevance arises from a convergence of indirect findings on curing efficiency, nitrite utilization, and redox control, which collectively influence nitrosation potential.

From a process standpoint, high-intensity ultrasound enhances mass transfer and accelerates physicochemical interactions within food matrices via cavitation. The collapse of cavitation bubbles generates localized microturbulence and transient high-energy sites, facilitating the penetration and dispersion of curing agents and bioactive compounds. In meat systems, this effect has been shown to accelerate curing reactions, improve the uniformity of salt and additive distribution, and reduce processing time (da Rosa et al., 2023; da Silva et al., 2020; Pinton et al., 2019). When applied to curing systems based on plant-derived nitrite sources, these effects become particularly relevant.

Plant-based curing systems, typically derived from nitrate-rich vegetables such as celery, beetroot, or other leafy greens, rely on microbial or enzymatic conversion of nitrate to nitrite followed by the formation of nitric oxide, which binds to myoglobin to produce the characteristic cured colour and contributes to microbial stability (Kim et al., 2025). Compared to conventional sodium nitrite, these systems often exhibit slower curing kinetics, greater variability, and lower curing efficiency under equivalent conditions, necessitating longer holding times or additional reducing agents to achieve comparable colour development. In this context, ultrasound-assisted processing has been explored as a tool to enhance mass transfer and improve the penetration and reactivity of nitrite and nitric oxide precursors within the muscle matrix. Experimental studies on pork loin marination have shown that ultrasound can influence physicochemical properties and microstructure when combined with naturally converted nitrite sources, such as pre-converted Angelica keiskei extracts. However, its effects on residual nitrite and curing efficiency vary depending on formulation and co-ingredients (Kim et al., 2025). These findings suggest that ultrasound may facilitate the development of cured colour in systems with reduced effective nitrite availability by improving diffusion and reaction dynamics, thereby supporting alternative curing strategies based on plant-derived nitrite precursors.

Mechanistically, synergy arises from the interaction between ultrasound-induced physical effects and the chemical characteristics of plant-derived curing systems. Ultrasound promotes rapid contact among nitrite, endogenous reductants, and myoglobin, favoring nitric oxide formation over accumulation of nitrosating species. At the same time, plant matrices often contain phenolic compounds and other antioxidants, further biasing reactive nitrogen chemistry toward non-nitrosating pathways. The combined effect is a shift in the nitrosation balance, achieved not by direct inhibition of nitrosamine formation but by precursor control and pathway redirection.

Notably, ultrasound has also been reported to induce cured-like pigment formation in meat systems even in the absence of added nitrite, potentially through activation of endogenous nitric oxide–related pathways (Leães et al., 2023). While the mechanistic basis of this phenomenon remains incompletely understood, it reinforces the notion that ultrasound can influence nitrogen chemistry, thereby reducing reliance on exogenous nitrite. From a nitrosamine mitigation perspective, this finding is conceptually significant, as it suggests alternative routes to cured color and functionality with inherently lower nitrosation potential.

Despite these promising indications, critical limitations must be acknowledged. The absence of studies directly quantifying nitrosamine formation under ultrasound-assisted, plant-based curing conditions precludes definitive conclusions regarding efficacy. Furthermore, ultrasound parameters, such as intensity, duration, and application mode, vary widely across studies, complicating cross-comparison and industrial translation. Excessive ultrasonic energy may also promote protein denaturation or lipid oxidation, potentially counteracting the benefits conferred by reduced nitrite levels if not properly controlled.

In summary, ultrasound-assisted processing combined with plant-derived curing systems constitutes a theoretically sound and technologically attractive synergistic approach, grounded in enhanced curing efficiency and reduced nitrite exposure rather than direct nitrosamine degradation. Its current status is best described as mechanistically supported but empirically incomplete. Targeted studies that explicitly measure nitrosamine formation under these combined conditions are needed to validate the proposed benefits and define operational boundaries. Nevertheless, this strategy exemplifies how process intensification and formulation design can be jointly leveraged to reshape nitrosation pathways in cured foods.

7.2.3 High-pressure processing combined with polyphenol-rich formulations

The combination of high-pressure processing (HPP) with polyphenol-rich formulations represents one of the most technologically mature and industrially feasible synergistic strategies for reducing the potential for nitrosamine formation in cured and processed foods. Unlike photonic or plasma-based approaches, this strategy integrates a well-established non-thermal preservation technology with formulation-level chemical control, enabling mitigation without introducing novel processing risks or significant sensory penalties.

From a mechanistic standpoint, synergy between HPP and polyphenol-rich ingredients arises from the complementary effects of kinetic control and chemical inhibition. As discussed in Section 5.3.1, HPP enables effective microbial inactivation at ambient or chilled temperatures, thereby reducing reliance on high nitrite concentrations and post-packaging thermal treatments, both known drivers of nitrosamine formation (Andrade et al., 2025; Honikel, 2008). Polyphenol-rich formulations, in turn, inhibit nitrosation through multiple chemical mechanisms, including scavenging of nitrosating agents, chelation of transition metals, and suppression of lipid oxidation (Jin & Jin, 2025; Tian et al., 2020).

Some studies have demonstrated that HPP can facilitate the development of reduced-nitrite or nitrite-free meat products by inactivating spoilage and pathogenic microorganisms while having minimal impact on sensory and nutritional characteristics and product safety (Hygreeva & Pandey, 2016). In parallel, the incorporation of polyphenol-rich plant extracts, such as rosemary, grape seed, or green tea, has been shown to significantly reduce nitrosation and oxidative reactions in cured meat systems, particularly during storage and thermal processing (Zhou et al., 2020).

The synergistic relevance of combining these approaches lies in their additive, yet non-redundant, effects. HPP reduces the need for nitrite by ensuring microbial safety and limiting thermal exposure, while polyphenols directly suppress nitrosation chemistry that may still occur at low nitrite concentrations. Polyphenols also help mitigate secondary effects associated with pressure-induced modifications of the food matrix. High-pressure processing can induce protein conformational changes and partially release iron from heme-containing proteins, potentially enhancing localized oxidative reactions. Polyphenolic compounds counteract these effects primarily through metal chelation and radical-scavenging mechanisms.

From an industrial perspective, the HPP–polyphenol strategy offers several practical advantages. Both components are already accepted within regulatory frameworks and consumer expectations, and their integration does not require fundamental changes in product formulation or processing flow. Moreover, because HPP operates uniformly across the entire product volume, it overcomes the spatial limitations of surface-restricted technologies such as pulsed light. This makes the approach particularly suitable for whole-muscle products and thick emulsions, where internal nitrosation processes are relevant.

Nevertheless, limitations must be acknowledged. The cost and availability of HPP infrastructure restrict its adoption to medium- and high-value product categories. Importantly, while the mechanistic rationale for nitrosamine mitigation is strong, targeted studies explicitly quantifying nitrosamine formation under combined HPP and polyphenol-rich formulations remain scarce, representing a clear research gap.

In summary, integrating high-pressure processing with polyphenol-rich formulations constitutes a robust, immediately translatable synergistic strategy for reducing the potential for nitrosamine formation. By simultaneously limiting nitrite dependence, suppressing nitrosation chemistry, and avoiding thermal drivers of nitrosamine formation, this approach exemplifies how mature processing technologies and natural inhibitors can be strategically combined to achieve meaningful risk reduction in real food systems.

7.3 Risk management perspective and trade-offs in integrated mitigation strategies

The development of synergistic strategies for nitrosamine mitigation inevitably introduces a risk management dimension that extends beyond chemical efficacy alone. As demonstrated in Sections 6.1 and 6.2, combining plant-based inhibitors with emerging processing technologies can enhance mitigation performance by targeting multiple control points within the nitrosation network. However, these benefits must be evaluated in the context of technological trade-offs, unintended chemical effects, and practical constraints that shape real-world applicability.

A central risk management challenge lies in balancing nitrosamine reduction against other food safety and quality objectives. Nitrite plays a multifunctional role in cured foods, contributing not only to nitrosamine formation but also to microbial safety, color stability, flavor development, and oxidative control. Strategies that aggressively suppress nitrosation without preserving these functions risk compromising product safety or identity. Integrated approaches mitigate this risk by redistributing control across formulation and processing steps, yet they do not eliminate the need for careful optimization.

One recurring trade-off involves oxidative stability. Several emerging technologies, including pulsed light, cold plasma, and ultrasound, can generate reactive species or induce structural changes that promote oxidation under certain conditions. While polyphenol-rich formulations and other natural antioxidants can buffer these effects, their capacity is finite and matrix-dependent. Excessive processing intensity or insufficient antioxidant protection may inadvertently amplify lipid oxidation, indirectly favoring nitrosation pathways rather than suppressing them. This highlights the importance of aligning processing parameters with formulation-level buffering capacity.

Sensory integrity represents another critical constraint. Although synergistic strategies often allow for lower concentrations of plant-based inhibitors, cumulative effects on color, flavor, and texture remain possible, particularly when multiple interventions are combined. High-pressure processing, for example, can alter texture and protein functionality, while polyphenols may introduce bitterness or color shifts. From a risk management perspective, mitigation strategies that achieve maximal chemical suppression at the expense of consumer acceptance are unlikely to succeed in commercial settings.

Regulatory considerations further shape the risk landscape. While most emerging technologies discussed in this chapter are permitted within existing regulatory frameworks, their combined use may raise interpretive challenges, particularly when plant-derived curing systems or plasma-generated nitrite sources are involved. Claims related to "natural" curing, nitrite reduction, or additive replacement must be substantiated not only chemically but also legally, and inconsistencies across regulatory jurisdictions can complicate technology transfer.

Importantly, risk management in this context is not limited to minimizing adverse outcomes; it also involves avoiding overinterpretation of limited evidence. As shown in Section 5.3, direct experimental evidence for nitrosamine reduction remains scarce for several emerging technologies. Synergistic effects demonstrated in specific matrices should not be generalized without validation across product categories and processing scenarios. Overgeneralization may lead to inappropriate application, undermining both safety and credibility.

From a systems perspective, effective risk management requires shifting from single-endpoint optimization toward multi-criteria decision-making. Nitrosamine mitigation must be evaluated alongside microbial safety, oxidative stability, sensory quality, cost, scalability, and regulatory compliance. Integrated strategies are most successful when they deliver incremental improvements across multiple dimensions rather than maximal suppression of a single hazard.

In this sense, synergistic mitigation strategies should be viewed as adaptive frameworks rather than fixed solutions. Their value lies in flexibility, the ability to tailor combinations of formulation and processing interventions to specific matrices, processing intensities, and market constraints. This adaptive mindset aligns closely with contemporary approaches to food risk management, which emphasize resilience and contextual optimization over prescriptive thresholds.

By explicitly acknowledging trade-offs and uncertainties, the integrated strategies discussed in this chapter provide a realistic pathway toward meaningful nitrosamine risk reduction. Rather than offering definitive solutions, they establish a structured decision space for evaluating future innovations. This perspective naturally leads to the forward-looking considerations addressed in the following section, where emerging research needs, technological gaps, and opportunities for translational impact are discussed.

8. Future perspectives and research needs

Despite decades of research on nitrosamines and sustained regulatory attention, the preceding sections of this chapter demonstrate that effective and transferable mitigation strategies remain an open scientific and technological challenge. Progress in this field has been uneven, with substantial mechanistic understanding at the chemical level but comparatively limited integration into processing design and industrial practice. Future advances will depend on addressing several interrelated research gaps that extend beyond incremental formulation adjustments.

A first and critical need concerns generating direct experimental evidence linking emerging processing technologies to nitrosamine modulation in real food matrices. As highlighted in Section 5, direct measurements of nitrosamine reduction are currently available for only a handful of technologies and product types. Many studies infer mitigation potential from surrogate indicators such as residual nitrite, oxidative markers, or cured color development. While informative, these proxies are insufficient to support robust risk assessment. Future research must prioritize the explicit quantification of nitrosamines using validated analytical methods across diverse matrices, processing intensities, and storage conditions.

Closely related to this is the need for mechanistic resolution beyond empirical observations. Emerging technologies such as pulsed light, cold plasma, and ultrasound create complex physicochemical environments where multiple reactive species coexist. Disentangling beneficial from adverse effects requires systematic investigation of reaction pathways, including the formation and decay of transient nitrosation intermediates. Integrating kinetic modeling, targeted analytical approaches, and controlled model systems will be essential to move from proof-of-concept toward predictive understanding.

Another priority area involves matrix specificity and transferability. Nitrosamine formation is inherently matrix-dependent, shaped by protein composition, lipid content, endogenous antioxidants, and microstructure. Strategies shown to be effective in a single product, such as sliced cured meats, cannot be assumed to perform similarly in fermented sausages, whole-muscle products, or non-meat foods. Comparative studies across matrices are therefore necessary to define boundaries of applicability and to avoid overgeneralization.

From a technological perspective, future research should emphasize integration rather than substitution. The results discussed in this chapter indicate that neither formulation-level nor process-level interventions alone are likely to deliver comprehensive mitigation. Instead, the most promising advances will arise from designed combinations that align chemical inhibition with environmental control of nitrosation. This requires interdisciplinary collaboration among food chemists, process engineers, and sensory scientists, as well as experimental designs that explicitly test interaction effects rather than isolated variables.

Industrial feasibility and scalability represent additional research imperatives. Many emerging technologies demonstrate encouraging effects under laboratory or pilot-scale conditions but face barriers related to energy efficiency, throughput, equipment cost, and process uniformity. Future work should incorporate techno-economic analysis and life-cycle considerations early in development, ensuring that proposed mitigation strategies are not only chemically effective but also practically viable.

Regulatory and risk communication aspects also warrant greater attention. As novel processing–formulation combinations blur traditional distinctions between additives and processes, regulatory frameworks may struggle to keep pace. Research that generates transparent, reproducible evidence linking mitigation strategies to reduced consumer exposure will be essential to support informed regulatory decisions and credible labeling claims.

Finally, a broader conceptual shift is needed in how nitrosamine mitigation is framed. Rather than pursuing absolute elimination, which may be neither achievable nor necessary, future strategies should focus on contextual risk reduction, balancing nitrosamine suppression with microbial safety, sensory quality, and nutritional value. This systems-based perspective aligns with contemporary food safety paradigms and offers a more realistic pathway toward safer, acceptable, and sustainable food products.

In summary, the future of nitrosamine mitigation lies not in isolated technological breakthroughs, but in integrated, mechanistically informed, and context-sensitive strategies. By addressing current gaps in evidence, mechanism, and translation, the field can move beyond precautionary constraints toward rational design of food systems that minimize nitrosamine risk without compromising product integrity.

9. Conclusions

Nitrosamine mitigation remains one of the most complex challenges in food processing, not because of insufficient chemical knowledge, but due to the multifunctional role of nitrite within food systems. As demonstrated throughout this chapter, nitrite cannot be viewed solely as a precursor of nitrosamines, but rather as a critical technological agent that contributes simultaneously to microbial safety, color stability, flavor development, and oxidative control, particularly in cured meat products. Consequently, indiscriminate nitrite reduction is neither realistic nor desirable from a food safety perspective.

Plant-based extracts provide chemically effective inhibition of nitrosation through antioxidant activity, metal chelation, and redox modulation. However, their application is constrained by sensory, technological, and regulatory boundaries that limit their standalone effectiveness. Emerging processing technologies offer additional leverage by modifying the physicochemical environment in which nitrosation occurs. Yet, their capacity to replace nitrite's protective functions remains limited and highly context-dependent.

A central contribution of this chapter is the reframing of nitrosamine control from ingredient elimination to reaction pathway management. Rather than targeting nitrite itself, effective mitigation strategies should focus on suppressing the conditions and pathways that favor nitrosamine formation while preserving nitrite's essential antimicrobial role. This distinction is critical for aligning chemical risk reduction with microbiological safety, a balance that is often overlooked in simplified mitigation narratives.

By organizing emerging technologies according to mechanistic action and evidentiary strength, this chapter highlights that no single intervention can simultaneously ensure safety, quality, and minimal nitrosamine formation. Instead, synergistic strategies combining formulation-level inhibitors with process-level controls offer the most promising pathway forward. Such approaches enable the redistribution of control across multiple intervention points, reducing the potential for nitrosamine formation without compromising nitrite's protective functions.

From a risk management perspective, the goal of nitrosamine mitigation should therefore be contextual risk reduction rather than absolute elimination. Integrated strategies enable incremental yet robust nitrosation suppression while maintaining microbial safety, sensory integrity, and industrial feasibility. This systems-based approach aligns with contemporary food safety paradigms and offers a scientifically grounded alternative to reductionist solutions.

In conclusion, sustainable control of nitrosamines in foods will depend not on minimizing nitrite at all costs, but on rational design of processing and formulation systems that decouple nitrosamine formation from nitrite's essential technological roles. By shifting the focus from additive reduction to pathway control, food science can advance toward safer products that meet regulatory, technological, and consumer expectations simultaneously.

Acknowledgments

JoséM. Lorenzo is member of the Competitive Reference Group "FunFood" (Functional Food), funded by GAIN (grant No. IN607A2023/01).

Declaration of AI and AI-assisted technologies in the writing process

During the preparation of this chapter, the authors used ChatGPT (OpenAI) as an AI-assisted language tool to support text organization, clarity, and linguistic refinement. The scientific content, interpretations, and conclusions were developed by the authors and were not generated by the AI tool. All text was critically reviewed and approved by the authors, who take full responsibility for the final manuscript.

References

Afzal, A., Kang, J., Choi, B.-M., & Lim, H.-J. (2016). Degradation and fate of N -nitrosamines in water by UV photolysis. *International Journal of Greenhouse Gas Control, 52*, 44–51. https://doi.org/10.1016/j.ijggc.2016.06.009.

Altomare, A., Baron, G., Gianazza, E., Banfi, C., Carini, M., & Aldini, G. (2021). Lipid peroxidation derived reactive carbonyl species in free and conjugated forms as an index of lipid peroxidation: Limits and perspectives. *Redox Biology, 42*, 101899. https://doi.org/10.1016/j.redox.2021.101899.

Andrade, B. F., Carmo, L. R. do, Tanaka, M. S., Torres Filho, R. de A., Ramos, A. de L. S., & Ramos, E. M. (2025). Evaluation of nonthermal technologies to reduce or replace nitrite in meat products. *Food Technology and Biotechnology, 63*(1), 94. https://doi.org/10.17113/ftb.63.01.25.8744.

Balasubramaniam, V. M. (Bala), Martínez-Monteagudo, S. I., & Gupta, R. (2015). Principles and application of high pressure–based technologies in the food industry. *Annual Review of Food Science and Technology, 6*(1), 435–462. https://doi.org/10.1146/annurev-food-022814-015539.

Bayne, A.-C. V., Misic, Z., Stemmler, R. T., Wittner, M., Frerichs, M., Bird, J. K., & Besheer, A. (2023). N-nitrosamine mitigation with nitrite scavengers in oral pharmaceutical drug products. *Journal of Pharmaceutical Sciences, 112*(7), 1794–1800. https://doi.org/10.1016/j.xphs.2023.03.022.

Beard, J. C., & Swager, T. M. (2021). An organic chemist's guide to *N*-nitrosamines: Their structure, reactivity, and role as contaminants. *The Journal of Organic Chemistry, 86*(3), 2037–2057. https://doi.org/10.1021/acs.joc.0c02774.

Bonifacie, A., Promeyrat, A., Nassy, G., Gatellier, P., Santé-Lhoutellier, V., & Théron, L. (2021). Chemical reactivity of nitrite and ascorbate in a cured and cooked meat model implication in nitrosation, nitrosylation and oxidation. *Food Chemistry, 348*, 129073. https://doi.org/10.1016/j.foodchem.2021.129073.

Broniowska, K. A., & Hogg, N. (2012). The chemical biology of S-nitrosothiols. *Antioxidants & Redox Signaling, 17*(7), 969–980. https://doi.org/10.1089/ars.2012.4590.

Castro, V., Carpena, M., Fraga-Corral, M., Lopez-Soria, A., Garcia-Perez, P., Barral-Martinez, M., ... Prieto, M. A. (2023). Sulfur-containing compounds from plants. *Natural secondary metabolites*. Springer International Publishing. 363–402. https://doi.org/10.1007/978-3-031-18587-8_11.

Chacha, J. S., Zhang, L., Ofoedu, C. E., Suleiman, R. A., Dotto, J. M., Roobab, U., ... Guiné, R. P. F. (2021). Revisiting non-thermal food processing and preservation methods—Action mechanisms, pros and cons: A technological update (2016–2021). *Foods, 10*(6), 1430. https://doi.org/10.3390/foods10061430.

Chemat, F., Zill-e-Huma, & Khan, M. K. (2011). Applications of ultrasound in food technology: Processing, preservation and extraction. *Ultrasonics Sonochemistry, 18*(4), 813–835. https://doi.org/10.1016/j.ultsonch.2010.11.023.

Chih, P.-S., Wang, W.-R., Chen, C.-Y., Yan, P.-R., Mahmudiono, T., Lee, C.-C., & Chen, H.-L. (2025). Influence of cooking and storage conditions on the formation of N-nitrosamines in processed meats and pickled fish. *LWT, 224*, 117822. https://doi.org/10.1016/j.lwt.2025.117822.

Choi, S. Y., Chung, M. J., Lee, S.-J., Shin, J. H., & Sung, N. J. (2007). N-nitrosamine inhibition by strawberry, garlic, kale, and the effects of nitrite-scavenging and N-nitrosamine formation by functional compounds in strawberry and garlic. *Food Control, 18*(5), 485–491. https://doi.org/10.1016/j.foodcont.2005.12.006.

Clayton-Cuch, D., Niklas, A. A., Duedahl-Olesen, L., Amlund, H., & Sloth, J. J. (2025). Nitrite, nitrate and N-nitrosamine occurrence in processed nitrate accumulating vegetables stored under different conditions. *Food Chemistry, 493*. https://doi.org/10.1016/j.foodchem.2025.145631.

da Rosa, J. L., Rios-Mera, J. D., Castillo, C. J. C., Lorenzo, J. M., Pinton, M. B., dos Santos, B. A., ... Campagnol, P. C. B. (2023). High-power ultrasound, micronized salt, and low KCl level: An effective strategy to reduce the NaCl content of Bologna-type sausages by 50. *Meat Science, 195*. https://doi.org/10.1016/j.meatsci.2022.109012.

da Silva, J. S., Voss, M., de Menezes, C. R., Barin, J. S., Wagner, R., Campagnol, P. C. B., & Cichoski, A. J. (2020). Is it possible to reduce the cooking time of mortadellas using ultrasound without affecting their oxidative and microbiological quality? *Meat Science, 159*, 107947. https://doi.org/10.1016/j.meatsci.2019.107947.

de La Pomélie, D., Santé-Lhoutellier, V., & Gatellier, P. (2018). Mechanisms and kinetics of heme iron nitrosylation in an in vitro gastro-intestinal model. *Food Chemistry, 239*, 86–93. https://doi.org/10.1016/j.foodchem.2017.06.092.

De Mey, E., De Klerck, K., De Maere, H., Dewulf, L., Derdelinckx, G., Peeters, M.-C., ... Paelinck, H. (2014). The occurrence of N-nitrosamines, residual nitrite and biogenic amines in commercial dry fermented sausages and evaluation of their occasional relation. *Meat Science, 96*(2), 821–828. https://doi.org/10.1016/j.meatsci.2013.09.010.

De Mey, E., De Maere, H., Paelinck, H., & Fraeye, I. (2017). Volatile *N*-nitrosamines in meat products: Potential precursors, influence of processing, and mitigation strategies. *Critical Reviews in Food Science and Nutrition, 57*(13), 2909–2923. https://doi.org/10.1080/10408398.2015.1078769.

del Pilar García Santos, M., Calle, E., & Casado, J. (2003). A method for the kinetic study of amino acid nitrosation reactions. *Polyhedron, 22*(8), 1059–1066. https://doi.org/10.1016/S0277-5387(03)00091-3.

Deng, S., Shi, S., & Xia, X. (2022). Effect of plant polyphenols on the physicochemical properties, residual nitrites, and N-nitrosamine formation in dry-fried bacon. *Meat Science, 191*, 108872. https://doi.org/10.1016/j.meatsci.2022.108872.

Domínguez, R., Pateiro, M., Gagaoua, M., Barba, F. J., Zhang, W., & Lorenzo, J. M. (2019). A comprehensive review on lipid oxidation in meat and meat products. *Antioxidants, 8*(10), 429. https://doi.org/10.3390/antiox8100429.

Drabik-Markiewicz, G., Dejaegher, B., De Mey, E., Kowalska, T., Paelinck, H., & Vander Heyden, Y. (2011). Influence of putrescine, cadaverine, spermidine or spermine on the formation of N-nitrosamine in heated cured pork meat. *Food Chemistry, 126*(4), 1539–1545. https://doi.org/10.1016/j.foodchem.2010.11.149.

Drabik-Markiewicz, G., Van den Maagdenberg, K., De Mey, E., Deprez, S., Kowalska, T., & Paelinck, H. (2009). Role of proline and hydroxyproline in N-nitrosamine formation during heating in cured meat. *Meat Science, 81*(3), 479–486. https://doi.org/10.1016/j.meatsci.2008.10.002.

Eisenbrand, G., Buettner, A., Diel, P., Epe, B., Först, P., Grune, T., ... Hengstler, J. G. (2024). Commentary of the SKLM to the EFSA opinion on risk assessment of N-nitrosamines in food. *Archives of Toxicology, 98*(6), https://doi.org/10.1007/s00204-024-03726-1.

European Commission. (2023). Commission Regulation (EU) 2023/2108 of 6 October 2023 amending Annex II to Regulation (EC) No 1333/2008 of the European Parliament and of the Council and the Annex to Commission Regulation (EU) No 231/2012 as regards food additives nitrites (E 249–250) and nitrates (E 251–252). Official Journal of the European Union, L, 2023/2108, 9 October 2023.

Fiddler, W., Pensabene, J. W., Piotrowski, E. G., Doerr, R. C., & Wasserman, A. E. (1973). Use of sodium ascorbate or erythorbate to inhibit formation of n-nitrosodimethylamine in frankfurters. *Journal of Food Science, 38*(6), 1084–1086. https://doi.org/10.1111/j.1365-2621.1973.tb02157.x.

Fracari, P. R., Tomasevic, I., Massia, A. G., Laroque, D. A., Balzan, M. M., dos Santos, B. A., ... Campagnol, P. C. B. (2025). Pulsed light and jabuticaba peel extract for nitrite reduction and quality enhancement in sliced mortadella. *Meat Science, 224*, 109777. https://doi.org/10.1016/j.meatsci.2025.109777.

Ghoshal, G. (2023). Comprehensive review on pulsed electric field in food preservation: Gaps in current studies for potential future research. *Heliyon, 9*(6), e17532. https://doi.org/10.1016/j.heliyon.2023.e17532.

Gilani, P. S., Fesahat, M., & Shariatifar, N. (2024). Nitrosamine in meat and meat products: A review. *Journal of Food Safety and Hygiene*. https://doi.org/10.18502/jfsh.v9i4.14998.

GÓmez-López, V. M., Ragaert, P., Debevere, J., & Devlieghere, F. (2007). Pulsed light for food decontamination: A review. *Trends in Food Science & Technology, 18*(9), 464–473. https://doi.org/10.1016/j.tifs.2007.03.010.

Gray, J. I., Irvine, D. M., & Kakuda, Y. (1979). Nitrates and N-nitrosamines in Cheese. *Journal of Food Protection, 42*(3), 263–272. https://doi.org/10.4315/0362-028X-42.3.263.

Guo, M., Xian, Y., Zhao, X., Qian, M., Li, J., Zeng, X., ... Dong, H. (2025). The latest advances on the formation, exposure level and control strategies of nitrosamines in meat, poultry and fish products. *Food Chemistry, 488*, 144837. https://doi.org/10.1016/j.foodchem.2025.144837.

Herrmann, S. S., Duedahl-Olesen, L., & Granby, K. (2015). Occurrence of volatile and non-volatile N-nitrosamines in processed meat products and the role of heat treatment. *Food Control, 48*, 163–169. https://doi.org/10.1016/j.foodcont.2014.05.030.

Hogg, N., & Kalyanaraman, B. (1999). Nitric oxide and lipid peroxidation. *Biochimica et Biophysica Acta (BBA) - Bioenergetics, 1411*(2–3), 378–384. https://doi.org/10.1016/S0005-2728(99)00027-4.

Homšak, M., Trampuž, M., Naveršnik, K., Kitanovski, Z., Žnidarič, M., Kiefer, M., & Časar, Z. (2022). Assessment of a diverse array of nitrite scavengers in solution and solid state: A study of inhibitory effect on the formation of alkyl-aryl and dialkyl n-nitrosamine derivatives. *Processes, 10*(11), 2428. https://doi.org/10.3390/pr10112428.

Honikel, K.-O. (2008). The use and control of nitrate and nitrite for the processing of meat products. *Meat Science, 78*(1–2), 68–76. https://doi.org/10.1016/j.meatsci.2007.05.030.

Hygreeva, D., & Pandey, M. C. (2016). Novel approaches in improving the quality and safety aspects of processed meat products through high pressure processing technology - A review. *Trends in Food Science & Technology, 54*, 175–185. https://doi.org/10.1016/j.tifs.2016.06.002.

International Agency for Research on Cancer (IARC). 2010. Ingested nitrate and nitrite, and cyanobacterial peptide toxins. Lyon (FR): International Agency for Research on Cancer. IARC Monographs on the Evaluation of Carcinogenic Risks to Humans 94.

Izham, I., Avin, F., & Raseetha, S. (2022). Systematic review: Heat treatments on phenolic content, antioxidant activity, and sensory quality of Malaysian mushroom: Oyster (Pleurotus spp.) and Black Jelly (Auricularia spp.). *Frontiers in Sustainable Food Systems, 6*. https://doi.org/10.3389/fsufs.2022.882939.

Jakobek, L. (2015). Interactions of polyphenols with carbohydrates, lipids and proteins. *Food Chemistry, 175*, 556–567. https://doi.org/10.1016/j.foodchem.2014.12.013.

Jin, D.-X., & Jin, Y.-X. (2025). Developing healthier meat products: Application of natural polyphenols to reduce hazardous compounds during high temperature processing and digestion. *Foods, 14*(22), 3952. https://doi.org/10.3390/foods14223952.

Jo, K., Lee, S., Jeong, S.-K.-C., Jeon, H., Eom, J.-U., Yang, H.-S., & Jung, S. (2025). Reduction of N-nitrosamine in cured ham using atmospheric cold plasma-treated cauliflower powder. *Meat Science, 219*, 109649. https://doi.org/10.1016/j.meatsci.2024.109649.

Kalogianni, A. I., Lazou, T., Bossis, I., & Gelasakis, A. I. (2020). Natural phenolic compounds for the control of oxidation, bacterial spoilage, and foodborne pathogens in meat. *Foods, 9*(6), 794. https://doi.org/10.3390/foods9060794.

Karwowska, M., & Kononiuk, A. (2020). Nitrates/nitrites in food—Risk for nitrosative stress and benefits. *Antioxidants, 9*(3), 241. https://doi.org/10.3390/antiox9030241.

Kim, M., Bae, S. M., Yoo, Y., Park, J., & Jeong, J. Y. (2025). Clean-label strategies for the replacement of nitrite, ascorbate, and phosphate in meat products: A review. *Foods, 14*(14), 2442. https://doi.org/10.3390/foods14142442.

Kim, Y.-J., Kim, T.-K., Cha, J. Y., Kim, J.-H., Park, M. K., Jung, S., & Choi, Y.-S. (2025). Influence of naturally converted nitrite and ultrasound marination on the curing efficiency and residual nitrite of pork loin. *Food Science of Animal Resources, 45*(4), 1014–1026. https://doi.org/10.5851/kosfa.2024.e67.

Knorr, D., Froehling, A., Jaeger, H., Reineke, K., Schlueter, O., & Schoessler, K. (2011). Emerging technologies in food processing. *Annual Review of Food Science and Technology, 2*(1), 203–235. https://doi.org/10.1146/annurev.food.102308.124129.

Lachenmeier, D. W., & FÜgel, D. (2007). Reduction of nitrosamines in beer-review of a success story. *BrewingScience, 60*(Numbers 5–6).

Laranjo, M., Elias, M., & Fraqueza, M. J. (2017). The use of starter cultures in traditional meat products. *Journal of Food Quality, 2017*, 1–18. https://doi.org/10.1155/2017/9546026.

Laroque, D. A., SeÓ, S. T., Valencia, G. A., Laurindo, J. B., & Carciofi, B. A. M. (2022). Cold plasma in food processing: Design, mechanisms, and application. *Journal of Food Engineering, 312*, 110748. https://doi.org/10.1016/j.jfoodeng.2021.110748.

Leães, Y. S. V., Lorenzo, J. M., Seibt, A. C. M. D., Pinton, M. B., Robalo, S. S., Mello, R. D. O., ... Cichoski, A. J. (2023). Do ultrasound form spontaneously nitrous pigments in nitrite-free pork meat batter? *Meat Science, 203*, 109231. https://doi.org/10.1016/j.meatsci.2023.109231.

Lee, C., Choi, W., & Yoon, J. (2005). UV photolytic mechanism of *N* -nitrosodimethylamine in water: Roles of dissolved oxygen and solution pH. *Environmental Science & Technology, 39*(24), 9702–9709. https://doi.org/10.1021/es051235j.

Lee, S., Lee, H., Kim, S., Lee, J., Ha, J., Choi, Y., ... Yoon, Y. (2018). Microbiological safety of processed meat products formulated with low nitrite concentration — A review. *Asian-Australasian Journal of Animal Sciences, 31*(8), 1073–1077. https://doi.org/10.5713/ajas.17.0675.

Lee, S. Y., Lee, D. Y., Kim, O. Y., Kang, H. J., Kim, H. S., & Hur, S. J. (2020). Overview of studies on the use of natural antioxidative materials in meat products. *Food Science of Animal Resources, 40*(6), 863–880. https://doi.org/10.5851/kosfa.2020.e84.

Li, L., Shao, J., Zhu, X., Zhou, G., & Xu, X. (2013). Effect of plant polyphenols and ascorbic acid on lipid oxidation, residual nitrite and N-nitrosamines formation in dry-cured sausage. *International Journal of Food Science & Technology, 48*(6), 1157–1164. https://doi.org/10.1111/ijfs.12069.

Li, Y., & Hecht, S. S. (2022). Metabolic activation and DNA interactions of carcinogenic N-nitrosamines to which humans are commonly exposed. *International Journal of Molecular Sciences, 23*(9), 4559. https://doi.org/10.3390/ijms23094559.

Lu, J., Li, M., Huang, Y., Xie, J., Shen, M., & Xie, M. (2022). A comprehensive review of advanced glycosylation end products and N- nitrosamines in thermally processed meat products. *Food Control, 131*, 108449. https://doi.org/10.1016/j.foodcont.2021.108449.

Mirvish, S. S. (1995). Role of N-nitroso compounds (NOC) and N-nitrosation in etiology of gastric, esophageal, nasopharyngeal and bladder cancer and contribution to cancer of known exposures to NOC. *Cancer Letters, 93*(1), 17–48. https://doi.org/10.1016/0304-3835(95)03786-V.

Misra, N. N., Tiwari, B. K., Raghavarao, K. S. M. S., & Cullen, P. J. (2011). Nonthermal plasma inactivation of food-borne pathogens. *Food Engineering Reviews, 3*(3–4), 159–170. https://doi.org/10.1007/s12393-011-9041-9.

Molina, J. R. G., Melios, S., Crofton, E., Frias Celayeta, J. M., Bolton, D. J., & Botinestean, C. (2025). Effects of nitrite level, packaging system and high-pressure processing treatment on physicochemical and sensory properties of cooked ham. *LWT, 231*, 118345. https://doi.org/10.1016/j.lwt.2025.118345.

Nabizadeh, S., Aeini, K., Barzegar, F., Arabameri, M., Hosseini, H., Kamankesh, M., & Mohammadi, A. (2024). Volatile N-nitrosamines in processed meat products: An approach for monitoring dietary exposure, assessing human risk, and evaluating variable correlations by principal component analysis and heat map. *Food and Chemical Toxicology, 188*, 114649. https://doi.org/10.1016/j.fct.2024.114649.

Niklas, A. A., Borge, G. I. A., RØdbotten, R., Berget, I., Müller, M. H. B., Herrmann, S. S., ... Kirkhus, B. (2023). Levels of nitrate, nitrite and nitrosamines in model sausages during heat treatment and in vitro digestion – The impact of adding nitrite and spinach (Spinacia oleracea L.). *Food Research International, 166*, 112595. https://doi.org/10.1016/j.foodres.2023.112595.

Nowosad, K., Sujka, M., Pankiewicz, U., & Kowalski, R. (2021). The application of PEF technology in food processing and human nutrition. *Journal of Food Science and Technology, 58*(2), 397–411. https://doi.org/10.1007/s13197-020-04512-4.

Ohshima, H. , & Bartsch, H. (1984). Monitoring endogenous nitrosamine formation in man. IARC Scientific Publications, (59).

PaiÉ-Ribeiro, J., Baptista, F., Gomes, M. J., Teixeira, A., Pinheiro, V., Outor-Monteiro, D., & Barros, A. N. (2024). Exploring the variability in phenolic compounds and antioxidant capacity in olive oil by-products: A path to sustainable valorization. *Antioxidants, 13*(12), 1470. https://doi.org/10.3390/antiox13121470.

PÉrez-GÁlvez, A., Viera, I., & Roca, M. (2020). Carotenoids and chlorophylls as antioxidants. *Antioxidants, 9*(6), 505. https://doi.org/10.3390/antiox9060505.

Pessego, M., Rosa da Costa, A., & Moreira, J. (2011). Importance of phenols structure on their activity as antinitrosating agents: A kinetic study. *Journal of Pharmacy And Bioallied Sciences, 3*(1), 128. https://doi.org/10.4103/0975-7406.76491.

Pinton, M. B., Correa, L. P., Facchi, M. M. X., Heck, R. T., Leães, Y. S. V., Cichoski, A. J., ... Campagnol, P. C. B. (2019). Ultrasound: A new approach to reduce phosphate content of meat emulsions. *Meat science, 152*, 88–95.

Rot, T., Kovačević, D., Habschied, K., & Mastanjević, K. (2025). N-Nitrosamines in meat products: Formation, detection and regulatory challenges. *Processes, 13*(5), 1555. https://doi.org/10.3390/pr13051555.

Sallan, S., Kaban, G., Şişik Oğraş, Ş., Çelik, M., & Kaya, M. (2020). Nitrosamine formation in a semi-dry fermented sausage: Effects of nitrite, ascorbate and starter culture and role of cooking. *Meat Science, 159*, 107917. https://doi.org/10.1016/j.meatsci.2019.107917.

Sallan, S., Yılmaz Oral, Z. F., & Kaya, M. (2023). A review on the role of lactic acid bacteria in the formation and reduction of volatile nitrosamines in fermented sausages. *Foods, 12*(4), 702. https://doi.org/10.3390/foods12040702.

Scanlan, R. A. (1983). Formation and occurrence of nitrosamines in food. *Cancer Research, 43*(5), https://doi.org/10.1201/9781351071383-9.

Schrenk, D., Bignami, M., Bodin, L., Chipman, J. K., del Mazo, J., Hogstrand, C., ... Grasl-Kraupp, B. (2023). Risk assessment of N-nitrosamines in food. *EFSA Journal, 21*(3), https://doi.org/10.2903/j.efsa.2023.7884.

Sen, N. P., Donaldson, Barbara, Charbonneau, Claudette, & Miles, W. F. (1974). Effect of additives on the formation of nitrosamines in meat curing mixtures containing spices and nitrite. *Journal of Agricultural and Food Chemistry, 22*(6), 1125–1130. https://doi.org/10.1021/jf60196a054.

Sen, N. P., Tessier, L., Seaman, S. W., & Baddoo, P. A. (1985). Volatile and nonvolatile nitrosamines in fish and the effect of deliberate nitrosation under simulated gastric conditions. *Journal of Agricultural and Food Chemistry, 33*(2), 264–268. https://doi.org/10.1021/jf00062a027.

Shakil, M. H., Trisha, A. T., Rahman, M., Talukdar, S., Kobun, R., Huda, N., & Zzaman, W. (2022). Nitrites in cured meats, health risk issues, alternatives to nitrites: A review. *Foods, 11*(21), 3355. https://doi.org/10.3390/foods11213355.

Shao, X., Zhu, M., Zhang, Z., Huang, P., Xu, B., Chen, C., & Li, P. (2021). N-nitrosodimethylamine reduction by Lactobacillus pentosus R3 in fermented cooked sausages. *Food Control, 124*, 107869. https://doi.org/10.1016/j.foodcont.2021.107869.

Sokal-Dembowska, A., Jarmakiewicz-Czaja, S., Tabarkiewicz, J., & Filip, R. (2025). Nitrate and nitrite in the diet: Protective and harmful effects in health and disease. *Current Nutrition Reports, 14*(1), 89. https://doi.org/10.1007/s13668-025-00678-5.

Spirić, D., Stefanović, S., Silađi, Č., Petronijević, R., Trbović, D., Borjan, N., & Videnović, D. (2025). Occurrence of nitrates in milk products — Review. *Meat Technology, 66*(3), 332–338. https://doi.org/10.18485/meattech.2025.66.3.55.

Steinberg, P. (2019). Red meat-derived nitroso compounds, lipid peroxidation products and colorectal cancer. *Foods, 8*(7), 252. https://doi.org/10.3390/foods8070252.

Sun, F., Kong, B., Chen, Q., Han, Q., & Diao, X. (2017). N -nitrosoamine inhibition and quality preservation of Harbin dry sausages by inoculated with Lactobacillus pentosus, Lactobacillus curvatus and Lactobacillus sake. *Food Control, 73*, 1514–1521. https://doi.org/10.1016/j.foodcont.2016.11.018.

Taheri, A., Ganjeali, A., Arefi-Oskouie, A., Çirak, C., & Cheniany, M. (2023). The variability of phenolic constituents and antioxidant properties among wild populations of Ziziphora clinopodioides Lam. *Physiology and Molecular Biology of Plants*. https://doi.org/10.1007/s12298-023-01283-y.

Tahmouzi, S., Alizadeh Salmani, B., Eskandari, S., & Arab, M. (2025). Effects of plant substitutes for nitrite on the technological characteristics of fermented sausages: A comprehensive review. *Food Science & Nutrition, 13*(4), https://doi.org/10.1002/fsn3.70186.

Tannenbaum, S., Wishnok, J., & Leaf, C. (1991). Inhibition of nitrosamine formation by ascorbic acid. *The American Journal of Clinical Nutrition, 53*(1), 247S–250S. https://doi.org/10.1093/ajcn/53.1.247S.

Tian, Y., Du, H., Wang, L., Li, S., Zhang, L., Zhang, & Li, W. (2020). Nitrite scavenging and inhibition of *N*-nitrosamines formation by phenolic extracts from diosp*yros lotus* L. leaves and active ingredients. *Natural Product Communications, 15*(9), https://doi.org/10.1177/1934578×20961186.

Toepfl, S., Heinz, V., & Knorr, D. (2006). Applications of pulsed electric fields technology for the food industry. In J. Raso, & V. Heinz (Eds.). *Pulsed electric fields technology for the food industry. Food engineering series* (pp. 197–221). Boston, MA: Springer. https://doi.org/10.1007/978-0-387-31122-7_7.

Wainright, T. (1986). The chemistry of nitrosamine formation: Relevance to malting and brewing. *Journal of the Institute of Brewing, 92*(1), https://doi.org/10.1002/j.2050-0416.1986.tb04373.x.

Wang, Y., Li, F., Zhuang, H., Chen, X., Li, L., Qiao, W., & Zhang, J. (2015). Effects of plant polyphenols and α-tocopherol on lipid oxidation, residual nitrites, biogenic amines, and N-nitrosamines formation during ripening and storage of dry-cured bacon. *LWT - Food Science and Technology, 60*(1), 199–206. https://doi.org/10.1016/j.lwt.2014.09.022.

Wu, H., Bak, K. H., Goran, G. V., & Tatiyaborworntham, N. (2024). Inhibitory mechanisms of polyphenols on heme protein-mediated lipid oxidation in muscle food: New insights and advances. *Critical Reviews in Food Science and Nutrition, 64*(15), 4921–4939. https://doi.org/10.1080/10408398.2022.2146654.

Xie, Y., Geng, Y., Yao, J., Ji, J., Chen, F., Xiao, J., Hu, X., & Ma, L. (2023). N-nitrosamines in processed meats: Exposure, formation and mitigation strategies. *Journal of Agriculture and Food Research, 13*, 100645. https://doi.org/10.1016/j.jafr.2023.100645.

Yin, X., Chen, K., Cheng, H., Chen, X., Feng, S., Song, Y., & Liang, L. (2022). Chemical stability of ascorbic acid integrated into commercial products: A review on bioactivity and delivery technology. *Antioxidants, 11*(1), 153. https://doi.org/10.3390/antiox11010153.

Yurchenko, S., & MÖlder, U. (2006). Volatile N-nitrosamines in various fish products. *Food Chemistry, 96*(2), 325–333. https://doi.org/10.1016/j.foodchem.2005.04.009.

Yurchenko, S., & MÖlder, U. (2007). The occurrence of volatile N-nitrosamines in Estonian meat products. *Food Chemistry, 100*(4), 1713–1721. https://doi.org/10.1016/j.foodchem.2005.10.017.

Zapata, J. E., Sepúlveda, C. T., & Álvarez, A. C. (2022). Kinetics of the thermal degradation of phenolic compounds from achiote leaves (Bixa orellana L.) and its effect on the antioxidant activity. *Food Science and Technology, 42*. https://doi.org/10.1590/fst.30920.

Zhang, H., Yu, D., Sun, J., Liu, X., Jiang, L., Guo, H., & Ren, F. (2014). Interaction of plant phenols with food macronutrients: Characterisation and nutritional–physiological consequences. *Nutrition Research Reviews, 27*(1), 1–15. https://doi.org/10.1017/S095442241300019X.

Zhang, L., Qu, H., Xie, M., Shi, T., Shi, P., & Yu, M. (2023). Effects of different cooking methods on phenol content and antioxidant activity in sprouted peanut. *Molecules (Basel, Switzerland), 28*(12), 4684. https://doi.org/10.3390/molecules28124684.

Zhang, Q. Q., Jiang, M., Rui, X., Li, W., Chen, X. H., & Dong, M. S. (2017). Effect of rose polyphenols on oxidation, biogenic amines and microbial diversity in naturally dry fermented sausages. *Food Control, 78*, 324–330. https://doi.org/10.1016/j.foodcont.2017.02.054.

Zhou, F., Xie, Y., Wang, Y., Zhang, H., Wang, J., Liao, X., & Chen, C. (2025). An overview of the formation mechanisms of endogenous and exogenous N-nitrosamines in human diets. *Journal of Environmental Sciences, 158*, 527–541. https://doi.org/10.1016/j.jes.2025.02.016.

Zhou, Y., Wang, Q., & Wang, S. (2020). Effects of rosemary extract, grape seed extract and green tea polyphenol on the formation of *N* -nitrosamines and quality of western-style smoked sausage. *Journal of Food Processing and Preservation, 44*(6), https://doi.org/10.1111/jfpp.14459.

Zou, X. N., Lu, S. H., & Liu, B. (1994). Volatile N-nitrosamines and their precursors in chinese salted fish—a possible etological factor for NPC in China. *International Journal of Cancer, 59*(2), 155–158. https://doi.org/10.1002/ijc.2910590202.

CHAPTER TWO

Processing innovations for tailoring eating quality and nutrition of meat-based foods

Renyu Zhang[a,*], Carolina E. Realini[a], Wangang Zhang[b], and Mustafa M. Farouk[a]

[a]Food Technology & Processing Team, Bioeconomy Science Institute, Palmerston North, New Zealand
[b]Jiangsu Synergetic Innovation Center of Meat Processing and Quality Control, College of Food Science and Technology, Nanjing Agricultural University, Nanjing, P.R. China
*Corresponding author. e-mail address: renyu.zhang@agresearch.co.nz

Contents

Abstract

Meat-based foods continue to play an important role in human diets, yet their future relevance will depend increasingly on how effectively eating quality, nutritional value, resource efficiency, and sustainability can be improved through processing innovation. This chapter examines how processing strategies can be deliberately designed to enhance the sensory and nutritional performance of meat-based foods across whole-muscle, minced, restructured, and hybrid systems. It first considers advances in post-mortem ageing, showing how conventional wet- and dry-ageing can be further refined through stepwise and tailored ageing approaches to better control tenderness, flavour development, yield, and consistency across muscles, species, and consumer segments. The chapter then explores the integration of freezing/thawing with ageing as a

Advances in Food and Nutrition Research, Volume 122
ISSN 1043-4526, https://doi.org/10.1016/bs.afnr.2026.07.001

practical route to improving the quality of frozen meat by managing structural and biochemical changes more effectively. In minced and restructured systems, attention is given to meat-to-meat fortification, nutrient enrichment with meat co-products, and hybrid meat production as strategies to increase nutritional density, valorise under-utilised resources, and broaden product offering. The chapter further discusses hybrid meat as an emerging processing concept and highlights the HYBRID meat framework as a systems-based approach for guiding ingredient selection, matrix compatibility, process design, and consumer-oriented product development. Innovative processing technologies, culinary developments and enabling tools are also considered as means to translate processing gains into consumer value. Collectively, these perspectives show how a more integrated, processing-led strategy can help deliver meat-based foods that are sensory optimised, nutritionally robust, economically viable, and responsive to evolving consumer preferences and sustainability expectations.

1. Introduction

Meat-based foods remain central to global dietary patterns due to their high-quality protein, bioavailable micronutrients, and distinctive sensory attributes. Recent national nutrition guidance, such as the Dietary Guidelines for Americans, continues to prioritise high-quality, nutrient-dense protein foods, including lean meats, poultry, eggs, seafood, and plant proteins as foundational elements of a healthy dietary pattern while recommending balanced consumption across food groups (U.S. Department of Agriculture and U.S. Department of Health and Human Services, 2026). This reflects a broader recognition in nutrition science that protein foods, with meat as a key contributor, occupy a core position among recommended food categories in many dietary guidelines. Beyond their nutritional contribution, meat products carry cultural, culinary, and economic significance across diverse food systems (Realini et al., 2022). However, contemporary meat production and consumption face increasing pressure to reconcile eating quality with nutritional optimisation, resource efficiency, and environmental sustainability. These pressures are driven by population growth, changing consumer expectations, and heightened scrutiny of the environmental footprint associated with animal-derived foods (Hutchings, Chheang, Hedderley, Realini, & Jaeger, 2024; Xu, Sharma, et al., 2021). Within this context, processing innovations through optimising post-mortem ageing, formulation and hybridisation of minced meat products have emerged as innovative pathways for tailoring the quality and nutritional performance of meat-based foods while making more effective use of animal resources (Zhang, Deb-Choudhury, et al., 2026; Zhang, Jacob, Agnew, Farouk, & Realini, 2025; Zhang, Jacob, Thum, Realini, & Farouk, 2026; Zhang, Ross, et al., 2023).

Traditionally, innovation in the meat sector has focused on improving processing efficiency, shelf-life, and safety. While these objectives remain essential, there has been a marked shift towards value-oriented innovation that prioritises sensory quality, nutritional functionality, and sustainability outcomes (Samad et al., 2026; Zhang, Farouk, Realini, & Thum, 2025; Zhang, Kallas, et al., 2024). Eating quality of meat, encompassing main attributes of tenderness, juiciness, flavour, and overall palatability, continues to be a primary driver of consumer acceptance (Realini et al., 2023). At the same time, nutritional considerations have expanded beyond macronutrient composition to include fatty acid profiles, micronutrient density, bioactive compounds, and the digestibility and bioaccessibility of nutrients, with increasing emphasis on the development of meat products tailored to specific consumer groups such as elderly and active adults (Egan, 2026; Farouk, Zhang, Jenkinson, & Realini, 2025; Jairath, Biswas, Mal, Suman, & Suman, 2024). These attributes are fundamentally shaped by biological and production-related factors (e.g., breed, diet, and muscle cuts), yet their functional expression in meat-based foods is substantially influenced by post-mortem processing and formulation strategies.

A defining feature of meat innovation is the recognition that different product formats offer distinct opportunities for quality and nutritional tailoring. In whole-muscle products such as steaks and roasts, the intact muscle structure constrains formulation options, placing greater emphasis on post-mortem biochemical processes as the primary drivers of eating quality improvement. Post-mortem ageing remains one of the most effective and widely applied natural processing techniques to enhance tenderness and flavour development through controlled proteolysis and metabolic transformations (Kim et al., 2018). Advances in ageing regimes, including variations in time, temperature, packaging, and integration of different ageing methods, have enabled more precise modulation of sensory outcomes while minimising quality losses and waste (Álvarez, Álvarez, Hamill, Mullen, & O'Neill, 2021; Li et al., 2022; Zhang, Yoo, Ross, & Farouk, 2022). These developments also enable the integration of emerging analytical tools to better understand muscle-specific responses to ageing and their implications for eating quality.

In contrast, minced meat products offer a fundamentally different scope for innovation. By disrupting muscle structure, mincing enables greater flexibility in formulation, portioning, and utilisation of diverse raw materials in reformed meat matrices. This product category plays a pivotal role in improving resource efficiency by valorising secondary cuts, trimmings, and co-products that may otherwise be underutilised. Moreover, minced meat matrices provide an effective vehicle for compositional optimisation through strategies such as meat-to-meat fortification (Farouk et al., 2025) and hybridisation (Zhang, Farouk, et al., 2025). Strategic blending of different muscles, species, or tissue types can be used to balance sensory quality, nutritional value, and cost, while also reducing reliance on premium cuts (Farouk et al., 2025). Moreover, hybrid meat systems, incorporating plant-based or other alternative protein ingredients, further extend this potential by allowing targeted reductions in environmental impact of meat consumption and improving resource-efficiency, while maintaining consumer acceptance through careful design of product format, structure, and sensory attributes (Zhang, Farouk, et al., 2025; Zhang, Jacob, et al., 2026).

This chapter integrates advances in post-mortem ageing, its integration with freezing and thawing, minced and restructured product design, and hybridisation within a unified, systems-based perspective. It examines how ageing strategies can be refined to deliver targeted sensory outcomes in whole-muscle products, how their integration with freezing/thawing processes can improve the quality of frozen meat, and how minced meat matrices can be leveraged for nutritional fortification, co-product valorisation, and the development of hybrid protein systems. Collectively, these approaches highlight practical processing strategies for tailoring meat-based foods while improving consistency, resource utilisation, and value recovery across the meat value chain.

2. Enhancing quality of meat products through post-mortem ageing

Post-mortem ageing is a well-recognised quality optimisation technique for whole-muscle meat and remains one of the most effective processing strategies for improving eating quality of meat products. Ageing leverages endogenous biochemical and structural changes that occur following slaughter, particularly the gradual weakening of the myofibrillar structure and the development of flavour-active compounds. Among conventional approaches, wet-ageing and dry-ageing are the most widely practiced, each offering distinct advantages and constraints in terms of quality outcomes, process control, and commercial feasibility.

The most widely recognised mechanism underpinning ageing-induced tenderisation is the action of endogenous proteolytic systems, particularly the calpain family. μ-Calpain is activated early post-mortem as calcium homeostasis is disrupted, while m-calpain contributes more gradually depending on ionic strength, temperature, and muscle characteristics (Kim et al., 2018). These enzymes selectively degrade key myofibrillar and cytoskeletal proteins, including desmin, titin, nebulin, and troponin-T, thereby weakening the structural integrity of the muscle fibre and reducing resistance to bite (Bischof et al., 2024). Alongside proteolysis, post-mortem ageing is accompanied by ongoing metabolic transformation. Residual glycolysis, nucleotide degradation, and amino acid turnover continues to alter eating quality through production of free amino acids, short peptides, and other low-molecular weight compounds that contribute directly to flavour perception, or act as precursors for Maillard reactions and lipid-derived aroma formation during cooking (Kim et al., 2018; Zhang et al., 2022). As a result, ageing influences not only texture but also the intensity, complexity, and balance of cooked meat flavour.

The extent to which these textural and flavour-related changes can develop is governed by the ageing potential of the meat, defined as the intrinsic capacity of a given muscle to respond positively to ageing. This potential is largely determined by biological factors such as muscle fibre type composition, connective tissue content, intramuscular fat level, and activity of endogenous enzymes (Kim et al., 2018). For example, meat cuts with different muscle fibre types exhibited distinct profiles of degraded proteins during ageing, indicating that ageing potential varies among cuts and requires tailored ageing to achieve desired outcomes (Kim, Lee, Jung, Song, & Jin Hur, 2021; Li, Leeming, Hastie, & Warner, 2026; Stafford, Alruzzi, Gagaoua, & Matarneh, 2025). Generally, muscles dominated by fast glycolytic fibres often exhibit more rapid proteolytic tenderisation but may be more susceptible to moisture loss and structural weakening, whereas oxidative muscles typically age more slowly yet develop more stable texture and flavour profiles over extended periods. Variations in connective tissue content, distribution and degree of collagen cross-linking further influence resistance to bite and the extent to which proteolysis contributes to tenderness (Purslow, 2018).

Ageing potential is further shaped by pre-slaughter conditions that influence the biochemical status of muscle at rigor onset and subsequent ageing process. Factors such as animal handling and stress exposure, nutritional status, and lairage management affect muscle energy reserves, pH decline kinetics, rigor temperature, electrical stimulation, immobilisation, and enzyme activity, thereby defining the biochemical environment in which post-mortem muscle transformations unfold (Álvarez, Koolman, Whelan, & Moloney, 2022; Terlouw et al., 2021). Together with inherent muscle properties, these factors influence the biochemical conditions that modulate tenderness development and the generation of flavour precursors during ageing.

While ageing potential establishes the biological boundaries of quality improvement, its realisation is strongly influenced by ageing regimes and processing conditions. Factors such as time, temperature, oxygen exposure, packaging environment, and water activity regulate the rate, direction, and stability of proteolytic, metabolic, and oxidative processes during ageing (Kim et al., 2018; Zhang et al., 2022). By modulating these conditions, it is possible to accelerate desirable transformations, mitigate undesirable reactions, and guide quality development towards defined product objectives. This interaction between intrinsic muscle characteristics and extrinsic process control positions ageing as a biologically driven yet deliberately adjustable process, providing the scientific basis for tailoring ageing regimes to muscle type, product format, and intended consumer use. The following sections provide an overview of the major ageing regimes applied to whole-muscle meat, including wet-ageing, dry-ageing, stepwise ageing, and introduce the concept of tailored ageing.

2.1 Wet-ageing

Wet-ageing is achieved by vacuum-packing primal or subprimal cuts in water- and oxygen-impermeable packaging and storing them under refrigerated conditions, typically between −1.5 and 4 °C, for periods ranging from several days to several weeks (e.g., 2–4 weeks). It remains the most widely adopted post-mortem ageing method in the meat industry, providing consistent tenderisation and extended shelf-life in fresh meat products (Zhang, Realini, Kim, & Farouk, 2023). The anaerobic environment inside the package limits evaporative moisture loss and restricts the growth of spoilage microorganisms, while allowing endogenous biochemical processes (e.g., proteolysis and lipolysis) to proceed in a controlled and more predictable manner. As a result, wet-ageing is highly compatible with contemporary cold-chain logistics and large-scale industrial operations.

Wet-ageing is typically associated with consistent improvements in meat tenderisation, moderate metabolic progression, and limited oxidative stress (Ramanathan et al., 2020). Water activity remains largely unchanged, and lipid oxidation is generally restrained, with progressive accumulation of low-molecular-weight flavour precursors contributing to more reliable eating quality outcomes across a wide range of product formats (Aaslyng & Meinert, 2017; Zhang, Yoo, Realini, Staincliffe, & Farouk, 2021). From a nutritional perspective, studies have suggested improved relative protein digestibility following an ageing process (Farouk, Wu, Frost, Staincliffe, & Knowles, 2019; Lee et al., 2020). However, prolonged ageing (e.g., >4 weeks) or compromised packaging (e.g., broken vacuum seal) can increase susceptibility to oxidation, particularly during subsequent frozen storage and post-thawing manipulation (Zhang, Realini, et al., 2023).

The main advantages of wet-ageing include high processing yield, consistent tenderisation and product quality, along with scalability and low operational complexity as it does not require dedicated ageing facilities. However, flavour development is limited during wet-ageing, and extended ageing periods are associated with increased purge loss and surface discolouration (Aaslyng & Meinert, 2017; Li et al., 2022). These changes negatively affect product appearance and may increase the potential risk of microbial proliferation.

2.2 Dry-ageing and in-bag dry-ageing

Dry-ageing is another post-mortem ageing method that has traditionally been used for meat preservation. It involves storing unpackaged meat under carefully controlled conditions of temperature, relative humidity, and air flow, allowing progressive moisture evaporation from the muscle surface. The core post-mortem biochemical processes, particularly calpain-mediated proteolysis, are broadly comparable to those occurring during wet-ageing, resulting in similar tenderisation effects (Zhang et al., 2022). Dry-ageing of meat is recognised for its contribution to the development of distinctive flavour profiles, becoming increasing popular among consumers seeking new and differentiated eating experiences. Over the last decade, a new variant of dry-ageing, called in-bag dry-ageing, has emerged. In-bag dry-ageing is a controlled adaptation of the traditional dry-ageing process by packaging meat cuts using moisture-permeable membranes (e.g., Tublin 10®) that allow dehydration while restricting microbial contamination and excessive oxidation. Compared with traditional dry-ageing, in-bag dry-ageing offers several advantages including greater control over microbial activity, reduced trimming losses, elimination of the need for dedicated dry-ageing facilities, and improved consistency in product quality, thereby supporting broader commercial adoption (Kim et al., 2018; Zhang et al., 2022).

The mechanisms underlying the development of distinctive dry-aged meat flavour have been recently elucidated by Zhang et al. (2022), involving the interplay among dehydration, oxidation, and microbial activity. Among the physical and biochemical processes occurring during dry-ageing, dehydration is the primary driver, concentrating meat constituents and fundamentally altering the biochemical environment governing post-mortem transformations. This concentration effect increases the availability of low molecular weight compounds, including peptides, free amino acids, nucleotides, and reducing sugars (Lee, Choe, et al., 2019; Zhang, Ross, et al., 2023). Some of these compounds (e.g., inosine 5'-monophophate, and glutamic acid) can contribute directly to the intensified umami/savoury taste of dry-aged meat products, while others can serve as critical precursors for Maillard reactions and Strecker degradation during cooking, amplifying flavour intensity without requiring fundamentally different metabolic pathways (Zhang et al., 2022).

Oxidative reactions (e.g., lipid oxidation) further differentiate the eating quality and flavour of dry-aged meat from its wet-aged counterpart. Thus, greater oxygen exposure during dry-ageing promotes moderate lipid oxidation, leading to the formation of aldehydes, ketones, and alcohols that contribute to the development of characteristic nutty, buttery, and roasted notes of dry-aged meat (Zhang, Maes, et al., 2024; Zhang, Ross, et al., 2023). When tightly controlled, this level of oxidation enhances flavour complexity and depth. However, oxidation remains a secondary driver relative to dehydration and must be carefully managed, as excessive oxidative reactions (e.g., during extended ageing periods) compromise lipid stability, nutritional quality, colour, and shelf-life (Domínguez et al., 2019).

Microbial activity also contributes to the differentiation of dry-ageing by influencing surface biochemistry and flavour development. Unlike wet-ageing, which intentionally mitigates microbial growth through processing under anaerobic vacuum conditions, dry-ageing supports the development of a selective surface microflora, typically dominated by yeasts and moulds such as *Thamnidium*, *Debaryomyces*, and *Penicillium* species, alongside limited bacterial populations. These microorganisms primarily proliferate on the meat surface and therefore do not contribute directly to the tenderisation of the underlying meat structure; however, they influence flavour differentiation through the secretion of extracellular proteases and lipases that act mainly at the surface (Gowda, De Zutter, Van Royen, & Van Damme, 2022; Xu, Liu, Cheng, & Qian, 2021). The resulting amino acids, peptides, and free fatty acids enrich the flavour precursor pool and generate volatile compounds that influence cooked aroma (Lee et al., 2022; Setyabrata et al., 2022). Therefore, microbial activity contributes to flavour differentiation, reinforcing the characteristic sensory profile of dry-aged meat. An increasing number of studies conducted over the last five years have focused on manipulating the microbial communities of dry-aged meat products by inoculation of selected strains from yeasts and moulds, such as *Mucor flavus* (Jaworska et al., 2025; Przybylski et al., 2024), *Yarrowia hollandica* and *Penicillium oxalicum* (Liu et al., 2024), and *Penicillium candidum* and *Penicillium nalgiovense* (Lee et al., 2022). Inoculation with these strains resulted in more significant improvements, such as increased levels of flavour precursors and enhanced proteolysis, in model systems using small meat pieces and untrimmed samples. However, analyses of trimmed samples showed only limited effects, including modest increases in proteolytic activity mainly involving sarcoplasmic proteins, with no clear impact on myofibrillar proteins. These observations support the findings of Zhang et al. (2022) that the core "ageing effects" driven by endogenous enzymatic activity are similar between wet- and dry-ageing processes. Following trimming, the differences between ageing methods are significantly reduced, reflecting the surface localisation of many dehydration- and microbially driven changes. As a result, the advantages of dry-ageing for improving eating quality need to be considered alongside its associated yield losses and processing costs. However, trimmings or crusts generated during dry-ageing have been shown to be valuable ingredients due to their concentrated flavour and bioactive (e.g., antioxidant) compounds (Kim et al., 2020; Setyabrata, Hernandez, Cooper, Legako, & Kim, 2024), and enhanced functional properties (e.g., emulsion activity index) (Xue et al., 2021).

Such paradoxical outcomes indicate that current dry-ageing practices require innovation to develop viable strategies that maximise quality benefits while reducing production costs and improving affordability of dry-aged meat products. To achieve these goals, an integrated understanding of the mechanisms underlying the dry-ageing process is needed. When dry-ageing is decoupled (Fig. 1), it becomes apparent that endogenous proteolysis proceeds similarly across ageing methods, while the distinctive attributes of dry-ageing arise primarily from dehydration-driven changes and associated surface phenomena. Consequently, the primary differences between the two ageing methods are attributable to "dry" versus "wet" effects rather than ageing per se. Dehydration is the key driver underpinning the unique quality profile of dry-aged meat products, leading not only to compositional changes but also to modifications in biochemical processes and microbial ecology (Zhang et al., 2022).

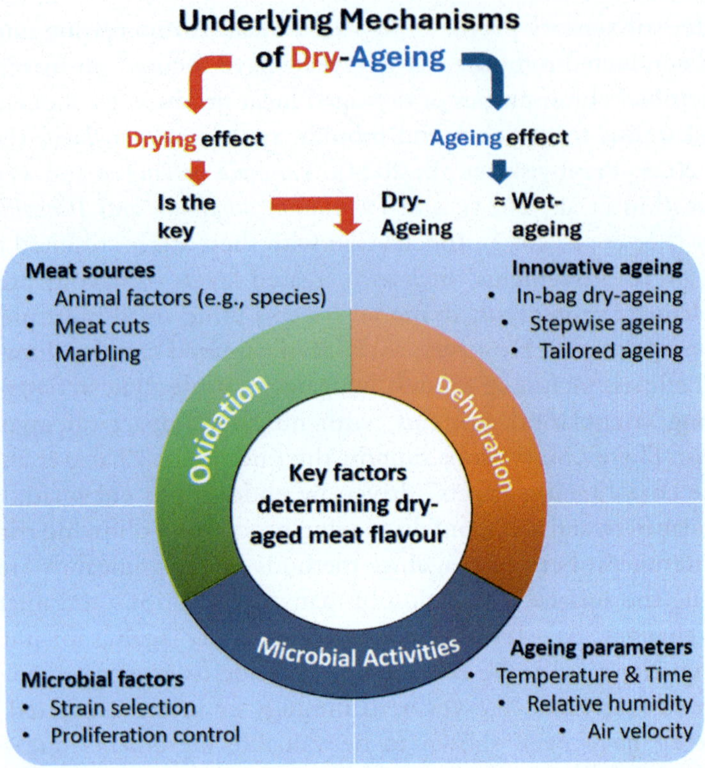

Fig. 1 Underlying mechanisms of dry-ageing and key factors influencing the development of dry-aged meat flavour.

There is a broad consensus that dry-ageing requires an extended duration, typically 6–8 weeks or longer, to achieve true dry-aged status. This extended timeframe is necessary because "dry" effects develop slowly at standard ageing temperatures (~2–4 °C). During the early stages of dry-ageing (up to 2–3 weeks), these effects occur primarily at the meat-air interface, resulting in crust formation without significant modification to the internal portion of the segment.

As dry-ageing progresses beyond approximately 4–8 weeks, moisture equilibrium occurs between the internal regions and the surface of meat, leading to progressive dehydration of the internal tissue, and allows the "dry" impacts to become sufficiently pronounced for consumer perception. This explains why many studies reporting short dry-ageing periods (3–4 weeks) fail to find significant differences between wet- and dry-aged products. Once the crust is removed, no significant concentration effect remains within the internal portion of the meat. Although oxidation levels may be higher than those of wet-aged equivalents due to oxygen exposure, the dominant biochemical process remains comparable to standard ageing.

Additionally, microbial proliferation on the crust proceeds relatively slowly during dry-ageing, particularly as water activity is progressively reduced. Under these conditions, the growth of yeasts and moulds is favoured (Lee, Yoon, et al., 2019; Mikami et al., 2021), as most spoilage bacteria and pathogens are inhibited at low water activities (e.g., <0.85–0.9). Most studies report increased yeast and mould development after approximately 3–4 weeks of dry-ageing, which coincides with enhanced sensory attributes of dry-aged meat (Zhang et al., 2022), reinforcing the contributions of "dry" effects to characteristic flavour development.

Optimisation of dry-ageing should therefore focus on maximising the beneficial outcomes of the drying process. However, many concentration effects and flavour-development mechanisms occur primarily within the outer crust. Unfortunately, this layer is commonly trimmed after ageing due to discolouration, excessive oxidation, and unappealing microbial growth. This trimming not only results in significant yield loss but also discards a substantial portion of the very "dry" effects that characterise a premium product.

Furthermore, the extended processing time required to develop these distinct characteristics significantly increases production costs, limiting both consumer affordability and broader industry adoption. Beyond these economic constraints, traditional dry-ageing is increasingly scrutinised for its environmental footprint, including high energy demands, the necessity for dedicated infrastructure, and the generation of significant waste. These limitations underscore the urgent need for alternative strategies that retain desirable sensory profiles while minimising yield losses, reducing processing times, and improving resource efficiency. Ultimately, this provides a strong reason for exploring novel approaches that decouple flavour development from excessive drying and the subsequent reliance on heavy trimming.

2.3 Tailoring eating quality through stepwise and tailored ageing

2.3.1 *Stepwise ageing*

The choice between wet- and dry-ageing is not simply a matter of preference but reflects fundamental trade-offs among eating quality attributes, nutrient stability, and processing efficiency. The advantages and limitations of different ageing strategies are summarised in Table 1.

Table 1 Comparative overview of post-mortem ageing strategies to improve eating quality of whole-muscle meat products.

Ageing strategy	Core principle	Dominant physicochemical mechanisms	Key quality outcomes	Advantages	Limitations
Wet-ageing	Vacuum storage under refrigerated conditions	Calpain-mediated proteolysis; limited oxidation	Improved tenderness; Mild flavour development	High yield; Process simplicity and consistency	Limited flavour complexity; Purge production
Traditional dry-ageing	Unpackaged cuts under controlled temperature, humidity, and airflow	Proteolysis + dehydration + oxidative flavour development	Concentrated beefy/nutty/buttery flavours; Improved tenderness	Premium sensory profile; Product differentiation	High weight and trimming losses; Microbial risk; Cost for dry-ageing facilities; Process variability
In-bag dry-ageing	Moisture-permeable vacuum bags enabling controlled dehydration	Proteolysis with moderated oxidation	Dry-age flavour with reduced loss and improved consistency	Faster drying + improved flavour; Better hygiene and consistency; Scalability	Lower flavour intensity vs. traditional dry-ageing
Stepwise ageing (wet → dry)	Sequential wet tenderisation followed by controlled dehydration	Early proteolysis → later flavour concentration	Balanced tenderness, flavour, and processing efficiency	Reduced dry-age losses; Flexible; Aligns with logistics; Reduced economic risk	Requires optimisation; not universally suitable

Stepwise ageing (dry → wet)	Short dry-ageing for dehydration and mild oxidation followed by controlled tenderisation	Proteolysis + dehydration + mild oxidation → proteolysis + moisture equilibration	Differentiated flavour profiles; Improved tenderness consistency	Supply-chain flexibility; More consistent quality	Requires optimisation; Microbial and oxidative risks
Tailored ageing	Ageing designed for species, muscle, and consumer group	Muscle-specific proteolysis and metabolism	Tailored texture and flavour	Improved consistency; demographic alignment	Requires deep muscle knowledge; higher R&D input

Information summarised in the table was based on previous publications (Kim et al., 2018; Zhang et al., 2022).

Wet-ageing is typically associated with greater tenderness consistency and nutrient preservation, whereas dry-ageing emphasises flavour intensity and textural differentiation. From a product-innovation perspective, improving the sensory and nutritional quality of meat products therefore requires moving beyond a binary comparison of ageing methods toward strategies that can selectively leverage the strengths of both approaches. In this context, stepwise ageing strategies that integrate wet- and dry-ageing have gained increasing attention over the past decade, demonstrating promising potential as alternatives to conventional single-mode ageing. By temporally separating tenderness development from drying-related flavour formation, combined wet- and dry-ageing approaches provide a practical strategy for enhancing eating quality while partially mitigating yield loss and processing inefficiencies associated with traditional dry-ageing. This reframing shifts ageing from a fixed, standardised practice to a deliberately designed process, providing the conceptual basis for the more adaptive approaches discussed in the following section.

Stepwise ageing involves the deliberate sequencing of two distinct ageing environments to exploit their complementary effects on eating quality. Unlike traditional ageing, which applies a single set of conditions throughout storage, stepwise ageing recognises that different environments promote different quality attributes and that these attributes can be developed sequentially rather than simultaneously.

Two principal stepwise configurations have been described in the literature. The first configuration involves wet-ageing followed by dry-ageing or in-bag dry-ageing (Fig. 2). In this configuration, early wet-ageing provides a stable, anaerobic environment that supports efficient calpain-mediated proteolysis and tenderisation while minimising oxidative stress and microbial activity (Kim et al., 2018; Zhang et al., 2022). Establishing baseline tenderness during this phase reduces the reliance on prolonged dry-ageing to achieve acceptable texture. Following wet-ageing, the product is subjected to controlled and accelerated dehydration, during which moisture loss concentrates flavour precursors and facilitates oxidation-driven flavour differentiation characteristic of dry-aged meat. Because tenderness has already been largely achieved, the subsequent dry-ageing phase can be substantially shortened, thereby improving processing efficiency while still delivering desirable sensory outcomes.

This strategy, also called "Age-and-Dry", demonstrates that dry-aged meat flavour intensities can be tailored by targeting specific levels of dehydration (Zhang, Ross, et al., 2023). In practice, accelerated dehydration can

Fig. 2 Representative images illustrating in-bag dry-ageing and stepwise ageing process of beef striploin sections in a commercial ageing cabinet, including in-bag dry ageing (28 days) and stepwise ageing (14 days wet-ageing followed by 14 days in-bag dry-ageing), and the resulting steaks.

be achieved within a few days using high-moisture permeable ageing bags, enabling the development of dry-aged flavour profiles with varying measurable intensities. This approach allows products to be customised to different consumer preferences while reducing time and yield losses associated with conventional dry-ageing. Importantly, the limited duration of the drying phase, combined with the protective function of moisture-permeable ageing bags, eliminates the need for trimming. These bags restrict cross-contamination from the ageing environment while enabling controlled moisture loss (Zhang et al., 2021). After dehydration, products are vacuum packaged for a short equilibration period (typically 2–3 days) to allow moisture redistribution between the dehydrated surface layer and the interior muscle. As a result, the surface crust formed during drying is softened or partially disappears, depending on the extent of dehydration. Overall, this "Age-and-Dry" strategy maximises flavour concentration effects derived from dehydration while maintaining high product yield and processing efficiency. Future research should focus on defining consumer preference thresholds for dry-aged flavour intensity across target markets, optimising the wet-ageing duration prior to dehydration, and developing innovative drying technologies to further enhance process efficiency and product consistency.

Additionally, a deeper understanding of drying kinetics during dry-ageing is required, particularly for meat cuts from different species (beef versus lamb), anatomical configurations (bone-in versus boneless), and cut sizes/weights.

Previous studies have established several models linking the initial sample weight with overall weight losses following a fixed dry-ageing period, demonstrating that the sample sizes/weights would significantly affect the drying rates and dehydration levels (Álvarez, Álvarez, Hamill, O'Neill, & Mullen, 2023; Zhang, Jacob, et al., 2025; Zhang et al., 2021; Zhang, Yoo, & Farouk, 2019; Zhang et al., 2022). In addition, the development of non-invasive tools to monitor moisture loss and drying dynamics would improve process control and reproducibility, thereby facilitating industry adoption of stepwise ageing strategies. A preliminary study by Dixit, Zhang, Farouk, dos Reis, and Realini (2024) showed that hyperspectral imaging can be used to predict dry-aged beef attributes including weight loss ($R^2 = 0.92$), moisture content ($R^2 = 0.78$), and instrumental colour traits (a* - $R^2 = 0.75$, chroma - $R^2 = 0.72$) during in-bag dry-ageing for up to 21 days.

The second configuration reverses this sequence, applying a moderate dry-ageing phase (e.g., approximately 2 weeks) followed by wet-ageing. This approach has been explored as a means of initiating dehydration-driven flavour concentration with mild oxidation at an early stage, after which wet-ageing is used to stabilise the product, limit further moisture loss, and reduce oxidative reactions and microbial activity (Zhang et al., 2022). Previous studies have shown that most of the weight loss during dry-ageing occurred within the initial two weeks, typically accounting for approximately 15–20 % moisture loss at drying rates of ~1–3 % per day. Beyond this initial phase, drying rates decrease substantially during extended dry-ageing, often to around 0.2 % per day (Zhang, Jacob, et al., 2025; Zhang et al., 2021). These findings indicate that applying dry-ageing early is particularly effective for achieving targeted dehydration levels, while prolonged dry-ageing yields diminishing returns in terms of moisture removal efficiency.

Once the desired level of dehydration has been attained, the remaining tenderisation and the continued generation of flavour precursors (e.g., free amino acids and short peptides) can be completed during the subsequent wet-ageing phase. Importantly, this wet-ageing step can be integrated into refrigerated storage or cold-chain distribution to destination markets, thereby decoupling flavour development from extended on-site ageing. As a result, this configuration offers opportunities to enhance processing flexibility and logistical efficiency while retaining key sensory attributes associated with dry-aged products.

Taken together, stepwise ageing represents a more holistic and adaptable processing strategy that integrates the complementary effects of wet- and dry-ageing to deliver consistent, high-quality meat products. By decoupling and selectively recombining the biochemical drivers of ageing, it enables tailored control of tenderness, flavour development, and yield, while mitigating the trade-offs associated with single-mode ageing processes. Importantly, reducing the duration of dry-ageing exposure improves yield and resource efficiency by limiting moisture loss and trimming, and the ability to integrate ageing steps within existing cold-chain systems enhances its practicality and scalability for industry application.

2.3.2 Tailored ageing

While stepwise ageing provides a promising strategy for decoupling and recombining the major drivers of different ageing methods, its effectiveness remains constrained by the inherent biological heterogeneity of meat. Conventional ageing regimes, whether wet, dry, or stepwise, often implicitly assume a degree of uniformity in muscle structure, composition, and ageing responsiveness that rarely exists across species, muscles, carcass characteristics, or production systems. Consequently, even well-designed stepwise protocols may not consistently deliver optimal tenderness, flavour development, and yield when applied uniformly across variable raw materials. For example, different tenderness levels and flavour profiles were observed among beef clod heart, brisket, and flat iron subjected to the same wet-ageing or in-bag dry-ageing process for 21 or 42 days (Leighton et al., 2023). In clod heart, in-bag dry-ageing improved some palatability attributes such as beef flavour, salty taste with decreased sour-dairy, stale/carboard flavours, and lipid oxidation-derived volatiles compared to wet-aged equivalents regardless of ageing periods. Enhanced salty taste and fatty aroma with reduced bloody/serumy flavour were also observed in brisket, however, loss of beef and buttery flavours and increase of unpleasant flavours were also detected. For flat iron, in-bag dry-ageing showed negative impacts on palatability regardless of ageing period, which was mainly due to its small size/volume, resulting in over-drying following 42 days of dry-ageing process.

To address this limitation, tailored ageing has been proposed as a further evolution of stepwise ageing, in which ageing conditions are deliberately designed to meet specific product attributes for defined raw materials, rather than applied as rigid processing protocols (Kim et al., 2018; Zhang et al., 2022). In this sense, tailored ageing shifts the emphasis from the ageing method (e.g., wet, dry, or stepwise ageing) to the intended outcome (e.g., tenderness-first, flavour-forward, high-yield, or nutrition-preserving), recognising that the optimal sequencing, intensity, and duration of ageing vary across species and meat cuts. Importantly, stepwise ageing provides the "toolkit" (i.e., separable process modules), whereas tailored ageing provides the "decision framework" for selecting and tuning these modules to achieve consistent and tailored quality outcomes.

At a practical level, tailored ageing needs to be operationalised by aligning process design with the biological attributes of specific muscles (e.g., species or cuts) and their intended end uses. For example, meat from older animals or meat with relatively higher connective tissue content (e.g., *semitendinosus* and *infraspinatus* muscles) may benefit from an ageing design that prioritises proteolysis primarily by extended wet-ageing (Roy & Bruce, 2024), and minimises unnecessary dehydration, particularly when the goal is improved ease of mastication. In contrast, cuts with higher intramuscular fat content may be more responsive to short, accelerated dehydration phases that concentrate flavour precursors and support mild oxidation-driven flavour differentiation under a more controlled environment to prevent excessive oxidative reactions (Zhang et al., 2022). Where improved flavour complexity is the principal objective, ageing strategies may prioritise controlled dehydration and moderated oxidative and microbial influences, recognising their roles in concentrating flavour precursors and driving aromatic development, while maintaining product stability and yield. The central principle is not that a given cut must follow a fixed pathway, but that processing conditions should be selected based on the primary limitations of the raw material to achieve defined quality outcomes.

Tailored ageing also enables clearer alignment with consumer segment requirements, reinforcing the broader sustainability and value proposition of stepwise processing. For example, products designed for young children may prioritise tenderness, ease of chewing, and overall palatability (Cichero, 2017), rather than intense dry-aged flavour notes; in such cases, processing strategies should favour proteolysis while minimising dehydration and oxidation changes. For elderly consumers, further refinement of processing strategies may be required to enhance flavour intensity (e.g., through moderate dehydration) while maintaining the tenderness and juiciness, given the increase in tasting thresholds associated with ageing (Farouk et al., 2025; Mojet et al., 2001). A recent study in beef striploin further supports this principle by showing that stepwise ageing (14 days wet-ageing + 14 days in-bag dry-ageing) produced similar tenderness and oxidative outcomes to 28 days of in-bag dry-ageing, but with lower ageing loss and higher moisture retention

(Zhang, Deb-Choudhury, et al., 2026). Combining stepwise ageing with tailored cooking methods (e.g., low-temperature oven precooking at 49 °C for 24 h) can generate distinct volatile fingerprints, highlighting the potential to co-optimise ageing and cooking to achieve specific balances of tenderness, yield, juiciness, and flavour expression according to product intent. Conversely, products targeting gourmet or specialist markets may justify ageing strategies that deliberately promote flavour differentiation (e.g., controlled dehydration, oxidation, and microbial inoculation) within defined limits for yield and consistency.

3. Combining freezing/thawing with ageing to produce high-quality frozen meat

Freezing remains the most effective large-scale approach for extending meat shelf-life and facilitating global distribution, yet the potential quality gap between fresh-never-frozen and frozen/thawed meat still influences consumer perception and market value of frozen meat products (Zhang, Realini, et al., 2023). The main issue is not that frozen meat is inherently low quality, but that freezing and thawing processes introduce a bundle of physical and biochemical changes, such as structural disruption due to ice crystal formation, solute concentration effects, protein denaturation, lipid/protein oxidation, purge formation, and colour instability (Du et al., 2022). These changes may intensify during frozen storage, post-thaw handling and shelf display, resulting in compromised meat juiciness, tenderness, and flavour (Zhang, Realini, et al., 2023). Over the past two decades, a growing body of evidence has demonstrated that integrating post-mortem ageing (mostly wet-ageing, and in some applications dry-ageing/stepwise ageing) with freezing/thawing can mitigate these effects. Such integrated approaches offer practical means of improving the eating quality of frozen meat to levels approaching fresh-never-frozen products, while leveraging processing strategies were already established at industrial scale.

3.1 Wet-ageing before freezing (WA-B-F)

Increasing evidence has shown that once a sufficient ageing period (e.g., 2–3 weeks) has been applied before exposure to freezing, i.e., WA-B-F, to establish baseline tenderness and modify muscle structure, some meat quality traits could be improved, including colour stability during post-thaw shelf display (Alvarenga, Hopkins, Ramos, Almeida, & Geesink, 2019), water holding capacity (Kim, Frandsen, & Rosenvold, 2011), and tenderness (Grayson, King, Shackelford, Koohmaraie, & Wheeler, 2014; Kim, Liesse, Kemp, & Balan, 2015). During the ageing phase, endogenous proteolytic enzymes degrade key cytoskeletal proteins, weakening structural linkages and increasing myofibrillar fragmentation (Kim et al., 2018). This process not only improves tenderness but also alters the microstructural organisation of the muscle. The breakdown of costameres and associated structural elements disrupts defined drip channels and modifies how water is distributed within the tissue, and as a result, the muscle develops an enhanced ability to physically retain water, which is referred to as the "sponge effect" (Farouk, Mustafa, Wu, & Krsinic, 2012). In this state, water becomes more effectively entrapped within the muscle matrix, supported by increased soluble proteins and localised increases in viscosity. When such pre-conditioned muscle is subsequently frozen, the structural stresses imposed by ice crystal formation are partially mitigated. Because the muscle fibres have already undergone controlled degradation, freezing-induced deformation is less localised and less catastrophic compared with unaged muscle (Setyabrata & Kim, 2019). This, in turn, could result in improved water retention upon thawing and reduced purge loss relative to meat frozen immediately post-mortem. On the other hand, the proteolysis induced by WA ensures that satisfactory improvement in tenderness has largely been achieved prior to freezing, making the final eating quality more consistent, predictable, and less dependent on post-thaw handling.

However, this approach is still constrained by the fundamental challenges associated with freezing. The same structural openness and increased solute concentration created during ageing can elevate susceptibility to lipid and protein oxidation and microbial spoilage, particularly during prolonged ageing and frozen storage (Holman, Coombs, Morris, Bailes, & Hopkins, 2018; Holman, Coombs, Morris, Kerr, & Hopkins, 2017). As a result, colour stability may decline more rapidly during post-thaw display, and the risk of flavour deterioration increases if storage duration and oxygen exposure are not carefully controlled (Kim, Kim, Seo, Setyabrata, & Kim, 2018). Furthermore, purge loss, although reduced relative to unaged frozen meat, remains higher than that of equivalently aged-never-frozen products (Schulte et al., 2019). These limitations highlight that while WA-B-F improves structural resilience and tenderness of frozen products, it does not fully eliminate the oxidative and physicochemical changes during frozen storage. Hence, successful application of WA-B-F process requires integrated optimisation of ageing and freezing conditions (e.g., temperature, time, freezing rate etc.) according to animal species and muscle type. This integration is critical to maximise the benefits of consistent tenderisation and aged flavour development while minimising quality deterioration related to oxidation and microbial spoilage. Based on reported evidence from literature, WA at low temperatures (e.g., −1.5–1 °C) for 2–3 weeks for beef, 2–4 weeks for lamb, and about 1 week for venison prior to frozen storage up to 1 year, is generally sufficient to achieve consistent tenderisation and acceptable overall quality (Coombs, Holman, Collins, Friend, & Hopkins, 2017; Coombs, Holman, Collins, et al., 2018; Coombs, Holman, Friend, & Hopkins, 2017; Coombs, Holman, Ponnampalam, et al., 2018; Kim et al., 2011; Wiklund et al., 2009).

3.2 Freezing before wet-ageing (F-B-WA)

In contrast, F-B-WA adopts an inverse logic, in which freezing-induced structural modifications are used to facilitate subsequent post-thaw WA. In this configuration, meat is typically frozen early post-mortem (24–48 h post-slaughter) and subsequently thawed and aged, a process that can be integrated into storage or distribution. Thus, the key biochemical processes governing eating quality in fresh (unaged) meat, such as enzymatic proteolysis and production of flavour precursors, are effectively inhibited under commercial freezing conditions (e.g., −16 to −20 °C) (Du et al., 2022). These processes can resume upon thawing, contributing to tenderisation of meat during post-thaw WA (Aroeira et al., 2016; Zhang & Ertbjerg, 2018). The key advantage of this approach arises from the way freezing alters both muscle structure and enzymatic regulation. Ice crystal formation during early freezing disrupts cellular membranes and expands the physical spacing between muscle fibres, increasing the accessibility of structural proteins to proteolytic enzymes (Grujić, Petrović, Pikula, & Amidžić, 1993). At the same time, freezing appears to affect the regulatory balance of proteolysis, as the endogenous inhibitor calpastatin is more susceptible to inactivation than calpain enzymes (Vieira, Diaz, Martínez, & García-Cachán, 2009). Combined with the release of Ca^{2+} from damaged cellular compartments, this creates conditions that favour enhanced enzymatic activity during thawing and subsequent ageing (Zhang & Ertbjerg, 2018). Through these mechanisms, F-B-WA may promote accelerated proteolysis during post-thaw storage, contributing to myofibrillar degradation and improved tenderness. In some studies, this has translated into more consistent tenderisation, particularly when compared with conventional ageing alone (Grayson et al., 2014; Sales et al., 2020). From a practical standpoint, the ability to freeze and distribute large volumes of product soon after slaughter, with ageing deferred to later stages of the cold chain, offers operational flexibility. This allows processors to decouple tenderisation from early post-mortem handling and align it with logistics, storage and market requirements.

However, the same mechanisms that facilitate proteolysis can also introduce quality deterioration. Because the muscle is structurally intact at the time of freezing (early post-mortem), ice crystal formation imposes greater mechanical stress, leading to more pronounced fibre shrinkage and disruption (Du et al., 2022; Zhang, Realini, et al., 2023). This promotes the migration of water into extracellular spaces and, ultimately, results in increased purge loss upon thawing (Aroeira et al., 2016; Setyabrata & Kim, 2019). Unlike WA-B-F, there is no prior ageing-induced structural weakening to buffer this effect. Additionally, the combination of structural damage and subsequent ageing enhances exposure of proteins and lipids to oxidative reactions, contributing to reduced colour stability and increased susceptibility to discolouration during post-thaw storage and display (Aroeira et al., 2017). These effects can become more pronounced when ageing is conducted under aerobic conditions or when freezing and storage conditions are not optimally controlled (e.g., slow freezing at higher temperatures or using aerobic packaging). Thus, while F-B-WA can enhance proteolysis and tenderness, this often occurs at the expense of water-holding capacity and visual quality, emphasising the need for careful control of freezing rate, storage conditions, and thawing protocols. Rapid freezing methods (e.g., air blast or cryogenic freezing to −30 to −40 °C or below) promote the formation of smaller, more uniform ice crystals, thereby reducing mechanical damage compared with conventional slow freezing (Du et al., 2022). Higher freezing rates have been associated with improved water-holding capacity (lower purge and drip losses) and more consistent tenderisation following post-thaw ageing (Balan et al., 2019).

Taken together, WA-B-F tends to favour structural stabilisation and water retention, resulting in more consistent eating quality with relatively lower purge, whereas F-B-WA capitalises on enhanced enzymatic accessibility and regulatory shifts to accelerate tenderisation, albeit with greater risk to reduce water-holding capacity and colour stability. These contrasting mechanisms highlight that ageing and freezing strategies should be designed as an integrated system rather than as independent steps. Looking forward, non-thermal technologies, such as ultrasound-assisted freezing or thawing, electrostatic field applications, radiofrequency tempering, and high-pressure processes, offer potential to improve ice crystal control, reduce thawing losses, or accelerate ageing processes (Du et al., 2022; Zhang, Realini, et al., 2023; Zhang, Liu, et al., 2023). These technologies may help address specific bottlenecks, including non-uniform thawing and excessive purge, particularly when applied in conjunction with ageing strategies. However, challenges related to scalability, energy efficiency, process uniformity, and cost remain significant, and broader industrial adoption will depend on demonstrating consistent benefits at commercially relevant scales.

4. Processing innovation for minced meat products

Minced and restructured meat systems provide a particularly versatile platform for processing innovation because their disrupted structure allows the matrix to be redesigned through blending, fortification, and hybridisation. In this context, innovations in the minced meat matrix include three complementary directions: meat-to-meat fortification, where secondary cuts or tissues are used to enhance targeted nutritional or functional attributes; co-product fortification, where offal and other meat co-products are valorised to improve micronutrient density to achieve some targeted nutritional functionalities; and hybridisation, where meat is combined with alternative protein sources to broaden nutritional diversity and create new product opportunities. As illustrated in Fig. 3, these strategies reposition the minced meat matrix from a structurally simple comminuted system to a flexible design platform capable of integrating both animal- and non-animal-derived ingredients aligned with defined product targets.

4.1 Meat-to-meat fortification in minced meat matrix

4.1.1 Minced meat matrix

Minced meat represents one of the most versatile and widely consumed forms of red meat and offers unique opportunities for developing foods that support health and wellbeing across the lifespan. Its popularity stems from a combination of attributes including affordability, convenience, ease of preparation, suitability for batch cooking and freezing, and its relatively soft texture compared with whole-muscle cuts. These characteristics make minced meat particularly attractive for populations

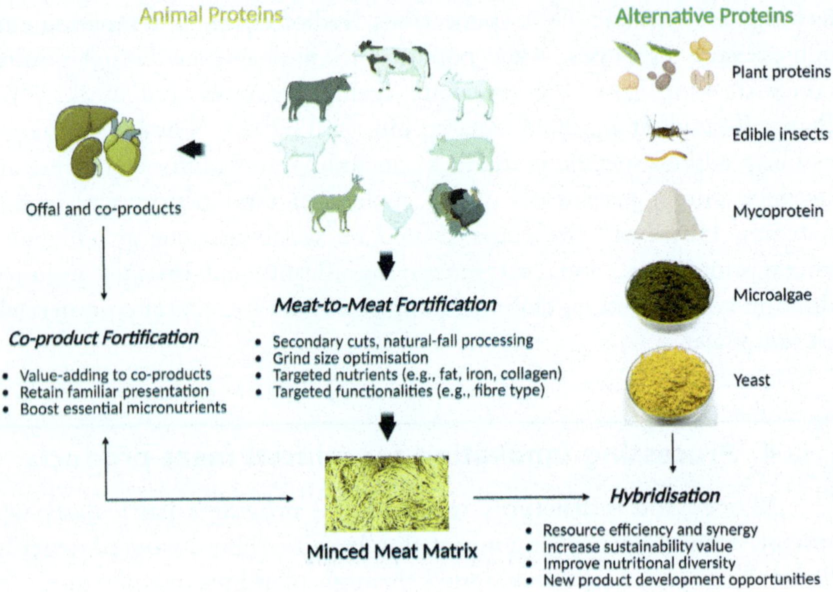

Fig. 3 Processing and formulation strategies to improve the eating quality and nutritional value of minced meat products. *Created in BioRender. https://BioRender.com/2f9aafc.*

with changing nutritional and physical needs, including older adults, children, and individuals requiring convenient, nutrient-dense meal options (Farouk et al., 2025).

Traditionally, mince has been marketed primarily based on fat content, leanness, and value for money (Fayet-Moore, Cunningham, Stobaus, & Droulez, 2014). However, its true potential lies in its ability to serve as a highly adaptable nutritional platform capable of delivering targeted health and wellness benefits. Unlike intact muscle cuts, which offer limited opportunities for modification, the minced meat matrix provides a flexible system that can be tailored to meet specific nutritional targets through the strategic selection and combination of ingredients derived entirely from animal sources (Farouk et al., 2025). The most significant advantage of a minced meat matrix is its amenability to reformulation. Different muscles within a carcass vary considerably in their nutritional composition, flavour, tenderness, and functional properties. By combining muscles from different anatomical locations, carcass classes, or production systems, it becomes possible to design products with enhanced levels of desirable nutrients while maintaining eating quality and consumer familiarity. More importantly, minced meat provides an ideal vehicle for incorporating edible organs and other nutrient-rich co-products that are often underutilised despite their exceptional nutritional value (Farouk, Jenkinson, Zhang, & Realini, 2024; Farouk et al., 2025).

Organs such as liver, heart, kidney, and other edible offal are naturally rich sources of bioavailable iron, vitamin B12, zinc, selenium, and a range of other micronutrients that are often lacking in modern diets. Incorporating these tissues into a minced meat matrix enables a "meat-to-meat fortification" approach, whereby nutritional enhancement is achieved using ingredients originating from the same animal rather than through the addition of external fortificants (Farouk et al., 2024). This strategy not only preserves the natural image of the product but also aligns with increasing consumer interest in minimally processed foods and whole-animal utilisation.

The versatility of minced meat allows nutrient composition to be tailored to achieve specific nutritional outcomes and, where appropriate, support regulated nutrition claims. For example, incorporation of selected organ meats can substantially increase iron content, creating opportunities for products targeted at groups with elevated iron requirements, such as women of reproductive age, adolescents, older adults, and young children. Similarly, mince formulations can be developed to enhance levels of protein, B vitamins, collagen, or other nutrients depending on the needs of the intended consumer group.

Beyond its nutritional flexibility, the minced meat matrix also provides opportunities to optimise texture, functionality, and eating quality (Farouk et al., 2025; Pennings et al., 2013). Particle size, fat content, moisture level, and ingredient selection can be manipulated to produce products with specific sensory characteristics. This is particularly important for older consumers and toddlers, for whom ease of chewing and swallowing may influence food choice and nutritional status (Farouk et al., 2018). Compared with whole-muscle cuts, minced meat products can be formulated to provide a softer texture while retaining desirable flavour and juiciness. The ability to control these attributes offers considerable scope for developing products tailored to the requirements of healthy ageing populations (Farouk et al., 2018). The minced meat matrix is also highly compatible with modern convenience food formats. Minced meat can be incorporated into a wide range of familiar dishes including burgers, meatballs, pasta sauces, pies, casseroles, ready meals, and snack products (Samarakoon Mudiyanselage, Farouk, Realini, Kantono, & Hamid, 2025). It can also be pre-cooked, frozen, and packaged in user-friendly formats that simplify meal preparation without compromising nutritional value. Such versatility increases consumer acceptance of nutrient-enhanced formulations because the nutritional benefits can be delivered through foods that are already familiar and widely consumed (Farouk et al., 2018).

From an industry perspective, the mince category provides an attractive platform for innovation because it combines high consumer acceptance with considerable formulation flexibility. It offers opportunities to increase the value of underutilised cuts and edible co-products while simultaneously addressing growing consumer interest in nutrition, health, sustainability, and whole-animal utilisation. As the red meat industry increasingly seeks to move beyond traditional commodity products, minced meat has the potential to evolve from a simple source of protein into a targeted nutritional delivery system capable of supporting a range of functional and health-related outcomes.

Viewed in this way, mince is much more than a convenient meat product. It is a versatile and adaptable food matrix that enables the integration of diverse animal-derived ingredients, supports the development of nutrient-enhanced foods, and provides a practical pathway for delivering tailored nutritional benefits to specific consumer groups. This combination of functionality, familiarity, and flexibility positions minced meat as one of the most promising vehicles for future innovation in value-added red meat products (Farouk et al., 2024; Farouk et al., 2025).

4.1.2 Minced meat matrix as a vehicle for fortification of nutrients

The approach for utilising the minced meat matrix for nutrient enhancement and product quality outcomes is depicted in Fig. 4, with emphasis on the incorporation of organ meats to enhance nutritional value and functionality. This strategy was used by Farouk et al. (2024) in a work motivated by the growing consumer interest in foods that provided nutritional benefits beyond basic nourishment and by industry interest in increasing whole-carcass utilisation through the development of higher-value products. Farouk et al. (2024) determined the technical feasibility, consumer acceptance, and commercial potential of enhancing conventional beef mince through the incorporation of nutrient-dense beef offal. The study sought to determine whether underutilised edible co-products such as liver, heart, and kidney could be incorporated into mince formulations in a manner that significantly enhanced nutritional value while maintaining consumer acceptability and commercial viability. The study integrated analytical evaluation, product formulation, consumer research,

Fig. 4 Strategies for utilising the minced meat matrix for nutrient fortification and product quality outcomes.

market assessment, and commercial evaluation to identify viable nutrient-enhanced mince concepts. A key finding was that consumers increasingly associated food choices with health and wellbeing and actively sought products that delivered meaningful, clearly communicated nutritional benefits. Within this context, iron emerged as the most influential nutrient driving consumer interest. However, consumer awareness of meat as a source of bioavailable iron remains limited. For example, Realini et al. (2022) reported that while consumers in Uruguay strongly associated meat with protein, it was less frequently recognised as a source of other key nutrients such as iron. This gap in perception is important, given that heme iron from meat is highly bioavailable and plays a critical role in preventing iron deficiency. Given the high global prevalence of iron deficiency, particularly among women, children, adolescents, and older adults, and its significant incidence in specific populations such as young children and pregnant women, meat-based products enriched with naturally bioavailable iron present a strong opportunity for value-added differentiation (Farouk et al., 2024).

The research evaluated several edible beef offal as potential ingredients for nutrient enhancement. Liver, heart, and kidney were selected due to their nutritional profiles and availability within existing processing systems.

Laboratory analyses demonstrated that all the three offal contributed meaningful levels of micronutrients, particularly iron and vitamin B12, when incorporated into mince formulations. However, liver was identified as the most effective fortification ingredient owing to its exceptionally high concentration of bioavailable iron and other essential micronutrients. Compared with conventional lean beef muscle, beef liver contained substantially higher concentrations of iron, allowing significant nutritional enhancement to be achieved with relatively modest inclusion levels (Farouk et al., 2024).

One of the most important findings of the project was the identification of an optimal formulation consisting of approximately 85 % beef mince and 15 % beef liver. This blend provided a balance between nutritional enhancement and consumer acceptability while meeting criteria for regulated nutrition claims under Australian food standards (https://www.foodstandards.gov.au/). At this inclusion level, the product could qualify as a source, good source, or complete source of iron depending on serving size and formulation, thereby providing a meaningful and marketable nutritional benefit. Importantly, the formulation delivered a substantial increase in nutrient density without fundamentally changing the product category or requiring consumers to adopt unfamiliar eating behaviours.

The project demonstrated that successful nutritional valorisation required more than simply adding nutrient-rich ingredients to a food product. Consumer research revealed that perceived nutritional benefits must be significant to justify product differentiation. Participants consistently indicated that minor nutritional improvements were unlikely to motivate purchase decisions. Instead, consumers expected meaningful enhancements that could be clearly communicated and readily understood. The findings suggested that a nutrient uplift of approximately fifty percent or greater was necessary to create a compelling value proposition. This observation has important implications for future product development initiatives involving meat co-products, as it highlights the need to generate substantial and measurable nutritional benefits rather than incremental improvements.

Another particularly noteworthy outcome of the study concerned consumer attitudes toward offal. While edible offal offered significant nutritional advantages, they also presented sensory and psychological challenges. Consumers frequently associated liver and other organ meats with strong flavours, unfamiliar aromas, undesirable textures, and negative perceptions from previous experiences. Consequently, the principal barrier to adoption was not technological feasibility but consumer perception (Farouk et al., 2024). Nevertheless, the research demonstrated that these challenges can be mitigated through appropriate product design and use contexts. Acceptance improved significantly when nutrient-enhanced mince was incorporated into familiar, flavourful meal formats (e.g., pasta sauces, curries, casseroles, chilli dishes), where the sensory impact of offal was moderated while nutritional benefits were retained. An important strategic insight emerging from the project was that transparent communication of nutritional benefits may be more effective than attempting to mask the inclusion of co-products. Consumers interested in health, wellness, nutrition, and sustainability responded positively to clear messaging around ingredient functionality and origin, suggesting that authenticity and purpose-driven positioning can enhance acceptance. Additionally, the project also highlighted broader consumer trends relevant to the red meat industry. Participants increasingly valued products that contributed to sustainability, minimise waste, and utilise more of the animal. This finding aligns with growing societal interest in circular food systems and whole-animal utilisation. Nutritional valorisation therefore offers dual benefits by simultaneously improving nutritional outcomes and increasing the value extracted from livestock resources. The incorporation of edible offal into mainstream food products represents a practical example of how sustainability objectives and nutritional enhancement can be achieved through the same innovation pathway.

From an industry perspective, nutrient-enhanced mince could create new opportunities for product differentiation within a highly competitive category. Conventional mince is often regarded as a commodity product competing primarily on price (Fayet-Moore et al., 2014). By contrast, nutrient-enhanced mince provides a mechanism for creating premium offerings supported by measurable nutritional claims. This approach enables processors and retailers to move beyond traditional price-based competition and instead focus on delivering added consumer value. The ability to communicate naturally occurring nutrients, rather than relying on artificial fortification, was identified by consumers as a particularly attractive aspect of the concept.

The findings have implications extending beyond mince products. The project illustrates broader principles that can be applied to future meat co-product valorisation initiatives. First, consumers respond most positively when nutritional enhancements are derived naturally from animal tissues rather than added through external fortification. Second, nutrient claims must be meaningful and supported by substantial increases in nutrient content. Third, product formats that are already familiar to consumers provide a more effective vehicle for introducing novel ingredients (Farouk et al., 2018; Farouk et al., 2025). Finally, successful commercialisation depends on communicating nutritional outcomes rather than focusing on the co-products themselves, and that consumers do not purchase co-products; they purchase solutions to nutritional needs. Consequently, the most successful valorisation strategies will be those that transform underutilised animal materials into products delivering meaningful, measurable, and easily communicated benefits for health and wellbeing (Farouk et al., 2024).

Overall, there is compelling evidence that nutritional valorisation can serve as a powerful mechanism for increasing the value of red meat co-products while responding to evolving consumer expectations regarding health, nutrition, sustainability, and food quality. Fig. 4 therefore represents not only a formulation strategy, but a broader design framework linking raw material selection, nutritional objectives, product formulation, and consumer acceptance.

4.2 Hybridisation with non-meat proteins: Designing dual-protein systems

4.2.1 Hybrid meat concepts

Meat-based hybrid foods, also called hybrid meat, dual protein foods or balanced protein foods, combining proteins from animal with alternative/non-animal sources (e.g., plant-based, microbial, and edible insects) within a single matrix, have emerged as one of the most dynamic areas of innovation in minced meat systems (Zhang, Farouk, et al., 2025). Unlike plant-based meat analogues, hybrid products retain meat as a central component, leveraging its sensory familiarity while using non-meat proteins to enhance sustainability, nutrition, and/or functionality.

From a processing perspective, minced meat provides an ideal environment for hybridisation. Mincing and blending with alternative protein sources disrupt meat structure and reduce the dominance of muscle fibre orientation, allowing alternative proteins to interact and contribute to water binding, gel formation, and texture development. From a consumer perspective, hybridisation offers both familiarity and flexibility, enabling the tailoring of product characteristics to meet diverse preferences while lowering the psychological barrier associated with fully replacing meat (Grasso, Rondoni, Bari, Smith, & Mansilla, 2022). This aligns closely with the flexitarian approach, where consumers intentionally reduce but do not eliminate meat intake (Dagevos, 2021; Melios & Grasso, 2024). Hybrid products enable gradual dietary shifts by combining meat with alternative proteins, supporting both health and environmental goals without requiring drastic changes in eating habits. By offering familiar and acceptable textures and flavours alongside enhanced nutritional and functional profiles, these products cater to flexitarians seeking balance between traditional meat consumption and more sustainable, plant-forward choices (Miao, Hastie, Ha, & Warner, 2023; Ryder, Jaworska, & Grasso, 2023; Salgaonkar & Nolden, 2024).

To date, a wide range of non-meat proteins has been explored in hybrid meat systems, such as legumes (e.g., soy, pea, lentil), cereals (e.g., wheat, rice, oat), fungi (mycoprotein), microalgae (*Chlorella*, *Spirulina*), yeast (*Saccharomyces cerevisiae*), and edible insects (e.g., superworm and mealworm). Each category has its own unique characteristics in terms of nutritional profiles, functional and sensorial properties, altering hybridisation performance with meat. Table 2 summarises the key advantages and disadvantages of different protein sources in hybrid meat systems.

Legume proteins have been extensively studied due to their large production volumes and relatively low prices compared to other protein sources. These plant-based proteins are mostly applied in the isolated powder form or have been further texturised through extrusion for replacing meat in burgers and sausages, demonstrating good water- and oil-holding capacity, emulsification performance and acceptable sensory properties when the substitution levels were relatively low (~10–20 % or below) (Barker & McSweeney, 2022; Grasso, Smith, Bowers, Ajayi, & Swainson, 2019; Taylor, Ahmed, Al-Juhaimi, & Bekhit, 2020). Increasing substitution levels (e.g., over 30–40 %) has often been found to introduce dry, beany flavours, altering colour (reduced a^* and increased b^*) and texture (soft, less cohesive) properties, resulting in low consumer acceptability (Barker & McSweeney, 2022; Dasiewicz, Szymanska, Opat, & Hac-Szymanczuk, 2024; dos Santos et al., 2023). Cereal proteins, particularly wheat gluten, provide structural support and chewiness, making them suitable for restructured products. Texturised wheat protein has been found to perform better than soy protein in preserving the texture properties of hybrid salami compared to traditional pork salami, likely due to the functional contributions of glutenin and gliadin for forming a dense and cohesive hybrid structure through strong intermolecular disulfide and hydrogen bonds (Li et al., 2024; Wang, Jin, & Xu, 2015).

Table 2 Comparative overview of different alternative protein sources used in meat-based hybrid systems.

Alternative proteins	Main ingredient formats	Advantages	Disadvantages	Tested formats
Legumes (soy, pea, lentil)	Protein isolate or concentrate Texturised proteins	Enriched protein and fibre. Good water- and oil-binding. Emulsification.	Beany flavour. Potential allergenicity. Variable solubility. Compromised texture at high replacement levels.	Model systems Patties Sausages Meatballs
Cereals (wheat, rice, oat)	Protein isolates Gluten	Structural support. Chewiness. Cost-effective	Gluten allergenicity (wheat). Variable protein quality.	Model systems Sausages Patties
Fungi (mycoprotein)	Fermented fungal biomass	Fibrous texture. Umami flavour. High-quality protein.	Moisture sensitivity. Requires binders or structuring aids for texture formation.	Sausages Meatballs Patties
Microalgae (*Chlorella*, *Spirulina*)	Biomass	High protein, micronutrient and bio-actives. Sustainability value. Emulsification. Water and oil-binding.	Strong colour and flavour. Limited consumer familiarity.	Sausages/ frankfurters Patties Pâté Cooked turkey products Soft jerky

Yeast (*Saccharomyces cerevisiae*)	Biomass Protein extract	Concentrated protein, and micronutrients. Rapid production. Umami/savoury flavour. Water and oil-binding. Good filler for texture formation.	Colour effects. Bitterness and yeasty notes from some yeast sources.	Emulsified sausages. Patties Meatballs Soft jerky
Edible insects (mealworm, superworm, cricket)	Protein extract Biomass	High-quality protein and useful micronutrients. Good digestibility. Water-binding.	Colour effects. Texture disruption/chitin-related graininess. Flavour deviations. Consumer neophobia.	Model systems Patties/burgers Sausages/ Frankfurters

Information summarised in the table was based on previous publications (Zhang, Jacob, et al., 2026; Zhang, Liu, et al., 2023).

However, allergen concerns, especially those related to wheat and gluten proteins, may limit their wider applications in hybrid food products (Senarathna, Mel, & Malalgoda, 2024).

Among emerging non-animal proteins, mycoprotein has attracted particular interest because it combines a relatively strong nutritional profile with functional properties that are favourable for meat analogue and hybrid meat design. Mycoprotein is mainly produced from *Fusarium venenatum* through fermentation and its filamentous fungal structure differs from many plant proteins due to its inherently fibrous architecture, which helps support a more meat-like texture with softness and chewiness in finished products. In addition, it exhibits favourable emulsifying, foaming, and rheological properties which are suitable for mince-style products, contributing to texture development (Khan et al., 2024). From a nutritional perspective, mycoprotein provides high-quality protein (11–13 %) with a complete essential amino acid profile, while remaining relatively low in fat (2.9 %) and contributing substantial dietary fibre (6–7 %), mainly from β-glucan and chitin in the fungal cell wall (Finnigan, Mach, & Edlin, 2024; Finnigan, Theobald, & Bajka, 2025). It also supplies micronutrients such as zinc, selenium, phosphorus, and B vitamins. Despite these advantages, mycoprotein is not a simple one-to-one substitute for meat. Product performance still depends strongly on formulation and processing design, and commercial systems often rely on complementary binders (e.g., egg white or other plant proteins) or structuring aids (e.g., moulding, heating, and freezing) to optimise texture (Khan et al., 2024). In addition, safety considerations such as RNA reduction during manufacture, control of mycotoxin risk, and possible intolerance in mould-sensitive individuals remain important (Finnigan et al., 2025).

Microalgae proteins or biomasses are emerging as promising ingredients in hybrid meat products because the high nutritional density and sustainability value with potential high production volumes and lower environmental impacts (Yang et al., 2024). *Chlorella* and *Spirulina* are two attractive species due to their high protein content (up to 70 % on dry basis), balanced amino acid profile, and the presence of bioactive compounds such as polyunsaturated fatty acids (e.g., omega-6), pigments (e.g., chlorophylls, carotenoids), and antioxidants (Bernaerts, Gheysen, Foubert, Hendrickx, & Van Loey, 2019). Microalgae are also of interest in hybrid meat design because their protein-rich biomass and polysaccharide-containing cell structures can interact with the meat matrix to affect water retention, cook loss, rheological behaviour, and texture development, although these effects are highly dependent on species, inclusion level, and product format (Siladji et al., 2024; Zhang, Farouk, et al., 2025). The main constraints of incorporating microalgae biomass in hybrid meat systems include strong colour and flavour deviations, especially the green-blue pigmentation and characteristic algal notes. Due to these constraints, relatively low inclusion levels (~1–3 %) are usually applied to achieve comparable or improved product performance compared to equivalent meat-only products, while reducing negative impacts on sensory quality (Abdel-Moatamed, El-Fakhrany, Elneairy, Shaban, & Roby, 2024; Bošković Cabrol et al., 2023).

In hybrid meat products, yeast biomass most commonly refers to *Saccharomyces cerevisiae*, used either as nutritional yeast produced through controlled fermentation or as brewer's spent yeast recovered from brewing industry and further processed for food applications. Both forms are relevant because yeast biomass typically contains ~40–60 % protein (on a dry basis), provides a broad amino acid profile with relatively high lysine, and also contributes B vitamins, minerals, and cell-wall components such as β-glucans and mannoproteins, which can support water binding, viscosity, and emulsification in complex food matrices (Jaeger, Arendt, Zannini, & Sahin, 2020; Ma et al., 2023). In addition, yeast is particularly attractive in hybrid meat design because it contains umami-active amino acids, peptides, nucleotides (e.g., 5′-nucleotides, glutamic acid, and cysteine) (Alim et al., 2019; Jaeger et al., 2020), as well as kokumi peptides (Liu et al., 2015), making it valuable not only as a protein ingredient but also as a functional flavour-enhancing component. This functionality is illustrated in a recent beef hybrid model study using nutritional yeast flakes (*S. cerevisiae*) alone or blended with *Chlorella* biomass at 20 % and 35 % total meat substitution (Zhang, Jacob, et al., 2026). Yeast-dominant formulations reduced crude fat and energy, lowered cooking loss, and produced stronger elastic gel behaviour during heating than microalgae-rich systems, while also maintaining colour closer to the meat control. In jerky-type products, yeast-containing hybrids improved drying performance and textural properties, while mitigating some of the colour and sensory deviations associated with *Chlorella* incorporation. Overall, the study indicates that nutritional yeast is particularly valuable at modest substitution levels (around 20 %), where it can improve processing performance and flavour functionality while enabling greater inclusion of other alternative proteins.

Edible insects represent another emerging protein category attracting growing interest in hybrid meat systems. The most commonly studied species include mealworms, superworms, and crickets, typically incorporated as flours, protein-rich fractions, or larvae-derived ingredients (Zhang, Farouk, et al., 2025). Across species, insect biomass typically contains around 35 %–61 % protein (dry basis), with values reported to be even higher in some crickets and grasshoppers (up to ~77 %), and their protein digestibility is generally high (~76 %–98 %) (Lange & Nakamura, 2021). They also contribute variable amounts of fat (~2 %–62 %), often with a relatively high proportion of unsaturated fatty acids, chitin-rich fibre, as well as useful micronutrients such as iron, zinc, magnesium, and B vitamins (Rumpold & Schlüter, 2013).

Incorporation of edible insect ingredients in hybrid meat matrices can provide useful functional properties, such as water binding, emulsification, and some gel-forming ability, although their performance is usually less uniform than meat proteins because of differences in protein structure and the presence of chitin-rich particles (Gravel & Doyen, 2020; Zhang, Farouk, et al., 2025). In practice, the most favourable outcomes are generally reported at modest inclusion levels of about 5 %–15 %, where insect ingredients can improve nutritional value and, in some formulations, water retention or cooking yield (Gomes Martins et al., 2024; Vlahova-Vangelova, Balev, & Kolev, 2023; Zhang, Li, et al., 2024). For example, cricket flour at around 10 % inclusion in burger-type products remained broadly acceptable, whereas 15 %–20 % more often produced darker colour, lower redness, off-aromas, and drier or sandy textures (Gomes Martins et al., 2024). Consumer acceptance therefore remains a major barrier, particularly in Western markets, where food neophobia, perceived disgust, and limited familiarity with edible insects as food continue to constrain adoption (Alhujaili, Nocella, & Macready, 2023).

Taken together, the development of hybrid meat products has largely followed a meat-analogue design paradigm, focusing on familiar product formats such as mince, patties, sausages, and nuggets, where consumers hold well-defined expectations regarding appearance, texture and flavour. Under these constrains, partial substitution of meat with alternative proteins often results in functional and sensory mismatches, including reduced structural integrity (e.g., softer or less cohesive textures), altered flavour profiles derived from the alternative ingredients, and visible colour deviations, all of which can negatively influence consumer acceptance.

To mitigate these formulation challenges, many studies have restricted the inclusion levels of alternative proteins to relatively low proportions (typically <10 %–20 %, and even lower for more disruptive ingredients such as microalgae at <3 %), thereby maintaining acceptable sensory characteristics. However, at these low inclusion levels, hybrid products function more as fortified meat systems than truly integrated dual-protein structures, limiting their potential contributions to sustainability, resource efficiency, and nutritional diversification. These observations indicate that the current meat-analogue approach imposes structural and sensory constraints that restrict the transformative potential of hybrid systems, particularly when evaluated against broader goals of dietary transition and protein diversification. Future innovation should therefore move beyond mimicking conventional meat products toward the development of format-specific hybrid designs, in which meat and alternative proteins are combined based on their complementary functional and sensory properties. Such approaches enable greater flexibility in product structure, flavour design, and consumer positioning, particularly in categories where expectations are less rigid (e.g., snack products, restructured formats, or novel convenience foods). Achieving this transition requires a systems-based design strategy that integrates ingredient functionality, matrix interactions, processing conditions, and consumer expectations, rather than relying solely on substitution-based formulation. This need provides the conceptual foundation for the hybrid meat framework introduced in the next section (Zhang, Farouk, et al., 2025), which supports the rational design of hybrid products beyond conventional meat analogues.

4.2.2 Hybrid meat framework

The HYBRID meat framework, established by Zhang, Farouk et al. (2025), provides a holistic systems-based roadmap for developing meat-based dual-protein foods by integrating raw material characteristics, hybridization mechanisms, processing conditions, targeted product attributes, and consumer insights into one connected design strategy. Rather than treating hybridisation as a simple replacement exercise, the framework recognises that successful product development depends on how meat and alternative proteins interact across multiple levels from their intrinsic composition and functionality to the final eating experience. At the ingredient level, both meat and alternative proteins bring variability in solubility, gelation behaviour, emulsification, flavour, colour, and nutritional profile, all of which influence their compatibility within a shared matrix. At the mechanistic level, hybrid product performance is shaped by protein-protein interactions such as hydrophobic interactions, hydrogen bonding, electrostatic attraction, and disulfide cross-linking, which are, in turn, strongly influenced by key physicochemical factors including pH, temperature, and salt content. These interactions determine the ability of the system to form cohesive structures, retain water and fat, and deliver the desired texture and stability.

Importantly, the framework emphasises that these mechanisms cannot be considered independently from processing design and consumer expectations. Processing technologies such as extrusion (Pöri, Aisala, Liu, Lille, & Sozer, 2023; Schmid, Farahnaky, Adhikari, & Torley, 2022), high-pressure treatments (Bernasconi, Szerman, Vaudagna, & Speroni, 2020; Zhao et al., 2024), ultrasound (Dong et al., 2024; Rahman & Lamsal, 2021), or targeted protein pre-modification (e.g., pH shifting) (Jiang, Wang, & Xiong, 2018; Niu, Xia, Wang, Kong, & Liu, 2018) can enhance ingredient compatibility and functionality; however, their relevance depends on alignment with the intended product format, processing intensity, and quality targets. From a consumer perspective, the framework reinforces that sensory quality remains the primary determinant of acceptance, while nutritional, sustainability, and transparency claims are most effective when they complement, rather than compensate for sensory attributes (Realini et al., 2023). Importantly, this aligns with increasing consumer preference for minimally processed foods, where products perceived as "closer to nature" are favoured over highly processed meat analogues. In this context, hybrid meat systems offer a distinct advantage by retaining a substantial fraction of conventional meat while incorporating alternative proteins, thereby enabling nutritional and sustainability improvements without requiring extensive structural modification or reliance on highly processed ingredients. In this way, the hybrid meat framework positions hybridisation not as a compromise between meat and alternatives, but as an integrated, systems-based product design strategy that aligns ingredient choice, mechanistic understanding, process design, and consumer relevance to guide the development of scalable, acceptable, and fit-for-purpose hybrid meat products.

5. Future outlooks

There is growing interest in processing technologies that can accelerate or partially replicate the effects of post-mortem ageing to address constraints associated with time, throughput, and cold-chain capacity. These technologies operate through different mechanisms and therefore serve distinct roles within meat processing systems. Some function as ageing accelerators, enhancing endogenous biochemical processes such as proteolysis by increasing enzyme accessibility, modifying cellular structures or water distribution within muscle. Others act as ageing alternatives, achieving improvements in tenderness through physical disruption or external energy input without reproducing the underlying biochemical pathways responsible for flavour development. Technologies such as pulsed electric fields (Gómez et al., 2019), ultrasound (Zhang et al., 2025), microwave (Horikoshi & Serpone, 2022), and other non-thermal or hybrid approaches (Carrillo-Lopez et al., 2024) have shown potential to shorten ageing time or improve process consistency, however, their effects are often selective, predominantly targeting the influences on tenderness with limited or inconsistent impacts on flavour and overall eating quality. In addition, their application remains constrained by challenges related to scalability, process control, product uniformity, and economic feasibility. Accordingly, these technologies are best viewed not as replacements for conventional ageing, but as complementary tools within integrated and tailored processing strategies, where their application is guided by raw material characteristics, desired quality attributes, and product positioning.

Post-mortem ageing, freezing, and minced meat formulation and hybridisation define the biochemical potential of meat products, while ultimately it is culinary transformation that determines how this potential is expressed, perceived, and valued by consumers. Culinary innovations should be treated as a delivery system for the quality potential created by processing, since even well-designed ageing (e.g., targeted strategies) and improved freeze-thaw management only translate into consumer value if they are expressed reliably at the point of cooking and consumption. This is particularly relevant for frozen/thawed meat, where water distribution, purge formation, and oxidative susceptibility can shift the way tenderness and flavour are perceived after thawing. The concept of stepwise cooking is well-known among chefs, offering a practical route to stabilise eating quality of meat by steering structural transitions more consistently than conventional high-heat cooking. In simple terms, it helps separate internal doneness and moisture management from surface browning and aroma generation, by applying gentle thermal treatment at lower temperatures (e.g., *sous-vide*) to reduce avoidable moisture loss and support tenderness development, especially in collagen-rich muscles (Naqvi et al., 2021). This is considered as a preparation step to allow the meat to be cooked to a more desirable and guaranteed doneness before a brief finishing step by searing or high-humidity convection cooking to restore roasted notes and surface colour that consumers associate with "fresh-cooked" meat. This two-stage culinary logic is particularly valuable for frozen/thawed products, where structural disruption and purge increase the risk of dryness under conventional high heat cooking (Alfaifi, Al-Ghamdi, Othman, Hobani, & Suliman, 2023). The cooking method (e.g., *sous-vide*, convection dry heat) and conditions (e.g., temperature, time) during the preparation step can also be tailored to target specific quality requirement (e.g., flavour, tenderness) (Sara et al., 2025; Subbaraj, Deb-choudhury, Pavan, & Realini, 2024). A recent study by Zhang, Deb-Choudhury, et al. (2026) provided a useful example of this principle, showing that low-temperature long-time oven precooking (49 °C for 24 h) reduced the shear force of unaged beef to levels comparable with well-aged samples, while also generating volatile fingerprint patterns that overlapped with those of dry-aged beef. Combination of stepwise ageing and tailored cooking also produced distinct volatile fingerprints while maintaining tenderness relative to dry-ageing straight, illustrating how culinary design can be used not only to express processing potential, but also to help compensate for, or strategically complement, the ageing regimes.

Looking forward, enabling technologies, such as digital and data-driven tools (e.g., artificial intelligence, AI, and machine learning), advanced analytical technologies (e.g., omics, rapid fingerprinting), and biotechnologies (e.g., precision fermentation), are expected to play a central role in advancing processing innovation to improve the sensory and nutritional quality of meat-based foods. These tools enable the modelling of complex, multivariate relationships among raw material variability, processing conditions, formulation strategies, and quality outcomes, supporting a shift from empirical, time-based rules to predictive and adaptive process design (Al-Sarayreh, Gomes Reis, Carr, & Reis, 2023; Khan, Sablani, Nayak, & Gu, 2022). Non-invasive sensing technologies (e.g., spectroscopy and imaging) strengthens process control by providing rapid, in-line indicators of composition and quality status, reducing operator subjectivity and enabling feedback loops during ageing, freezing/thawing, and downstream preparation (Dixit & Reis, 2025). Where deeper biochemical insights are required, direct analysis in real-time mass spectrometry (e.g., REIMS, DART-MS) offers a complementary capability by generating rapid molecular "fingerprints" linked to lipid profiles and metabolites, which can help tracking of flavour-precursor formation, formulation consistency, or product authentication without lengthy sample preparation (Ross et al., 2020; Zhang, Pavan, et al., 2023).

Such high-throughput phenotyping is particularly valuable when the goal is to connect mechanistic changes to sensory outcomes and ensuring product quality consistency across heterogeneous raw materials.

However, the implementation and wider industry uptake of these enabling technologies remain constrained in practice, especially by data interoperability across platforms, scaling from pilot to industrial settings, capital and operating costs, and workforce capability requirements spanning meat science, engineering, and analytics. Therefore, the current application of these enabling technologies is mainly as supporting infrastructure for outcome-oriented processing to improve reproducibility, strengthen quality assurance, and accelerate learning cycles as processors tailor ageing and product design to different species, cuts, and consumer expectations.

6. Conclusions

A central message of this chapter is that the future of meat processing innovation lies in the development of consumer-centric, systems-based processing strategies that more deliberately shape eating quality, nutritional value, and resource efficiency in response to evolving consumer demands. In whole-muscle products, innovations in stepwise and tailored ageing show how tenderness, flavour development, yield, and process efficiency can be more precisely controlled and aligned with intrinsic raw material variability, consumer expectations, and industrial constraints. The integration of ageing with freezing/thawing further illustrates that improvements in frozen meat quality depend on treating these processes as interconnected components within a unified processing system, rather than as independent operations. In minced and restructured meat systems, innovations in meat-to-meat fortification and co-product utilisation highlight the potential to improve nutritional density, increase whole-animal utilisation, and create higher value products from underutilised resources. At the same time, hybrid meat systems demonstrate that successful innovation depends not simply on replacing part of the meat fraction, but on understanding and designing the structural, functional, and sensory interactions between meat and alternative proteins within a shared matrix. In this context, the HYBRID meat framework provides a valuable systems-based approach for integrating ingredient selection, protein compatibility, processing design, product format, and consumer expectations into a coherent development strategy.

More broadly, future progress of the meat sector will depend less on single technological interventions and more on the integration and coordination of multiple processing strategies across the value chain. Within this context, the applications of advanced processing technologies, culinary development and enabling technologies can help translate processing gains into consumer value while improving process control, consistency, and scalability. Overall, a more integrated, processing-led approach offers a realistic pathway for the meat sector to deliver products that are sensorially satisfying, nutritionally relevant, economically viable, and increasingly aligned with resource efficiency and sustainability goals.

Declaration of generative AI and AI-assisted technologies in the writing process

During the preparation of this manuscript, Microsoft Copilot was used to assist in language refinement. All content was reviewed and verified by the authors, who take full responsibility for the integrity and accuracy of the work.

References

Aaslyng, M. D., & Meinert, L. (2017). Meat flavour in pork and beef – From animal to meal. *Meat Science, 132*, 112–117. https://doi.org/10.1016/j.meatsci.2017.04.012.

Abdel-Moatamed, B. R., El-Fakhrany, A.-E. M., Elneairy, N. A., Shaban, M. M., & Roby, M. H. (2024). The impact of Chlorella vulgaris fortification on the nutritional composition and quality characteristics of beef burgers. *Foods, 13*(12), 1945. https://doi.org/10.3390/foods13121945.

Al-Sarayreh, M., Gomes Reis, M., Carr, A., & Reis, M. M. D. (2023). Inverse design and AI/Deep generative networks in food design: A comprehensive review. *Trends in Food Science & Technology, 138*, 215–228. https://doi.org/10.1016/j.tifs.2023.06.005.

Alfaifi, B. M., Al-Ghamdi, S., Othman, M. B., Hobani, A. I., & Suliman, G. M. (2023). Advanced red meat cooking technologies and their effect on engineering and quality properties: A review. *Foods, 12*(13), 2564. https://doi.org/10.3390/foods12132564.

Alhujaili, A., Nocella, G., & Macready, A. (2023). Insects as food: Consumers' acceptance and marketing. *Foods, 12*(4), 886. https://doi.org/10.3390/foods12040886.

Alim, A., Yang, C., Song, H., Liu, Y., Zou, T., Zhang, Y., & Zhang, S. (2019). The behavior of umami components in thermally treated yeast extract. *Food Research International, 120*, 534–543. https://doi.org/10.1016/j.foodres.2018.11.002.

Alvarenga, T. I., Hopkins, D. L., Ramos, E. M., Almeida, A. K., & Geesink, G. (2019). Ageing-freezing/thaw process affects blooming time and myoglobin forms of lamb meat during retail display. *Meat Science, 153*, 19–25. https://doi.org/10.1016/j.meatsci.2019.02.016.

Álvarez, C., Koolman, L., Whelan, M., & Moloney, A. (2022). Effect of pre-slaughter practises and early post-mortem interventions on sheep meat tenderness and its impact on microbial status. *Foods, 11*(2), 181. https://doi.org/10.3390/foods11020181.

Álvarez, S., Álvarez, C., Hamill, R., Mullen, A. M., & O'Neill, E. (2021). Drying dynamics of meat highlighting areas of relevance to dry-aging of beef. *Comprehensive Reviews in Food Science and Food Safety, 20*(6), 5370–5392. https://doi.org/10.1111/1541-4337.12845.

Álvarez, S., Álvarez, C., Hamill, R., O'Neill, E., & Mullen, A. M. (2023). Influence of meat sample geometry on dehydration dynamics during dry-aging of beef. *Meat Science*, 109216. https://doi.org/10.1016/j.meatsci.2023.109216.

Aroeira, C. N., Torres Filho, R. d A., Fontes, P. R., Gomide, L. A. d M., Ramos, A. d L. S., Ladeira, M. M., & Ramos, E. M. (2016). Freezing, thawing and aging effects on beef tenderness from bos indicus and Bos taurus cattle. *Meat Science, 116*, 118–125. https://doi.org/10.1016/j.meatsci.2016.02.006.

Aroeira, C. N., Torres Filho, R. d A., Fontes, P. R., Ramos, A. d L. S., Gomide, L. A. d M., Ladeira, M. M., & Ramos, E. M. (2017). Effect of freezing prior to aging on myoglobin redox forms and CIE color of beef from Nellore and Aberdeen Angus cattle. *Meat Science, 125*, 16–21. https://doi.org/10.1016/j.meatsci.2016.11.010.

Balan, P., Kim, Y. H. B., Stuart, A. D., Kemp, R., Staincliffe, M., Craigie, C., & Farouk, M. M. (2019). Effect of fast freezing then thaw-aging on meat quality attributes of lamb M. longissimus lumborum. *Animal Science Journal, 90*(8), 1060–1069. https://doi.org/10.1111/asj.13216.

Barker, S., & McSweeney, M. B. (2022). Sensory characterization of yellow pea and ground chicken hybrid meat burgers using static and dynamic methodologies. *Journal of Food Science, 87*(12), 5390–5401. https://doi.org/10.1111/1750-3841.16380.

Bernaerts, T. M., Gheysen, L., Foubert, I., Hendrickx, M. E., & Van Loey, A. M. (2019). The potential of microalgae and their biopolymers as structuring ingredients in food: A review. *Biotechnology Advances, 37*(8), 107419. https://doi.org/10.1016/j.biotechadv.2019.107419.

Bernasconi, A., Szerman, N., Vaudagna, S. R., & Speroni, F. (2020). High hydrostatic pressure and soybean protein addition to beef patties: Effects on the formation of mixed aggregates and technological parameters. *Innovative Food Science & Emerging Technologies, 66*, 102503. https://doi.org/10.1016/j.ifset.2020.102503.

Bischof, G., Witte, F., Terjung, N., Heinz, V., Juadjur, A., & Gibis, M. (2024). Metabolic, proteomic and microbial changes postmortem and during beef aging. *Critical Reviews in Food Science and Nutrition, 64*(4), 1076–1109. https://doi.org/10.1080/10408398.2022.2113362.

Bošković Cabrol, M., Glišić, M., Baltić, M., Jovanović, D., Silađi, Č., Simunović, S., ... Raymundo, A. (2023). White and honey Chlorella vulgaris: Sustainable ingredients with the potential to improve nutritional value of pork frankfurters without compromising quality. *Meat Science, 198*, 109123. https://doi.org/10.1016/j.meatsci.2023.109123.

Carrillo-Lopez, L. M., Suleman, R., Garcia-Galicia, I. A., Huerta-Jimenez, M., Felix-Portillo, M., & Alarcon-Rojo, A. D. (2024). *Ultrasound combined with other technologies to accelerate meat processing. Sonochemistry*. CRC Press197–244.

Cichero, J. A. Y. (2017). Unlocking opportunities in food design for infants, children, and the elderly: Understanding milestones in chewing and swallowing across the lifespan for new innovations. *Journal of Texture Studies, 48*(4), 271–279. https://doi.org/10.1111/jtxs.12236.

Coombs, C. E., Holman, B. W., Collins, D., Friend, M. A., & Hopkins, D. L. (2017). Effects of chilled-then-frozen storage (up to 52 weeks) on lamb M. longissimus lumborum quality and safety parameters. *Meat Science, 134*, 86–97. https://doi.org/10.1016/j.meatsci.2017.07.017.

Coombs, C. E., Holman, B. W., Collins, D., Kerr, M. J., Friend, M. A., & Hopkins, D. L. (2018). Effects of chilled-then-frozen storage (up to 52 weeks) on an indicator of protein oxidation and indices of protein degradation in lamb M. longissimus lumborum. *Meat Science, 135*, 134–141. https://doi.org/10.1016/j.meatsci.2017.09.013.

Coombs, C. E., Holman, B. W., Friend, M. A., & Hopkins, D. L. (2017). Long-term red meat preservation using chilled and frozen storage combinations: A review. *Meat Science, 125*, 84–94. https://doi.org/10.1016/j.meatsci.2016.11.025.

Coombs, C. E., Holman, B. W., Ponnampalam, E. N., Morris, S., Friend, M. A., & Hopkins, D. L. (2018). Effects of chilled and frozen storage conditions on the lamb M. longissimus lumborum fatty acid and lipid oxidation parameters. *Meat Science, 136*, 116–122. https://doi.org/10.1016/j.meatsci.2017.10.013.

Dagevos, H. (2021). Finding flexitarians: Current studies on meat eaters and meat reducers. *Trends in Food Science & Technology, 114*, 530–539. https://doi.org/10.1016/j.tifs.2021.06.021.

Dasiewicz, K., Szymanska, I., Opat, D., & Hac-Szymanczuk, E. (2024). Development and characterization of hybrid burgers made from pork and multi-ingredient plant mixtures and protected with lactic acid bacteria. *Applied Sciences, 14*, 6272. https://doi.org/10.3390/app14146272.

Dixit, Y., & Reis, M. M. (2025). Chapter Three - From farm to fork: Spectroscopy in meat quality and safety assurance. In D. Cozzolino (Vol. Ed.), *Advances in food and nutrition research: 115*, (pp. 131–164). Academic Press.

Dixit, Y., Zhang, R., Farouk, M. M., dos Reis, M. M., & Realini, C. E. (2024). *Hyperspectral imaging can be a rapid tool to monitor weight loss and quality changes during dry-aging of beef.* Paper presented at the 70th International Congress of Meat Science and Technology, Foz do Iguaçu, Brazil. https://digicomst.ie/wp-content/uploads/2025/05/244.pdf.

Domínguez, R., Pateiro, M., Gagaoua, M., Barba, F. J., Zhang, W., & Lorenzo, J. M. (2019). A comprehensive review on lipid oxidation in meat and meat products. *Antioxidants, 8*(10), 429. https://doi.org/10.3390/antiox8100429.

Dong, C., Zhao, J., Wang, L., Wang, X., Jiang, J., & Bi, J. (2024). Understanding the textural enhancement of low-salt myofibrillar protein gels filled with pea protein pre-emulsions through interfacial behavior: Effects of structural modification and oil phase polarity. *Food Chemistry, 460*, 140632. https://doi.org/10.1016/j.foodchem.2024.140632.

dos Santos, M., Ribeiro, W. O., Monteiro, J. D. S., dos Santos, B. A., Campagnol, P. C. B., & Pollonio, M. A. R. (2023). Effect of transglutaminase treatment on the structure and sensory properties of rice- or soy-based hybrid sausages. *Foods, 12*(23), 4226. https://doi.org/10.3390/foods12234226.

Du, X., Wang, B., Li, H., Liu, H., Shi, S., Feng, J., ... Xia, X. (2022). Research progress on quality deterioration mechanism and control technology of frozen muscle foods. *Comprehensive Reviews in Food Science and Food Safety, 21*(6), 4812–4846. https://doi.org/10.1111/1541-4337.13040.

Egan, B. (2026). Meat in the diet of athletes and active adults. *Meat Science, 233*, 110005. https://doi.org/10.1016/j.meatsci.2025.110005.

Farouk, M. M., Jenkinson, D. I., Zhang, R., & Realini, C. E. (2024). Mincing words: Valorization of mince beef through nutrient/functional formulation. *Nutritional Valorisation of Mince Beef.*

Farouk, M. M., Mustafa, N. M., Wu, G., & Krsinic, G. (2012). The "sponge effect" hypothesis: An alternative explanation of the improvement in the waterholding capacity of meat with ageing. *Meat Science, 90*(3), 670–677. https://doi.org/10.1016/j.meatsci.2011.10.012.

Farouk, M. M., Wu, G., Frost, D. A., Staincliffe, M., & Knowles, S. O. (2019). Factors affecting the digestibility of beef and consequences for designing meat-centric meals. *Journal of Food Quality, 2019*(1), 2590182. https://doi.org/10.1155/2019/2590182.

Farouk, M. M., Yoo, M. J. Y., Hamid, N. S. A., Staincliffe, M., Davies, B., & Knowles, S. O. (2018). Novel meat-enriched foods for older consumers. *Food Research International, 104*, 134–142. https://doi.org/10.1016/j.foodres.2017.10.033.

Farouk, M. M., Zhang, R., Jenkinson, D. I., & Realini, C. E. (2025). Tailoring meat products for the elderly: A comprehensive review. *Meat Science, 219*, 109669. https://doi.org/10.1016/j.meatsci.2024.109669.

Fayet-Moore, F., Cunningham, J., Stobaus, T., & Droulez, V. (2014). Fat content and composition in retail samples of Australian beef mince. *Nutrients, 6*(6), 2217–2228. https://doi.org/10.3390/nu6062217.

Finnigan, T., Mach, K., & Edlin, A. (2024). *Mycoprotein: A healthy new protein with a low environmental impact. Sustainable protein sources.* Elsevier 539–566.

Finnigan, T. J. A., Theobald, H. E., & Bajka, B. (2025). Mycoprotein: A healthy and sustainable source of alternative protein-based foods. *Annual Review of Food Science and Technology, 16*(16), 105–125. https://doi.org/10.1146/annurev-food-111523-121802.

Gomes Martins, V. M., Milano, P., Rodrigues Pollonio, M. A., dos Santos, M., de Oliveira, A. P., Savay-da-Silva, L. K., ... de Souza Paglarini, C. (2024). Adding cricket (Gryllus assimilis) flour in hybrid beef patties: Physicochemical, technological and sensory challenges. *International Journal of Food Science and Technology, 59*(1), 450–459. https://doi.org/10.1111/ijfs.16826.

Gómez, B., Munekata, P. E. S., Gavahian, M., Barba, F. J., Martí-Quijal, F. J., Bolumar, T., ... Lorenzo, J. M. (2019). Application of pulsed electric fields in meat and fish processing industries: An overview. *Food Research International, 123*, 95–105. https://doi.org/10.1016/j.foodres.2019.04.047.

Gowda, T. K. G. M., De Zutter, L., Van Royen, G., & Van Damme, I. (2022). Exploring the microbiological quality and safety of dry-aged beef: A cross-sectional study of loin surfaces during ripening and dry-aged beef steaks from commercial meat companies in Belgium. *Food Microbiology, 102*, 103919. https://doi.org/10.1016/j.fm.2021.103919.

Grasso, S., Rondoni, A., Bari, R., Smith, R., & Mansilla, N. (2022). Effect of information on consumers' sensory evaluation of beef, plant-based and hybrid beef burgers. *Food Quality and Preference, 96*. https://doi.org/10.1016/j.foodqual.2021.104417.

Grasso, S., Smith, G., Bowers, S., Ajayi, O. M., & Swainson, M. (2019). Effect of texturised soy protein and yeast on the instrumental and sensory quality of hybrid beef meatballs. *Journal of Food Science and Technology, 56*(6), 3126–3135. https://doi.org/10.1007/s13197-018-3552-9.

Gravel, A., & Doyen, A. (2020). The use of edible insect proteins in food: Challenges and issues related to their functional properties. *Innovative Food Science & Emerging Technologies, 59*, 102272. https://doi.org/10.1016/j.ifset.2019.102272.

Grayson, A., King, D., Shackelford, S., Koohmaraie, M., & Wheeler, T. (2014). Freezing and thawing or freezing, thawing, and aging effects on beef tenderness. *Journal of Animal Science, 92*(6), 2735–2740. https://doi.org/10.2527/jas.2014-7613.

Grujić, R., Petrović, L., Pikula, B., & Amidžić, L. (1993). Definition of the optimum freezing rate - 1. Investigation of structure and ultrastructure of beef M. longissimus dorsi frozen at different freezing rates. *Meat Science, 33*(3), 301–318. https://doi.org/10.1016/0309-1740(93)90003-Z.

Holman, B. W., Coombs, C. E., Morris, S., Bailes, K., & Hopkins, D. L. (2018). Effect of long term chilled (up to 5 weeks) then frozen (up to 12 months) storage at two different sub-zero holding temperatures on beef: 2. Lipid oxidation and fatty acid profiles. *Meat Science, 136*, 9–15. https://doi.org/10.1016/j.meatsci.2018.01.028.

Holman, B. W., Coombs, C. E., Morris, S., Kerr, M. J., & Hopkins, D. L. (2017). Effect of long term chilled (up to 5 weeks) then frozen (up to 12 months) storage at two different sub-zero holding temperatures on beef: 1. Meat quality and microbial loads. *Meat Science, 133*, 133–142. https://doi.org/10.1016/j.meatsci.2017.06.015.

Horikoshi, S., & Serpone, N. (2022). Stimulating the aging of beef with microwaves. In S. Horikoshi, G. Brodie, K. Takaki, & N. Serpone (Eds.). *Agritech: Innovative agriculture using microwaves and plasmas: Thermal and non-thermal processing* (pp. 91–110). Singapore: Springer Singapore.

Hutchings, S. C., Chheang, S. L., Hedderley, D. I., Realini, C. E., & Jaeger, S. R. (2024). Investigating UK consumer perception of food sustainability and consumer strategies for improving the sustainability of their most recent meal. *Journal of Sensory Studies, 39*(5), e12948. https://doi.org/10.1111/joss.12948.

Jaeger, A., Arendt, E. K., Zannini, E., & Sahin, A. W. (2020). Brewer's spent yeast (BSY), an underutilized brewing by-product. *Fermentation, 6*(4), 123. https://doi.org/10.3390/fermentation6040123.

Jairath, G., Biswas, A. K., Mal, G., Suman, S. P., & Suman, S. (2024). Bioactive compounds in meat: Their roles in modulating palatability and nutritional value. *Meat and Muscle Biology, 8*(1), https://doi.org/10.22175/mmb.16992.

Jaworska, D., Pawłowska, J., Kostyra, E., Piotrowska, A., Płecha, M., Ostrowski, G., ... Przybylski, W. (2025). Dry-aged beef quality with the addition of Mucor flavus – Sensory, chemosensory and fatty acid analysis. *Meat Science, 220*, 109691. https://doi.org/10.1016/j.meatsci.2024.109691.

Jiang, J., Wang, Q., & Xiong, Y. L. (2018). A pH shift approach to the improvement of interfacial properties of plant seed proteins. *Current Opinion in Food Science, 19*, 50–56. https://doi.org/10.1016/j.cofs.2018.01.002.

Khan, M. I. H., Sablani, S. S., Nayak, R., & Gu, Y. (2022). Machine learning-based modeling in food processing applications: State of the art. *Comprehensive Reviews in Food Science and Food Safety, 21*(2), 1409–1438. https://doi.org/10.1111/1541-4337.12912.

Khan, R., Brishti, F. H., Arulrajah, B., Goh, Y. M., Abd Rahim, M. H., Karim, R., ... Saari, N. (2024). Mycoprotein as a meat substitute: Production, functional properties, and current challenges-a review. *International Journal of Food Science and Technology, 59*(1), 522–544. https://doi.org/10.1111/ijfs.16791.

Kim, G. D., Lee, S. Y., Jung, E. Y., Song, S., & Jin Hur, S. (2021). Quantitative changes in peptides derived from proteins in beef tenderloin (psoas major muscle) and striploin (longissimus lumborum muscle) during cold storage. *Food Chemistry, 338*, 128029. https://doi.org/10.1016/j.foodchem.2020.128029.

Kim, H. C., Baek, K. H., Ko, Y.-J., Lee, H. J., Yim, D.-G., & Jo, C. (2020). Characteristic metabolic changes of the crust from dry-aged beef using ^{2}D NMR spectroscopy. *Molecules (Basel, Switzerland), 25*(13), 3087. https://doi.org/10.3390/molecules25133087.

Kim, H. W., Kim, J. H., Seo, J. K., Setyabrata, D., & Kim, Y. H. B. (2018). Effects of aging/freezing sequence and freezing rate on meat quality and oxidative stability of pork loins. *Meat Science, 139*, 162–170. https://doi.org/10.1016/j.meatsci.2018.01.024.

Kim, Y. H. B., Frandsen, M., & Rosenvold, K. (2011). Effect of ageing prior to freezing on colour stability of ovine longissimus muscle. *Meat Science, 88*(3), 332–337. https://doi.org/10.1016/j.meatsci.2010.12.020.

Kim, Y. H. B., Liesse, C., Kemp, R., & Balan, P. (2015). Evaluation of combined effects of ageing period and freezing rate on quality attributes of beef loins. *Meat Science, 110*, 40–45. https://doi.org/10.1016/j.meatsci.2015.06.015.

Kim, Y. H. B., Ma, D., Setyabrata, D., Farouk, M. M., Lonergan, S. M., Huff-Lonergan, E., & Hunt, M. C. (2018). Understanding postmortem biochemical processes and post-harvest aging factors to develop novel smart-aging strategies. *Meat Science, 144*, 74–90. https://doi.org/10.1016/j.meatsci.2018.04.031.

Lange, K. W., & Nakamura, Y. (2021). Edible insects as future food: Chances and challenges. *Journal of Future Foods, 1*(1), 38–46. https://doi.org/10.1016/j.jfutfo.2021.10.001.

Lee, H. J., Choe, J., Kim, M., Kim, H. C., Yoon, J. W., Oh, S. W., & Jo, C. (2019). Role of moisture evaporation in the taste attributes of dry- and wet-aged beef determined by chemical and electronic tongue analyses. *Meat Science, 151*, 82–88. https://doi.org/10.1016/j.meatsci.2019.02.001.

Lee, H. J., Yoon, J. W., Kim, M., Oh, H., Yoon, Y., & Jo, C. (2019). Changes in microbial composition on the crust by different air flow velocities and their effect on sensory properties of dry-aged beef. *Meat Science, 153*, 152–158. https://doi.org/10.1016/j.meatsci.2019.03.019.

Lee, S., Jo, K., Lee, H. J., Jo, C., Yong, H. I., Choi, Y.-S., & Jung, S. (2020). Increased protein digestibility of beef with aging in an infant in vitro digestion model. *Meat Science, 169*, 108210. https://doi.org/10.1016/j.meatsci.2020.108210.

Lee, Y. E., Lee, H. J., Kim, C. H., Ryu, S., Kim, Y., & Jo, C. (2022). Effect of Penicillium candidum and Penicillium nalgiovense and their combination on the physicochemical and sensory quality of dry-aged beef. *Food Microbiology, 107*, 104083. https://doi.org/10.1016/j.fm.2022.104083.

Leighton, P. L. A., Barragán-Hernández, W., López-Campos, Ó., Segura, J., Aalhus, J. L., & Prieto, N. (2023). Effects of in-the-bag dry-ageing on meat quality, palatability and volatile compounds of low-value beef cuts. *Meat Science*, 109219. https://doi.org/10.1016/j.meatsci.2023.109219.

Li, X., Zhang, R., Hassan, M. M., Cheng, Z., Mills, J., Hou, C., ... Hicks, T. M. (2022). Active packaging for the extended shelf-life of meat: Perspectives from consumption habits, market requirements and packaging practices in China and New Zealand. *Foods, 11*(18), 2903. https://doi.org/10.3390/foods11182903.

Li, Y., Liu, X., Zhang, J., Yang, Z., Zhou, C., Wu, P., ... Liu, Y. (2024). Textured vegetable protein as a partial replacement for lean meat in salami analogues: Perspectives on physicochemical properties, flavour and proteome changes. *Food Chemistry, 463*, 140844. https://doi.org/10.1016/j.foodchem.2024.140844.

Li, Z., Leeming, M. G., Hastie, M., & Warner, R. D. (2026). Identification of protein markers of muscle fibre type and ageing progress in beef cutaneous trunci and masseter. *Food Chemistry, 503*, 147747. https://doi.org/10.1016/j.foodchem.2025.147747.

Liu, J., Song, H., Liu, Y., Li, P., Yao, J., & Xiong, J. (2015). Discovery of kokumi peptide from yeast extract by LC-Q-TOF-MS/MS and sensomics approach. *Journal of the Science of Food and Agriculture, 95*(15), 3183–3194. https://doi.org/10.1002/jsfa.7058.

Liu, Y., Sun, D., Peng, A., Li, T., Li, H., Mu, B., ... Li, G. (2024). Hydrolysis of beef sarcoplasmic protein by dry-aged beef-isolated penicillium oxalicum and Its associated metabolic pathways. *Foods, 13*(7), 1038. https://doi.org/10.3390/foods13071038.

Ma, J., Sun, Y., Meng, D., Zhou, Z., Zhang, Y., & Yang, R. (2023). Yeast proteins: The novel and sustainable alternative protein in food applications. *Trends in Food Science & Technology, 135*, 190–201. https://doi.org/10.1016/j.tifs.2023.04.003.

Melios, S., & Grasso, S. (2024). Meat fans' and meat reducers' attitudes towards meat consumption and hybrid meat products in the UK: A cluster analysis. *International Journal of Food Science and Technology, 59*, 9394–9401. https://doi.org/10.1111/ijfs.17586.

Miao, X., Hastie, M., Ha, M., & Warner, R. (2023). Consumer response to blended beef burgers and chicken nuggets is influenced by ingredient and nutrition claims - qualitative assessment. *Future Foods, 8*, 100247. https://doi.org/10.1016/j.fufo.2023.100247.

Mikami, N., Toyotome, T., Yamashiro, Y., Sugo, K., Yoshitomi, K., Takaya, M., ... Shimada, K. (2021). Dry-aged beef manufactured in Japan: Microbiota identification and their effects on product characteristics. *Food Research International, 140*, 110020. https://doi.org/10.1016/j.foodres.2020.110020.

Mojet, J., Christ-Hazelhof, E., & Heidema, J. (2001). Taste perception with age: Generic or specific losses in threshold sensitivity to the five basic tastes? *Chemical Senses, 26*(7), 845–860. https://doi.org/10.1093/chemse/26.7.845.

Naqvi, Z. B., Thomson, P. C., Ha, M., Campbell, M. A., McGill, D. M., Friend, M. A., & Warner, R. D. (2021). Effect of sous vide cooking and ageing on tenderness and water-holding capacity of low-value beef muscles from young and older animals. *Meat Science, 175*, 108435. https://doi.org/10.1016/j.meatsci.2021.108435.

Niu, H., Xia, X., Wang, C., Kong, B., & Liu, Q. (2018). Thermal stability and gel quality of myofibrillar protein as affected by soy protein isolates subjected to an acidic pH and mild heating. *Food Chemistry, 242*, 188–195. https://doi.org/10.1016/j.foodchem.2017.09.055.

Pennings, B., Groen, B. B., van Dijk, J.-W., de Lange, A., Kiskini, A., Kuklinski, M., ... Van Loon, L. J. (2013). Minced beef is more rapidly digested and absorbed than beef steak, resulting in greater postprandial protein retention in older men. *The American Journal of Clinical Nutrition, 98*(1), 121–128. https://doi.org/10.3945/ajcn.112.051201.

Pöri, P., Aisala, H., Liu, J., Lille, M., & Sozer, N. (2023). Structure, texture, and sensory properties of plant-meat hybrids produced by high-moisture extrusion. *LWT, 173*, 114345. https://doi.org/10.1016/j.lwt.2022.114345.

Przybylski, W., Jaworska, D., Kresa, P., Ostrowski, G., Płecha, M., Korsak, D., ... Pawłowska, J. (2024). Fungal biostarter and bacterial occurrence of dry-aged beef: The sensory quality and volatile aroma compounds after 21 days of aging. *Applied Sciences, 14*(19), 9053. https://doi.org/10.3390/app14199053.

Purslow, P. P. (2018). Contribution of collagen and connective tissue to cooked meat toughness; some paradigms reviewed. *Meat Science, 144*, 127–134. https://doi.org/10.1016/j.meatsci.2018.03.026.

Rahman, M. M., & Lamsal, B. P. (2021). Ultrasound-assisted extraction and modification of plant-based proteins: Impact on physicochemical, functional, and nutritional properties. *Comprehensive Reviews in Food Science and Food Safety, 20*(2), 1457–1480. https://doi.org/10.1111/1541-4337.12709.

Ramanathan, R., Mafi, G. G., Yoder, L., Perry, M., Pfeiffer, M., VanOverbeke, D. L., & Maheswarappa, N. B. (2020). Chapter 5 - Biochemical changes of postmortem meat during the aging process and strategies to improve the meat quality. In A. K. Biswas, & P. K. Mandal (Eds.). *Meat quality analysis* (pp. 67–80). Academic Press.

Realini, C. E., Ares, G., Antúnez, L., Brito, G., Luzardo, S., del Campo, M., ... Montossi, F. M. (2022). Meat insights: Uruguayan consumers′ mental associations and motives underlying consumption changes. *Meat Science, 192*, 108901. https://doi.org/10.1016/j.meatsci.2022.108901.

Realini, C. E., Driver, T., Zhang, R., Guenther, M., Duff, S., Craigie, C. R., ... Farouk, M. M. (2023). Survey of New Zealand consumer attitudes to consumption of meat and meat alternatives. *Meat Science, 203*, 109232. https://doi.org/10.1016/j.meatsci.2023.109232.

Ross, A., Brunius, C., Chevallier, O., Dervilly, G., Elliott, C., Guitton, Y., ... Vanhaecke, L. (2020). Making complex measurements of meat composition fast: Application of rapid evaporative ionisation mass spectrometry to measuring meat quality and fraud. *Meat Science*, 108333. https://doi.org/10.1016/j.meatsci.2020.108333.

Roy, B. C., & Bruce, H. L. (2024). Contribution of intramuscular connective tissue and its structural components on meat tenderness-revisited: A review. *Critical Reviews in Food Science and Nutrition, 64*(25), 9280–9310. https://doi.org/10.1080/10408398.2023.2211671.

Rumpold, B. A., & Schlüter, O. K. (2013). Nutritional composition and safety aspects of edible insects. *Molecular Nutrition & Food Research, 57*(5), 802–823. https://doi.org/10.1002/mnfr.201200735.

Ryder, C., Jaworska, S., & Grasso, S. (2023). Hybrid meat products and co-creation: What do consumers say, feel and think? *Frontiers in Nutrition, 10*, 1106079. https://doi.org/10.3389/fnut.2023.1106079.

Sales, L. A., Rodrigues, L. M., Silva, D. R. G., Fontes, P. R., Torres Filho, R. D. A., Ramos, A. D. L. S., & Ramos, E. M. (2020). Effect of freezing/irradiation/thawing processes and subsequent aging on tenderness, color, and oxidative properties of beef. *Meat Science, 163*. https://doi.org/10.1016/j.meatsci.2020.108078.

Salgaonkar, K., & Nolden, A. A. (2024). Exploring consumer preferences and challenges in hybrid meat products: A conjoint analysis of hotdogs. *Foods, 13*(10), 1460. https://doi.org/10.3390/foods13101460.

Samad, A., Muazzam, A., Alam, A. M. M. N., Kim, S., Hwang, Y.-H., & Joo, S.-T. (2026). A comprehensive review of technological advances in meat safety, quality, and sustainability for public health. *Foods, 15*(1), 47. https://doi.org/10.3390/foods15010047.

Samarakoon Mudiyanselage, C. S. S., Farouk, M. M., Realini, C. E., Kantono, K., & Hamid, N. (2025). Determinants of purchase intention for meat-based chilled ready meals in New Zealand: A consumer behaviour perspective. *Foods, 14*(6), 1038. https://doi.org/10.3390/foods14061038.

Sara, V. G., Tyler, W. T., Ifigenia, G., Robert, J. D., John, A. S., & Mahesh, N. N. (2025). Optimizing sous vide cooking temperature and time for beef palatability. *Meat and Muscle Biology, 9*(1), https://doi.org/10.22175/mmb.18928.

Schmid, E. M., Farahnaky, A., Adhikari, B., & Torley, P. J. (2022). High moisture extrusion cooking of meat analogs: A review of mechanisms of protein texturization. *Comprehensive Reviews in Food Science and Food Safety, 21*(6), 4573–4609. https://doi.org/10.1111/1541-4337.13030.

Schulte, M., Johnson, L., Zuber, E., Patterson, B., Outhouse, A., Fedler, C., ... Lonergan, S. (2019). Influence of postmortem aging and post-aging freezing on pork loin quality attributes. *Meat and Muscle Biology, 3*, 313. https://doi.org/10.22175/mmb2019.05.0015.

Senarathna, S., Mel, R., & Malalgoda, M. (2024). Utilization of cereal-based protein ingredients in food applications. *Journal of Cereal Science, 116*, 103867. https://doi.org/10.1016/j.jcs.2024.103867.

Setyabrata, D., Hernandez, S., Cooper, B. R., Legako, J. F., & Kim, Y. H. B. (2024). Inclusion of dry-aged beef trimmings as a quality and flavor enhancer for ground beef. *Meat and Muscle Biology, 8*(1), https://doi.org/10.22175/mmb.17794.

Setyabrata, D., & Kim, Y. H. B. (2019). Impacts of aging/freezing sequence on microstructure, protein degradation and physico-chemical properties of beef muscles. *Meat Science, 151*, 64–74. https://doi.org/10.1016/j.meatsci.2019.01.007.

Setyabrata, D., Xue, S., Vierck, K., Legako, J., Ebner, P., Zuelly, S., & Kim, Y. H. B. (2022). Impact of various dry-aging methods on meat quality and palatability attributes of beef loins (M. longissimus lumborum) from cull cow. *Meat and Muscle Biology, 6*(1), https://doi.org/10.22175/mmb.13025.

Siladji, C., Djordjevic, V., Milijasevic, J. B., Heinz, V., Terjung, N., Sun, W., & Tomasevic, I. (2024). Micro-and macroalgae in meat products. *Foods, 13*(6), 826.

Stafford, C. D., Alruzzi, M. A., Gagaoua, M., & Matarneh, S. K. (2025). Postmortem proteolysis and its indicators vary within bovine muscles: Novel insights in muscles that differ in their contractile, metabolic, and connective tissue properties. *Meat Science, 221*, 109718. https://doi.org/10.1016/j.meatsci.2024.109718.

Subbaraj, A. K., Deb-choudhury, S., Pavan, E., & Realini, C. E. (2024). Volatile fingerprints of beef cooking methods using sol–gel-based solid-phase microextraction (SPME) and direct analysis in real-time mass spectrometry (DART-MS). *Rapid Communications in Mass Spectrometry, 38*(1), e9655. https://doi.org/10.1002/rcm.9655.

Taylor, J., Ahmed, I. A. M., Al-Juhaimi, F. Y., & Bekhit, A. E.-D. A. (2020). Consumers' perceptions and sensory properties of beef patty analogues. *Foods, 9*(1), 63. https://doi.org/10.3390/foods9010063.

Terlouw, E. M. C., Picard, B., Deiss, V., Berri, C., Hocquette, J.-F., Lebret, B., ... Gagaoua, M. (2021). Understanding the determination of meat quality using biochemical characteristics of the muscle: Stress at slaughter and other missing Keys. *Foods, 10*(1), 84. https://doi.org/10.3390/foods10010084.

U.S. Department of Agriculture and U.S. Department of Health and Human Services. (2026). *Dietary Guidelines for Americans, 2025–2030*. U.S. Department of Agriculture and U.S. Department of Health and Human Services. Retrieved from . https://cdn.realfood.gov/DGA_508.pdf#page=2.08.

Vieira, C., Diaz, M., Martínez, B., & García-Cachán, M. (2009). Effect of frozen storage conditions (temperature and length of storage) on microbiological and sensory quality of rustic crossbred beef at different states of ageing. *Meat Science, 83*(3), 398–404. https://doi.org/10.1016/j.meatsci.2009.06.013.

Vlahova-Vangelova, D., Balev, D., & Kolev, N. (2023). Cricket powder (Acheta domesticus) as a lean pork meat replacer in cooked sausages. *Future of Food: Journal on Food, Agriculture and Society, 11*(4), 1–12. https://doi.org/10.17170/kobra-202210056951.

Wang, P., Jin, Z., & Xu, X. (2015). Physicochemical alterations of wheat gluten proteins upon dough formation and frozen storage – A review from gluten, glutenin and gliadin perspectives. *Trends in Food Science & Technology, 46*(2, Part A), 189–198. https://doi.org/10.1016/j.tifs.2015.10.005.

Wiklund, E., Farouk, M. M., Stuart, A., & Dobbie, P. (2009). *Consumer evaluation of chilled-never-frozen versus chilled-frozen-thawed beef and venison.* Paper presented at the 55th International Congress of Meat Science and Technology, Copenhagen, Denmark.

Xu, L., Liu, S., Cheng, Y., & Qian, H. (2021). The effect of aging on beef taste, aroma and texture, and the role of microorganisms: A review. *Critical Reviews in Food Science and Nutrition,* 1–12. https://doi.org/10.1080/10408398.2021.1971156.

Xu, X., Sharma, P., Shu, S., Lin, T.-S., Ciais, P., Tubiello, F. N., ... Jain, A. K. (2021). Global greenhouse gas emissions from animal-based foods are twice those of plant-based foods. *Nature Food, 2*(9), 724–732. https://doi.org/10.1038/s43016-021-00358-x.

Xue, S., Setyabrata, D., Bonham, C. C., & Brad Kim, Y. H. (2021). Evaluation of functional and chemical properties of crust from dry-aged beef loins as a novel food ingredient. *Meat Science, 173*, 108403. https://doi.org/10.1016/j.meatsci.2020.108403.

Yang, S., Wang, Y., Wang, J., Cheng, K., Liu, J., He, Y., ... Sun, H. (2024). Microalgal protein for sustainable and nutritious foods: A joint analysis of environmental impacts, health benefits and consumer's acceptance. *Trends in Food Science & Technology, 143*, 104278. https://doi.org/10.1016/j.tifs.2023.104278.

Zhang, F., Li, X., Liang, X., Kong, B., Sun, F., Cao, C., ... Liu, Q. (2024). Feasibility of Tenebrio molitor larvae protein to partially replace lean meat in the processing of hybrid frankfurters: Perspectives on quality profiles and in vitro digestibility. *Food Research International, 176*, 113846. https://doi.org/10.1016/j.foodres.2023.113846.

Zhang, J., Toldrá, F., Kang, D., Zhou, L., Wang, J., Zhang, W., & Hu, Y. (2025). Benefits of ultrasonic technology application in meat field and its influential mechanism: A review. *Critical Reviews in Food Science and Nutrition, 65*(28), 5726–5751. https://doi.org/10.1080/10408398.2024.2429002.

Zhang, R., Deb-Choudhury, S., Jacob, N., Subbaraj, A. K., Hefer, C. A., Ellies-Oury, M.-P., ... Realini, C. E. (2026). Tailoring dry-ageing and cooking to enhance technological quality and volatile fingerprints in beef striploin. *Meat Science, 239*, 110136. https://doi.org/10.1016/j.meatsci.2026.110136.

Zhang, R., Farouk, M. M., Realini, C. E., & Thum, C. (2025). Hybridization in meat-based dual protein foods: Mechanisms, challenges, and consumer insights. *Comprehensive Reviews in Food Science and Food Safety, 24*(4), e70216. https://doi.org/10.1111/1541-4337.70216.

Zhang, R., Jacob, N., Agnew, M., Farouk, M. M., & Realini, C. E. (2025). Dry-ageing impacts on meat quality, oxidative stability, and release of free amino acids in striploins from dairy crossbred yearling and 2-year-old steers. *Meat Science, 219*, 109674. https://doi.org/10.1016/j.meatsci.2024.109674.

Zhang, R., Jacob, N., Thum, C., Realini, C. E., & Farouk, M. M. (2026). Hybridisation of yearling beef mince with Chlorella biomass and nutritional yeast in patty and soft jerky model systems. *Meat Science, 232*, 109977. https://doi.org/10.1016/j.meatsci.2025.109977.

Zhang, R., Kallas, Z., Conner, T. S., Loeffen, M. P., Lee, M., Day, L., ... Realini, C. E. (2024). Factors influencing the willingness to pay a price premium for red meat with potential to improve consumer wellness in Australia and the United States of America. *Meat science, 213*, 109495. https://doi.org/10.1016/j.meatsci.2024.109495.

Zhang, R., Maes, E., Hefer, C., Noble, A., Thomas, A., Yoo, M. J., ... Deb-Choudhury, S. (2024). A peptidomic approach to discover the potential biomolecular signatures discriminating in-bag dry-and wet-aged lamb. *Food Chemistry Advances, 4*, 100664. https://doi.org/10.1016/j.focha.2024.100664.

Zhang, R., Pavan, E., Ross, A. B., Deb-Choudhury, S., Dixit, Y., Mungure, T. E., ... Farouk, M. M. (2023). Molecular insights into quality and authentication of sheep meat from proteomics and metabolomics. *Journal of Proteomics, 276*, 104836. https://doi.org/10.1016/j.jprot.2023.104836.

Zhang, R., Realini, C. E., Kim, Y. H. B., & Farouk, M. M. (2023). Challenges and processing strategies to produce high quality frozen meat. *Meat Science, 205*, 109311. https://doi.org/10.1016/j.meatsci.2023.109311.

Zhang, R., Ross, A. B., Jacob, N., Agnew, M., Staincliffe, M., & Farouk, M. M. (2023). Rapid evaporative ionisation mass spectrometry fingerprinting can discriminate lamb meat due to different ageing methods and levels of dehydration. *Journal of Proteomics, 272*, 104771. https://doi.org/10.1016/j.jprot.2022.104771.

Zhang, R., Yoo, M. J., Realini, C. E., Staincliffe, M., & Farouk, M. M. (2021). In-bag dry- vs. wet-aged lamb: Quality, consumer acceptability, oxidative stability and in vitro digestibility. *Foods, 10*(1), 41. https://doi.org/10.3390/foods10010041.

Zhang, R., Yoo, M. J. Y., & Farouk, M. M. (2019). Quality and acceptability of fresh and long-term frozen in-bag dry-aged lean bull beef. *Journal of Food Quality, 2019*, 1975264. https://doi.org/10.1155/2019/1975264.

Zhang, R., Yoo, M. J. Y., Ross, A. B., & Farouk, M. M. (2022). Mechanisms and strategies to tailor dry-aged meat flavour. *Trends in Food Science & Technology, 119*, 400–411. https://doi.org/10.1016/j.tifs.2021.12.023.

Zhang, Y., & Ertbjerg, P. (2018). Effects of frozen-then-chilled storage on proteolytic enzyme activity and water-holding capacity of pork loin. *Meat Science, 145*, 375–382. https://doi.org/10.1016/j.meatsci.2018.07.017.

Zhang, Y., Liu, G., Xie, Q., Wang, Y., Yu, J., & Ma, X. (2023). A comprehensive review of the principles, key factors, application, and assessment of thawing technologies for muscle foods. *Comprehensive Reviews in Food Science and Food Safety, 22*(1), 107–134. https://doi.org/10.1111/1541-4337.13064.

Zhao, S., Yang, L., Chen, X., Zhao, Y., Ma, H., Wang, H., & Su, A. (2024). Modulation of the conformation, water distribution, and rheological properties of low-salt porcine myofibrillar protein gel influenced by modified quinoa protein. *Food Chemistry, 455*. https://doi.org/10.1016/j.foodchem.2024.139902.

CHAPTER THREE

Hybrid food design as a way to enhance the nutritional and functional properties of animal-based products

L. Théron, M. Meurillon, S. Portanguen, P.-S. Mirade, and T. Astruc*
INRAE, UR370 Qualité des Produits Animaux (QuaPA), Saint-Genès-Champanelle, France
*Corresponding author. e-mail address: thierry.astruc@inrae.fr

Contents

Abstract

This chapter examines hybrid food systems as a strategic response to the challenges of rising global protein demand, population aging, environmental constraints, and changing dietary expectations. Hybrid foods, defined as formulations combining animal-derived ingredients with plant, fungal, insect, or other novel protein sources, are presented as a pragmatic pathway to reconcile nutritional adequacy, technological functionality, sustainability, and consumer acceptance within the ongoing protein transition.

The chapter reviews the wide range of raw materials used in hybrid food design, emphasizing their complementary roles. Plant and fungal ingredients contribute

Advances in Food and Nutrition Research, Volume 122
ISSN 1043-4526, https://doi.org/10.1016/bs.afnr.2026.03.007

dietary fiber, bioactive compounds, and reduced environmental impact, while animal-derived components ensure high-quality protein, essential micronutrients, and techno-functional properties. These combinations enable nutritionally complete products while promoting resource efficiency and circularity through the valorization of underutilized by-products.

Processing technologies are identified as central to hybrid food performance. Conventional and emerging processes, such as mixing, cooking, extrusion, non-thermal treatments, and 3D food printing, govern structure, sensory quality, digestibility, and nutrient bioavailability by shaping interactions between heterogeneous protein sources. Evidence reviewed in the chapter indicates that well-designed hybrid matrices can maintain, or even enhance, protein digestibility and mineral bioavailability compared to purely animal- or plant-based products.

The chapter further addresses health implications, consumer acceptance, and environmental performance. Hybrid foods can reduce saturated fat intake and environmental footprints while maintaining sensory appeal, particularly for flexitarian consumers. Life cycle assessments and market analyses suggest that hybrid foods align with dietary guidelines and sustainability policies, positioning them as a scalable and socially acceptable lever for future food systems.

1. Introduction

The global agri-food system is currently facing unprecedented challenges that call for a profound transformation of production and consumption models. In parallel with global population growth, the demand for dietary proteins is expected to increase by up to 70 % by 2050 (FAO, 2018), placing unprecedented pressure on food systems that are already resource-intensive and environmentally constrained. At the same time, global population aging is accelerating: the number of individuals aged 65 years and older is projected to double between 2020 and 2050, reaching approximately 1.6 billion worldwide (United Nations, 2022). Aging populations are associated with increased protein requirements to counteract sarcopenia, anabolic resistance, and age-related declines in appetite and masticatory efficiency (Bauer et al., 2013; Cruz-Jentoft et al., 2019; Morley, 2012). At the same time, animal-based proteins, while nutritionally dense and technologically functional, are associated with disproportionately high greenhouse gas emissions and land use compared to plant-derived proteins (Gerber, Food and Agriculture Organization of the United Nations, 2013; Poore & Nemecek, 2018). In this context, hybrid foods combining animal, plant, and/or fungal protein sources represent a promising lever to reconcile high protein quality, technological functionality, and environmental sustainability within the ongoing protein transition (Aiking & De Boer, 2020;

McClements & Grossmann, 2021). These dynamics profoundly reshapes global food demand and supply, compelling both producers and consumers to reconsider food choices considering sustainability, health, and system resilience (Grasso, 2024; Willett et al., 2019).

This ongoing transformation, often described as a global food transition, reflects a collective effort to move toward more diversified and sustainable diets, a reduction in environmental footprints, and greater efficiency in the use of biological resources. Governments, industries, and research institutions are increasingly seeking innovative pathways to secure food availability while mitigating the ecological and social costs associated with conventional animal-based production. Adapting to these pressures requires not only technological innovation but also a fundamental shift in consumption patterns, with diets evolving from animal-dominant models toward more balanced approaches that integrate plant-based alternatives. A central aspect of this evolution is the need to limit the overconsumption of red meat, which is associated with both environmental (Migliore et al., 2015; Parlasca & Qaim, 2022; Roozen & Raedts, 2023) and health-related concerns (Parlasca & Qaim, 2022; Willett et al., 2019). From an ecological perspective, livestock production contributes substantially to greenhouse gas emissions and water use, but animal breeding also helps to fertilize the soil by providing organic carbon, and biological activity (Conant, 2010). Moreover, the shift towards a consumption of exclusively plant-based food, whose protein yield is significantly lower, would be accompanied by intense deforestation, resulting in less CO_2 recycling and large modification of our ecosystem. From a health standpoint, moderate consumption of red meat is beneficial to health because it is rich in high-quality protein and easily absorbed iron, but an overconsumption of red and processed meats has been linked to an increased risk of cardiovascular diseases, certain cancers, and metabolic disorders WHO (2023). Reducing meat consumption, therefore, has become a key objective in the pursuit of both planetary and human health, encouraging consumers and industries alike to seek more sustainable sources of animal protein.

Beyond consumption habits, the optimization of resource use within the meat production chain also represents a crucial step toward sustainability. In conventional processing systems, a large share of animal by-products, such as offal's, blood, and connective tissues, remains underexploited despite their high nutritional potential (Llauger et al., 2021). Enhancing their recovery and valorization not only reduces waste but also increases the total protein yield per animal, contributing to a circular bioeconomy in which every fraction of the carcass is used efficiently and responsibly (Llauger et al., 2021). These evolving priorities have prompted extensive research into the development of new foods that reconcile sensory appeal, nutritional quality, and environmental responsibility. Technological innovation in food science, such as advances in protein extraction, plant-animal ingredient texturization, and functionalization, will enable the design of products that are simultaneously nutritious, palatable, and low in carbon emissions. The goal is to create foods that satisfy consumer expectations for taste and convenience while addressing the urgent need to reduce environmental impacts and reliance on resource-intensive ingredients. Furthermore, more and more consumers are interested in the health properties of the food that they eat. Raw materials from various sources contain micronutrients and compounds that could help prevent certain deficiencies and metabolic diseases. Food fortification strategies leveraging the properties of these raw materials, are being explored to meet consumer needs and produce healthy foods (Ashraf, 2025; Marcìa-Fuentes et al., 2026).

Within this context, the emergence of hybrid foods represents a significant milestone in the reconfiguration of global food systems. Positioned at the intersection of sustainability, nutrition, and innovation, these products offer a tangible and scalable solution to the challenge of feeding a growing global population while preserving both environmental integrity and the sensory quality that drives consumer acceptance.

2. Main raw materials used in the design of hybrid food

The design of hybrid foods relies on the careful selection and combination of raw materials from both plant and animal origins, with the possible inclusion of fungal or insect sources to further enhance nutritional, functional, and sensory properties (Fig. 1). The goal is to create good sensory and nutritional properties while incorporating sustainable, high-quality plant and fungal ingredients that reduce the overall environmental impact and improve health outcomes.

2.1 Plant source

From a plant-based perspective, the most relevant sources used in hybrid formulations are legumes, cereals, algae and oilseeds, which offer complementary nutritional and functional attributes (Table 1). Legumes such as

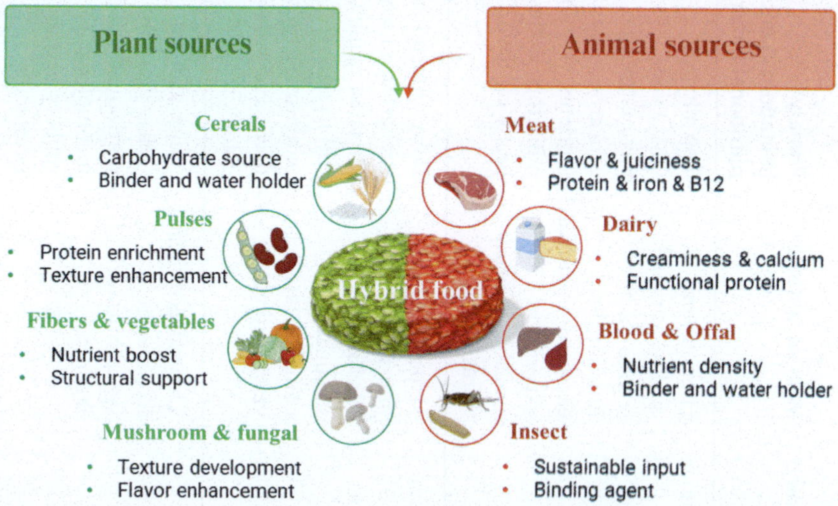

Fig. 1 Conceptual overview of plant, fungal, insect, and animal raw materials used in the design of hybrid foods. *Created in BioRender. (Theron, 2026) https://BioRender.com/prs4wvk.*

soy, pea, lentil, faba bean, and chickpea are particularly valuable due to their high protein content (typically 20–35 %), balanced amino acid profile, and technological versatility (Tkaczewska et al., 2025). Isolated or concentrated plant proteins derived from these sources can improve water- and fat-binding capacities, gel formation, and emulsion stability, making them suitable for integration into comminuted meat products such as burgers, sausages, or *pâtés* (Boukid et al., 2024).

Cereal proteins, especially from wheat (gluten) and rice, contribute to texture and structural cohesion, while starch fractions from corn or potato may enhance binding and juiciness (Kaur et al., 2026). Oilseeds like sunflower and rapeseed also provide lipid components and bioactive compounds (e.g., tocopherols, polyunsaturated fatty acids) that contribute to the nutritional balance of hybrid formulations (Grompone, 2011). Moreover, plant-derived fibers and polysaccharides (e.g., inulin, pectin, or cellulose derivatives) are often incorporated to improve water retention, reduce cooking losses, and increase the dietary fiber content of the final product (Zinina et al., 2019).

2.2 Fungal/mushroom source

In parallel, fungal or mushroom-based ingredients are gaining increasing attention as innovative raw materials for hybrid food design. Edible mushrooms

Table 1 Comparison of plant protein sources for hybrid food design.

Plant protein source	Protein content (%)	Nutritional profile & PDCAAS	Key functional properties	Limiting amino acids	Main advantages	Primary limitations & anti-nutritional factors
Soy Protein	Approximately 40 % (flour), more than 80–90 % (isolates)	Complete EAA profile; high digestibility (>90 %); PDCAAS close to 1.0	Excellent gelling, emulsifying, foaming, water and fat binding; forms firm elastic textures	Sulfur amino acids (methionine, cysteine)	High protein quality; versatile techno-functionality; extensive industrial experience; available in many forms	Top allergen; GMO and phytoestrogen concerns; anti-nutrients (trypsin inhibitors, lectins); beany off flavors
Pea Protein	From 20 to 25 % (seed), and 80–85 % (concentrates)	Rich in lysine and threonine; high digestibility (>90 %); PDCAAS of 0.78–0.91	Good emulsification; moderate foaming; gels and texturizes but less robust than soy	Sulfur amino acids (methionine, cysteine)	Hypoallergenic; non-GMO; clean-label image; sustainable; complements cereal proteins	Lower functional strength than soy; phytates, tannins, and trypsin inhibitors; earthy flavor and yellow color
Wheat Gluten	From 75 to 85 %	Low PDCAAS (0.25−0.40) due to lysine deficiency; high digestibility	Unique viscoelasticity; critical for fibrous, meat-like chew; excellent binding and water-holding	Lysine	Unparalleled meat-like texture; synergistic with legumes	Major allergen (celiac disease); nutritionally incomplete; rubbery texture if overused

Faba Bean Protein	From 26 to 30 % (seed)	High protein yield; high digestibility (processed); PDCAAS of 0.7–0.8	Excellent for extrusion; creates fibrous layered structures; good for emulsion gels (3D printing)	Sulfur amino acids (methionine, cysteine) and tryptophan	Highly suitable for meat analogues; high protein content; low allergenicity; safe if vicine-free	Vicine and convicine (favism risk); tannins and phytates; bitter taste; emerging supply chain
Chickpea Protein	Approximately 20 % (seed)	Balanced EAA profile; high amino acid score; PDCAAS of 0.7–0.8 (boiled)	Exceptional water and oil binding; stable foams; moderate gelling	Sulfur amino acids (methionine), marginal tryptophan/threonine	Hypoallergenic; no phytoestrogens; consumer familiarity; improves juiciness and yield in hybrids	Low solubility at neutral pH; low foaming capacity; phytic acid and raffinose-family oligosaccharides
Lentil Protein	Approximately 25 % (seed)	Rich in lysine; includes bioactive polyphenols; PDCAAS from 0.68 to 0.80	Good emulsification; high gel strength at specific pH; moderate water-holding	Sulfur amino acids (methionine, cysteine), leucine, or valine	Nutrient-dense (fiber, iron); mild flavor; versatile generalist; sustainable and non-allergenic	High tannins and phytates; less texturization potential than soy; flatulence factors; color issues

PDCAAS: Protein Digestibility Corrected Amino Acid Score; EAA: Essential Amino Acids; GMO: Genetically Modified Organism.
Adapted from Hertzler et al. (2020), Ma et al. (2022), Benković et al. (2023), and Zhang et al. (2025).

such as *Pleurotus ostreatus* (oyster mushroom), *Agaricus bisporus* (button mushroom), *Lentinula edodes* (shiitake), and *Flammulina velutipes* (enoki) provide a unique combination of umami flavor, fibrous texture, and high moisture content, which can enhance the palatability and mouthfeel of hybrid products. Rich in protein (20–30 % dry weight), dietary fiber (notably β-glucans), phenolic compounds and minerals such as selenium, zinc, iron and potassium, mushrooms offer both nutritional and functional benefits (Pérez-Montes et al., 2021). Their natural flavor compounds, such as glutamic acid and 5'-nucleotides, serve as natural flavor enhancers, allowing for a reduction in added salt and artificial additives (Pérez-Montes et al., 2021). In addition to whole mushroom tissues, mycoprotein ingredients derived from fungal biomass (e.g., *Fusarium venenatum*, used in Quorn™ products) represent a promising avenue for incorporation into hybrid foods. Mycoproteins combine high-quality protein with low environmental impact and a fibrous texture that mimics muscle tissue, making them ideal for structuring and texturizing hybrid formulations.

2.3 Animal source

Raw materials of animal origin remain essential in hybrid food systems, as they provide nutritional and functional properties that plant-based components alone cannot fully replicate. These include lean meat, mechanically separated meat, blood plasma, collagen, and edible offal such as liver, heart, or kidney. Animal proteins are characterized by their high biological value, complete amino acid profile, and strong heat-induced gelation capacity, which underpins the texture of traditional meat products. In addition, animal ingredients ensure the presence of the "meat factor", which enhances the bioavailability of key micronutrients such as iron, zinc, and vitamin B12, including whose of plant origins (Consalez et al., 2022). The incorporation of underutilized animal tissues, particularly edible offal, contributes simultaneously to nutritional enrichment and improved resource efficiency. Offal are among the most nutrient-dense animal-derived foods, containing high levels of high-quality proteins, B-complex vitamins, fat-soluble vitamins, and essential minerals. Liver and kidney, for instance, are especially rich in vitamin B12, folate, vitamin A, iron, zinc, and selenium in highly bioavailable forms, with 100 g of liver covering more than half of the daily requirements for several micronutrients and exceeding vitamin A recommendations (Latoch et al., 2024). This exceptional nutrient density makes offal an effective means of naturally fortifying hybrid foods, reducing reliance on synthetic additives. From a sustainability perspective, the valorization of offal supports whole-animal utilization, reduces waste, and lowers the environmental footprint of meat processing. Integrating organ meats into human diets decreases disposal burdens while providing affordable, nutrient-rich foods (Jayathilakan et al., 2012). Offal can partially replace prime muscle meat in hybrid formulations, thereby reducing costs and resource use without compromising nutritional quality. Functionally, offal offers additional advantages: liver contributes to a smooth texture and emulsifying capacity, heart provides a firm, meat-like structure, and kidney can enhance flavor complexity when used in small amounts. Although sensory challenges related to strong flavors may arise, appropriate preprocessing and formulation strategies can mitigate these effects. Overall, the use of offal in hybrid foods represents a "win-win" strategy, combining enhanced nutrient density with improved sustainability and circularity (Jayathilakan et al., 2012; Latoch et al., 2024).

Animal connective tissue-derived ingredients such as collagen and gelatin play a key role as functional binders in hybrid food systems. Collagen, the main structural protein of animal skin, bones, and tendons, yields gelatin upon partial hydrolysis and is widely used for its gelation, water-binding, emulsifying, foaming, and film-forming properties (Ahmad et al., 2024). In hybrid formulations, collagen and gelatin improve texture, stability, and juiciness by reinforcing protein networks and enhancing water and fat retention, thereby preventing dryness and crumbliness (Gómez-Guillén et al., 2011). These effects are particularly valuable in reduced-meat products, where gelatin can compensate for lower meat content by restoring elasticity and a meat-like bite. In practice, the addition of small amounts of gelatin in hybrid sausages or nuggets increases yield and firmness, resulting in textures comparable to full-meat products (Suurs & Barbut, 2020). From a nutritional perspective, collagen is rich in glycine, proline, and hydroxyproline but limited in essential amino acids; thus, its primary contribution in hybrid foods is functional rather than nutritional, although it can still contribute to protein content and is often regarded as a natural, clean-label ingredient (Ahmad et al., 2024).

Slaughterhouse blood, particularly the plasma fraction, is another animal-derived resource that can be harnessed in hybrid foods as a potent binder and emulsifier. Blood plasma is rich in albumin, globulins, and fibrinogen proteins (Stožinić et al., 2025), which collectively exhibit excellent functional properties: high water-holding capacity, strong gelation, foaming, and emulsifying ability. In food processing, spray-dried plasma powder has long been used in processed meat products (like frankfurters and pâtés) to improve texture and stability (Appiah & Peggy, 2012). Incorporating plasma proteins can dramatically enhance water retention and fat binding in a hybrid formulation, yielding a juicier, firmer product. For example, replacing soy or other binders with 1–2 % plasma protein in an emulsified pork sausage was shown to increase the emulsion stability and produce a firmer gel structure (Stožinić et al., 2025). Beyond functionality, nutritional advantages come with blood plasma use. Plasma is sometimes dubbed "liquid protein" because it contains high-quality proteins with all essential amino acids (Aung et al., 2023). Plasma has been incorporated into baked goods and pasta as well, for instance, as an egg white replacer in cakes or a protein fortifier in noodles (Toldrà et al., 2019). Plasma has the advantage of being colorless, which prevents it from darkening the food. On the other hand, plasma does not contain red blood cells, unlike whole blood which therefore provides a significant amount of bioavailable heme iron to combat anemia.

Beyond conventional muscle meats and offal, other animal-derived ingredients, including milk proteins, eggs, fish, krill, and even insects, are increasingly recognized for their complementary roles in hybrid formulations. Dairy proteins (e.g. casein and whey) are often incorporated to bolster both nutrition and functionality in plant-animal blends. They exemplify proteins of high biological value, with excellent digestibility and a complete essential amino acid profile (Domić et al., 2025). Functionally, caseins and their derivatives offer exceptional surface activity and water-binding capacity, enabling emulsification, foaming, and gelation in food matrices (Hammam et al., 2021). The addition of milk proteins can thus improve texture and stability (for instance, forming heat-induced gels or emulsions) while also enhancing flavor and mouthfeel. Egg proteins similarly provide all essential amino acids with one of the highest protein quality scores in human nutrition (Dubin et al., 1994). Egg whites and yolks contribute versatile techno-functional properties: ovalbumin in egg white forms firm gels and foams upon heating, and yolk phospholipids act as natural emulsifiers (Li et al., 2024). Incorporating eggs into hybrid recipes can improve binding, moisture retention, and emulsification, helping to create the cohesive texture of traditional meat products. In both cases, the inclusion of dairy or egg components helps overcome the sensory and structural limitations of plant proteins (Lima Nascimento et al., 2023), leveraging well-established food ingredients to achieve a more meat-like experience in hybrids.

Marine and novel animal proteins (from fish, krill, and insects) are also being explored in hybrid food design for their unique nutritional and functional benefits as well as sustainability advantages. Fish-derived proteins, such as those from lean fish muscle or refined surimi, possess remarkable gel-forming and water-holding abilities (Singh et al., 2025). Myofibrillar fish proteins (rich in actin and myosin) can create the firm, elastic network characteristic of products like fish balls or imitation crab and similarly can enhance the texture of hybrid meat-plant products. Nutritionally, fish offers high-quality complete protein alongside long-chain omega-3 fatty acids and important micronutrients (Noreen et al., 2025). The inclusion of even a modest portion of fish in a hybrid formulation boosts its content of cardioprotective omega-3. From a sustainability perspective, partially substituting terrestrial meat with sustainable seafood or fish protein can improve resource efficiency. For instance, combining surimi with plant proteins has been proposed to alleviate overfishing pressures on high-value fish stocks while still delivering acceptable taste and nutrition and iodine, for example, which purely terrestrial ingredients may lack (Singh et al., 2025). In practice, this means a hybrid product can retain the desirable gel texture and savory flavor contributed by fish, but with a lower ecological footprint than a 100 % meat or seafood item.

Edible insects have recently gained attention as sustainable protein inputs for hybrid foods. Insects such as mealworms (*Tenebrio molitor* larvae) or crickets are inherently high in protein (typically 50–70 % protein by dry weight) with a well-balanced amino acid profile, and they are dense in micronutrients like iron, zinc, and vitamin B12 (Zhang et al., 2025). From a functional standpoint, insect ingredients (usually added as a defatted powder or paste) can influence the texture of hybrid products. Research has shown that replacing a portion of meat with cricket or mealworm flour (on the order of 5–15 %) in products like sausages or patties can improve water-binding, reduce cooking losses, and even increase yield, without markedly detracting from taste or texture. Rearing insects for protein entails far lower land and water use and generates fewer greenhouse emissions than traditional livestock farming. For example, beef patties enriched with 10–20 % cricket flour retained more moisture and had less shrinkage upon cooking compared to all-meat controls (Gomes Martins et al., 2024). These functional benefits are thought to arise from the proteins and fiber (chitin) in insect meals, which can hold water and interact with the meat proteins. That said, very high substitution levels (e.g. more than 20–25 % of the meat replaced with insect protein) may disrupt the protein network: studies noted that excessive insect inclusion (higher than 25 %) led to softer, less cohesive gels, likely due to the dilution of myofibrillar proteins and the presence of insoluble insect components (Vlahova-Vangelova et al., 2023). Optimizing the level and processing of insect ingredients is therefore crucial to maintain the desired bite and structure in hybrids.

Beyond this diversity of animal-derived ingredients used in hybrid food formulation, cellular agriculture is an emerging technology that allows the production of animal-derived agricultural products using cells cultivated in bioreactors rather than directly from animals. This approach has strong potential as a sustainable and environmentally friendly alternative for producing meat and meat products, as it enables efficient production without the need to grow supporting tissues or functional systems such as skeletal or digestive structures. The design of hybrid foods using cultured meat is currently being investigated (Maseko et al., 2025).

Overall, the diversity of animal-derived ingredients used in hybrid food formulations highlights their complementary nutritional, functional, and sustainability-related roles; these aspects are synthesized in Table 2, which provides a comparative overview of their key properties, technological applications, and associated challenges in hybrid food design.

2.4 Examples of hybrid foods

A growing number of innovative hybrid food products have already demonstrated the potential of these material combinations. For instance, hybrid burgers and sausages containing 30–50 % plant proteins (e.g., pea or soy) mixed with beef or pork have reached commercial markets, offering lower environmental footprints and reduced saturated fat content while maintaining consumer-accepted sensory qualities. Similarly, poultry-vegetable blends and fish-legume formulations have been developed to meet the demand for healthier, more sustainable convenience foods. Research on hybrid food design has explored the integration of mycoproteins or mushroom fibers with minced meat to improve texture and flavor retention while reducing meat content. Emerging innovations also include fortified hybrid products enriched with plant-derived omega-3 fatty acids, dietary fibers, or phytochemicals to enhance functional value through their anti-oxidant activity (Kumar et al., 2015). These developments illustrate the versatility of hybrid foods as an adaptable technological and nutritional platform capable of addressing sustainability, health, and sensory challenges simultaneously.

Table 2 Animal-derived ingredients in hybrid foods.

Animal source	Key nutritional properties	Functional properties	Application in hybrid foods	Sustainability and circularity benefits	Processing challenges
Insect	Protein (dry weight); well-balanced amino acids; rich in iron, zinc, and vitamin; "meat factor" effect.	Water-binding; reduces cooking loss; increases yield; chitin/protein interaction holds water.	Replacement of meat in sausages or patties to improve moisture retention and reduce shrinkage.	Low land/water use; fewer greenhouse emissions compared to traditional livestock.	Substitution levels disrupt protein network, leading to softer, less cohesive gels due to myofibrillar dilution.
Liver, heart, kidneys, spleen	High biological value protein; rich in folate, choline, vitamin A, iron, zinc, selenium, and B-complex vitamins; high bioavailability.	Liver: emulsifying capacity; Heart: solid meaty texture; Kidney: flavor complexity and protein enrichment.	Added to blended meat spreads or sausage formulations to increase micronutrient content while diluting strong flavors.	Promotes whole-animal utilization; reduces waste of edible by-products; provides affordable nutrition.	Strong taste or odor; cultural acceptance issues; requires preprocessing like soaking or blanching.
Slaughterhouse blood (plasma fraction)	High-quality protein with all essential amino acids; highly bioavailable heme iron.	High water-holding capacity, strong gelation, foaming, and emulsifying ability; improves emulsion stability.	Plasma protein in emulsified sausages; used as egg white replacer in baked goods or protein fortifier in noodles.	Harnessing underutilized slaughterhouse resources to improve food yield and nutrition.	Whole blood can darken products (mitigated by using plasma fraction).
Fish (lean muscle/ surimi)	High-quality complete protein; rich in long-chain omega-3 fatty acids and iodine.	Remarkable gel-forming and water-holding abilities via myofibrillar proteins (actin and myosin).	Combining surimi with plant proteins to create firm, elastic networks in hybrid seafood-plant products.	Alleviates overfishing pressures on high-value stocks; lower ecological footprint than seafood items.	NA

(continued)

Table 2 Animal-derived ingredients in hybrid foods. *(cont'd)*

Animal source	Key nutritional properties	Functional properties	Application in hybrid foods	Sustainability and circularity benefits	Processing challenges
Animal skin, bones, and tendons (collagen/gelatin)	Source of bioactive peptides; rich in glycine, proline, and hydroxyproline; lacks some essential amino acids.	Gelation, water-binding, emulsifying, foaming, and film-forming; improves texture, stability, and juiciness.	Used in hybrid sausages or nuggets (at a few percent) to increase yield and provide a firmer, snappier texture.	Valorization of structural proteins from processing by-products; considered a "clean label" natural additive.	Mainly used for texture rather than complete nutrition due to incomplete amino acid profile.
Milk (casein and whey)	High biological value; excellent digestibility; complete essential amino acid profile.	Surface activity, water-binding, emulsification, foaming, and gelation.	Incorporated to bolster nutrition and improve texture, flavor, and mouthfeel in plant-animal blends.	NA	NA
Eggs (White and Yolk)	Highest protein quality scores; contains all essential amino acids.	Ovalbumin forms firm gels/foams; yolk phospholipids act as natural emulsifiers; improves binding and moisture retention.	Used to create cohesive texture and overcome sensory or structural limitations of plant proteins.	NA	NA

Adapted from Ahmad et al., (2024), Appiah and Peggy (2012), Aung et al. (2023), Consalez et al. (2022), Domić et al. (2025), Dubin et al. (1994), Gomes Martins et al. (2024), Gómez-Guillén et al. (2011), Hammam et al. (2021), Jayathilakan et al. (2012), Latoch et al. (2024), Li et al., (2024), Lima Nascimento et al. (2023), Maseko et al. (2025), Noreen et al. (2025), Singh et al. (2025), Stožinić et al. (2025), Suurs and Barbut (2020), Toldrà et al. (2019), Vlahova-Vangelova et al. (2023), and Zhang et al., (2025).

Le et al. (2024) explored the potential applications of bioactive peptides derived from algae, insects, and food waste, which could be used in the manufacture of hybrid foods. Bioactive peptides have various health benefits, particularly antioxidant, anti-inflammatory, and antimicrobial properties. By incorporating these bioactive compounds into hybrid foods, it is possible to create products with enhanced nutritional and functional benefits. This approach also has the advantage of promoting the use of underutilized resources, such as algae and insects, to create high value-added food products. At the same time, the design of hybrid food can also be oriented towards fortified foods to the extent that natural components rich in nutrients and molecules of nutritional interest are added (Ashraf, 2025; Marcìa-Fuentes et al., 2026).

Currently, although the market remains marginal, hybrid products are being marketed in various countries. This is particularly the case for the "BOTH burger" from the American company 50–50 Foods Inc., Danish Crown's 50/50 hybrid mince, the Maple Leaf 50/50™ product range in Canada, and products from the FiftyFifty brand (Boukid et al., 2024). In the United Kingdom, 38 hybrid meat products were launched between 2016 and 2020 (Grasso & Jaworska, 2020). In France, the Triballat Noyal group has launched a new brand called "Pâquerette et Compagnie," which sells hybrid milk made from 50 % cow's milk and 50 % plant-based juice (oats, hazelnuts, almonds). In charcuterie and delicatessen products, Madrange has launched the "Ô P'tits Légumes" range, producing hams, sausages, and meat specialties combining pork with at least 30 % vegetables.

3. Hybrid food processing

The development of hybrid foods not only depends on the selection of appropriate raw materials but also on the application of suitable processing technologies capable of achieving desirable structural, sensory, and nutritional properties (Fig. 2). Processing plays a decisive role in defining the final quality of hybrid products by influencing the interactions between animal, plant, and fungal components, as well as their stability and digestibility. The technological challenge lies in combining ingredients of diverse physicochemical nature in a way that produces homogeneous, safe, nutritionally efficient and palatable foods with a reduced environmental footprint.

Fig. 2 Schematic overview of the main processing technologies involved in the manufacture of hybrid foods. *Created in BioRender. (Theron, 2026) https://BioRender.com/pxy26ba. HHP: High Hydrostatic Pressure; PEF: Pulsed Electric Field; UV-C: Light-based technology.*

3.1 Main process applied for hybrid food manufacturing

A variety of process can be applied in the manufacture of hybrid foods, depending on the type of product and the intended functionality of the components. Generally speaking, foods from different sources are broken down before being blended and cooked. Mixing involves mechanically combining comminuted animal tissues (such as minced meat or fish) with plant or fungal ingredients (protein isolates, fibers, or powders) to form an integrated matrix. Mixing must ensure proper dispersion and compatibility between hydrophilic and lipophilic phases, often requiring the use of binders or emulsifiers. This method is particularly suited to products such as hybrid burgers, meatballs, or pâtés, where a heterogeneous but cohesive texture is acceptable. The presence of raw materials of animal origin, and in particular meat and offal, necessitates the almost systematic cooking of meat-offal-hybrid foods. Cold processes such as marinating, high-pressure processing, pulsed electric field (PEF), shockwaves, ultrasound, or other process (Ağagündüz et al., 2025) may be applied to raw materials, most often to sterilize them by destroying bacteria. These processes have the advantage of preserving heat-sensitive compounds, such as vitamin C, E, K2 and phenolic compounds. While this is beneficial for foods consumed raw, such as fruits and vegetables, it is less advantageous for animal products, which are often cooked before consumption (aside from extending their shelf life when fresh). However, it has been shown that, beyond the destruction of bacteria, most of these processes also modify the ultrastructure of the muscle, resulting in an increased rate of meat digestion (Bhat et al., 2021, 2022). These processes could therefore also find application to improve the digestibility of hybrid foods.

Marinade is commonly used to improve texture and flavor of meat and fish flesh. Although the scientific literature does not yet report on the effect of marinating on hybrid foods, this process could be used to correct sensory defects and promote consumer acceptance. Acidic marinades can improve the microbiological safety of meat products by reducing the pH of the food which can counteract the effect of the plant ingredients which tend to cause an increase in pH (Gargi & Sengun, 2021). Marinades can be enriched with probiotics designed to inhibit the growth of pathogenic bacteria through a competitive phenomenon. Marinades can also improve the chemical safety of meat products by limiting the formation of process-induced toxicants, such as heterocyclic aromatic amines, during the cooking process (Meurillon & Engel, 2016).

Ultrasound can be used to functionalize plant-based raw materials as they modify protein structure or functionality (O'Sullivan et al., 2017; Zakki et al., 2024). Indeed, ultrasonication, allows the disruption of aggregated pulse proteins into individual subunits and partially change the protein macromolecular structure through cavitation, improving their functionality (McCarthy et al., 2016). In a recent study, incorporating ultrasound treated peas protein isolate into pork sausages reduced hardness and breaking force while increasing the deformability of the matrix as the substitution rate increased (Nie et al., 2025). In combination with a reduction in salt content (from 0.6 to 0.3 M), a denser protein network associated with lower water mobility was observed, which resulted in reduced cooking losses.

With the exception of salted and dried meat and fish, which do not undergo a cooking process (dry-cured ham, dry sausages, smoked fish, etc.), meat and offal are generally cooked to achieve microbiological stabilization of the products and improve their sensory properties. Therefore, most of meat-based hybrid food must be cooked. Cooking process, including cooking temperature, cooking time and the cooking process (boiling, sous vide, roasting …) are important parameter to control health (microbiology, process-induced toxicants), sensory (palatability, flavor, texture), and nutritional (preservation of nutrients, impact on digestibility) properties. While ensuring the destruction of pathogenic bacteria, soft cooking treatments are preferable to limit the destruction of molecules of nutritional interest. Laser cooking, which allows the texture of 3D-printed foods to be controlled, is also being studied (Le Tan, 2025). This method also has the advantage of controlling the temperature through highly localized heating (Blutinger et al., 2026).

3.2 3D printing as innovative process for hybrid food manufacturing

More sophisticated structuring can be achieved through extrusion processing, a thermomechanical treatment that subjects a mixture of proteins, starches, and fibers to controlled heat, pressure, and shear. During extrusion, molecular unfolding and realignment occur, leading to the formation of fibrous structures that can replicate the texture of muscle tissue. High-moisture extrusion allows for the creation of anisotropic structures in which plant and animal proteins are interlaced, giving rise to hybrid products with improved chewiness, juiciness, and bite. Extrusion also offers advantages in terms of microbial stability, product uniformity, and scalability, making it a key technology for large-scale production of hybrid foods.

In parallel, emerging technologies such as 3D food printing are opening new perspectives in the design of hybrid foods (Portanguen, Tournayre, Gibert, et al., 2022; Portanguen, Tournayre, Sicard, et al., 2022). This additive manufacturing technique allows for precise spatial deposition of plant and animal ingredients, enabling the creation of customized textures, nutrient distributions, and aesthetic forms (Portanguen et al., 2019, 2026; Portanguen, Tournayre, Gibert, et al., 2022). For instance, 3D printing can be used to build layered structures that mimic the architecture of whole muscle cuts or to design portion-controlled products tailored to specific nutritional requirements. Beyond its creative potential, this technology also supports sustainability through ingredient optimization and reduced waste generation. 3D printing is generally based on extrusion, which mechanical stress impacts the microstructure of the food (Le Tohic et al., 2018) and can degrade the digitally designed geometry, as well as the final texture. This stress effect can be enhanced in hybrid foods where the different components have different viscoelastic properties.

3.3 Effect of processing on hybrid food structure and composition

Processing conditions exert a profound impact on the structure and composition of hybrid foods. Heat and mechanical stress can induce protein denaturation, aggregation, and cross-linking, which affect the network formation between plant and animal proteins. These interactions determine textural attributes such as firmness, cohesiveness, and water-holding capacity. A recent study demonstrated the benefit of heating liver and lentil proteins to their denaturation temperature to increase their interaction strength, with the prospect of better controlling cohesion at the interface of layers in a 3D-printed liver/lentil hybrid food (Schiell et al., 2026).

The balance between plant and animal fractions, as well as their processing compatibility, influences the stability of emulsions and gels, the retention of fat and water during cooking, and the uniformity of the matrix. Furthermore, the structural organization resulting from processing determines how easily digestive enzymes can access proteins and lipids, directly influencing the digestibility and bioavailability of nutrients.

3.4 Effect of processing on the quality attributes of hybrid foods

3.4.1 *Sensory properties*

From a sensory standpoint, processing technologies play a dual role: they can enhance desirable attributes while simultaneously generating potential defects.

Texture-wise, processing governs the formation of fibrous or gel-like structures that determine mouthfeel and juiciness. Well-optimized processing can reduce cooking losses and improve tenderness (Argel et al., 2022), whereas over-processing may cause dryness, excessive hardness, or phase separation. Adding plant matter to meat products affects their functional properties (Tkaczewska et al., 2025). This is particularly true in terms of protein solubility, particle size, and thermogelation when the pH increases. These changes also occur during high-temperature manufacturing processes, where free metal ions can increase lipid oxidation and alter food texture (Chen et al., 2022; Pérez-Montes et al., 2021). More generally, alternative proteins added to meat can modify its texture by altering its structural properties and water holding capacity (Argel et al., 2022), which can be useful for certain vulnerable populations seeking personalized foods that are less difficult to chew (Xia et al., 2024). For example, partially replacing pork with more than 20 % rice protein, as an alternative to legumes, reduced the stability of emulsions during cooking and decreased firmness (Hsiao et al., 2025). In another research on meat based hybrid food, increasing the proportion of peas relative to beef reduced cooking losses, fat content, and food hardness without negatively impacting acceptability (Nagatsuka & Koriyama, 2026).

Such changes are also observed in milk based hybrid foods proteins where combining pea protein with casein led to the formation of weaker and less structured gels (Xia et al., 2024). However, despite certain differences in terms of properties, the sensory acceptability of these products remains equivalent to traditional yogurts, particularly in terms of texture (Logan et al., 2025). These results are confirmed for a Cheddar-type cheese following a combination of pea, soybean, or fava bean proteins with milk proteins (Sarker et al., 2026).

Processing also affects flavor and color of food. For example, controlled heating during extrusion or mixing promotes Maillard reactions and lipid oxidation, which contribute to the development of flavor and color but may also produce off flavors if not properly managed (Chen et al., 2022; Pérez-Montes et al., 2021). Hybrid products often rely on a delicate equilibrium between plant and animal flavors, and excessive thermal treatment can lead to undesirable notes such as bitterness or beany flavors from plant proteins.

Color attributes are similarly influenced by the degree of protein denaturation, pigment stability (e.g., myoglobin oxidation), and ingredient composition, all of which affect consumer perception of freshness and quality. In the case of hybrid foods, some alternative protein sources contain components that can have a negative effect on the flavor, color or texture of food after processing. Plant protein flours, concentrates and isolates often contain phytochemicals that cause undesirable color or flavor, such as dark pigmentation or beany, bitter or astringent notes (Wang et al., 2022; Leonard et al., 2023; Saffarionpour, 2024). Similarly, insect-derived ingredients can lead to off-flavors due to volatile compounds production during cooking (Cheseto et al., 2020), off-colors, and unpleasant textures due to the high amount of chitin in insect exoskeletons (Sipponen et al., 2018). Different strategies are being investigated to improve the sensory properties, such as fermentation (Kim et al., 2021) and deodorisation to reduce off-flavors (Tzompa-Sosa et al., 2021) or grinding to improve texture (Sipponen et al., 2018). Therefore, identifying and mitigating undesirable components is essential when formulating hybrid foods with alternative proteins and more generally alternative raw matter including lipids, carbohydrates and nutrients.

3.4.2 Preservation properties

The formulation of hybrid foods leads to new products for which questions arise regarding their shelf life. Previous studies have shown that hybrid sausages containing different varieties of insect meal (Han et al., 2023; Kim et al., 2016) oxidized more quickly than the control based on meat alone, despite their antioxidant activity, which is lost when these hybrid foods are cooked (Vanqa et al., 2022). Similar results were obtained for hybrid foods based on shiitake mushrooms, which are rich in easily oxidizable unsaturated fatty acids and have low antioxidant activity (Park et al., 2023). The preservation process, particularly storage in a modified atmosphere, may therefore be an interesting alternative for limiting lipid oxidation in hybrid products and thus increasing their shelf life (Schiell et al., 2023). However, the design and monitoring of hybrid foods have only recently been studied, and few results on the effects of storage methods are available in the scientific literature (Tkaczewska et al., 2025). The need to control the microbiological status remains absolute and mandatory to protect consumer health.

3.4.3 Nutritional properties

Hybrid food can leverage the complementary nutritional attributes of each component to improve overall protein quality and micronutrient availability. Amino acid deficiencies in plant proteins are compensated by the amino acids from animal sources in hybrid formulations, yielding a complete essential amino acid profile (Tkaczewska et al., 2025). Animal proteins also break down into peptides during digestion that bind to non-heme iron from plant ingredients, thereby enhancing iron bioavailability from the alternative protein fraction. This synergy suggests hybrid products may offer a previously unattainable nutritional composition with a more comprehensive amino acid profile and improved mineral uptake. For example, Broucke et al. (2022) found that replacing up to 50 % of pork with pea protein in a hybrid sausage did not introduce any limiting essential amino acids. Beyond composition, recent findings indicate that hybrid matrices do not necessarily hinder protein digestibility. Peptidomic analysis revealed that the presence of plant proteins in a hybrid matrix can significantly increase the digestion of meat proteins by altering the kinetics of peptide release, potentially improving the enzymatic degradation of certain muscle proteins despite the plant proteins' lower solubility (Tkaczewska et al., 2025). In fact, hybrid sausages have shown extensive protein hydrolysis during *in vitro* digestion, one formulation saw roughly 90–95 % of its protein mass converted into peptides and amino acids after gastric and intestinal phase digestion (Nie et al., 2025). Such results directly challenge the assumption that hybrid products are less digestible. Moreover, in a comparison of purely plant-based vs. meat proteins, Zhou et al., (2021) observed that ground beef's proteins were more digestible (about 85 %) than those of a textured soy-based patty (about 70 %) under identical protein content. The same study noted that the soy-protein patty (formulated with plant ingredients like canola oil) formed a distinct structure in gastrointestinal fluids compared to pure beef, which led to faster protein breakdown in the stomach phase. However, this was accompanied by slower digestion in the small intestine, illustrating how ingredient composition can modulate the overall digestion kinetics. Taken together, these insights show that well-designed hybrid foods can achieve a complete amino acid profile and enhance certain nutrient bioavailability, all while maintaining high protein digestibility comparable to, and in some aspects even improving upon, their purely animal or plant counterparts (Broucke et al., 2022).

3.4.4 Consequence perspective on human health

Hybrid foods offer potential nutritional benefits through partial meat replacement with plant ingredients, while also presenting challenges associated with processing intensity and ingredient selection (Fig. 3).

Hybrid food products have emerged as a strategy to reduce red meat consumption without sacrificing familiar sensory attributes (Stanišić et al., 2025). This approach aligns with dietary recommendations, as high intake of red and processed meat has been linked to increased risks of chronic diseases, and partially replacing meat with plant ingredients can help

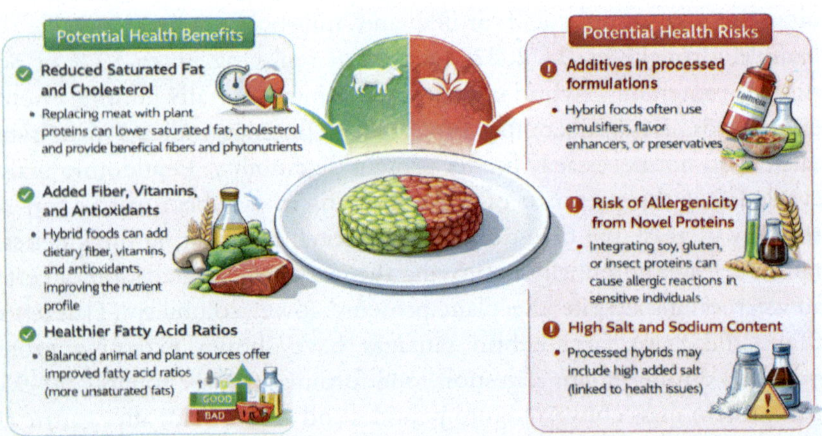

Fig. 3 Overview of the key positive and negative health implications of hybrid foods.

mitigate these health risks. Notably, incorporating legumes, grains, fungi, or other plant proteins into meat products can improve the overall nutrient profile. For example, blending meat with textured vegetable proteins significantly lowers saturated fat and cholesterol content while adding dietary fiber and micronutrients. Such hybrids often retain a high-quality protein profile, the animal component provides complete essential amino acids, complemented by plant proteins' amino acid diversity – and can offer healthier fatty acid ratios (e.g. more unsaturated fats) than all-meat equivalents (Tkaczewska et al., 2025). Plants contain numerous phytochemicals such as phenolic compounds, carotenoids, phytosterols, and glucosinolates that may help prevent certain diseases (Marcìa-Fuentes et al., 2026) which are unavailable in animal products.

However, alongside these potential benefits, there are important health considerations.

The plants that make up hybrid foods contain antinutritional factors which interfere with the absorption and utilization of essential nutrients (Ali et al., 2022; Okaiyeto et al., 2025). These substances include phytates, oxalates, tannins, polyphenols, alkaloids, lectins, saponins, etc. Phytates are found in high quantities in whole grains and legumes. They can bind to minerals such as calcium, iron, and zinc, thus limiting their absorption. Oxalates can also reduce the bioavailability of certain minerals and can contribute to the formation of kidney stone and tannins can inhibit the absorption of non-heme iron. Lectins, found in legumes and some vegetables, can disrupt the integrity of the intestinal wall if consumed in large quantities and saponins interfere with the absorption and utilization of essential nutrients (Okaiyeto et al., 2025). Although some of these substances also have beneficial effects on human health (Salim et al., 2023), it is necessary to take into account the antinutritional characteristics in the design of hybrid foods, especially when they are intended for deficient populations.

Hybrid foods (like many plant-based meat alternatives) are typically highly processed, and their formulations may include numerous additives for texture and flavor, leading to classification as ultra-processed foods (Messina et al., 2025). Consequently, they can contain high levels of sodium and certain additives (e.g. emulsifiers or flavor enhancers), which raise concerns given associations between excess salt/ultra-processed food intake and adverse health outcomes. Additionally, the inclusion of novel protein sources (such as soy, wheat gluten, or insect protein) introduces the risk of allergenicity in susceptible individuals (Stanišić et al., 2025).

Therefore, the health impact of hybrid foods depends on prudent formulation and consumption: they can indeed contribute to a more nutritious, plant-forward diet. For instance, trials have shown that hybrid formulations can achieve improved nutrient density and protein quality, but realizing these benefits without unintended negative effects requires minimizing added salt/additives and ensuring a balanced nutritional composition.

4. Consumer acceptability - target populations

The ultimate success of hybrid foods depends not only on technological feasibility and nutritional value, but also on consumer acceptance. Even the most innovative hybrid products will fail to penetrate the market if they do not meet the expectations, preferences, and cultural values of their target populations. Thus, understanding consumer perception and acceptance is essential, as it is shaped by a complex interplay of sensory experience, psychological attitudes, social norms, and environmental awareness. Studies consistently show that consumers are unwilling to compromise on taste, even if aware of environmental or health advantages. For example, one European survey found that while about half of consumers were willing to try hybrid meat products, they still perceived 100 % meat versions as tastier and more enjoyable, viewing the hybrids primarily as healthier or more ethical choices (Asioli et al., 2023). This highlights that if a hybrid food has noticeable "off" flavors, an altered texture, or an unusual appearance, its sustainability credentials alone are unlikely to sway consumers. On the other hand, if sensory quality is well executed, hybrid foods can leverage their partial meat content to deliver familiar satisfactions. Blending meat with palatable plant ingredients (such as mushrooms) can even enhance flavor; indeed, U.S. consumers found beef-mushroom burgers "new, tasty, and interesting," illustrating how culinary innovation can generate enthusiasm when it aligns with taste expectations (Lang, 2020; Prusaczyk et al., 2021).

Beyond the sensory domain, cognitive and affective factors also strongly influence consumer acceptability of hybrid foods. Many people are still unfamiliar with what "hybrid" foods exactly entail and how they differ from fully plant-based meat analogues. Clear labeling and transparent communication are therefore essential to build understanding and trust and to avoid consumer confusion. If hybrid foods are perceived as overly processed, artificial, or "engineered," they may provoke skepticism or rejection, especially among consumers who prioritize natural, minimally processed foods. A perceived lack of naturalness is known to be one of the most critical barriers to accepting novel food technologies such as 3D printing (Lupton & Turner, 2018; Portanguen et al., 2025; Ross et al., 2022). This suggests that hybrid food developers must mitigate notions of unnaturalness through education and the use of familiar ingredients.

Socio-demographic and cultural factors also shape the acceptance of hybrid foods, helping identify key target populations. For example, Chinese people were open to hybrid foods rich in sustainable proteins such as legumes and mushrooms while Danish people were more reluctant despite these behaviors also depended on the consumption scenario such as type of meal and social contexts (Jakobsen et al., 2026). Consumer acceptance is strongly influenced by factors such as taste, health benefits, sustainability, cost, and the novelty of this type of food. However, this acceptability remains affected by individual differences in food values, lifestyle, and culinary innovation (Lang, 2020). Hybrid food could play a decisive role in the successful transition to a more plant-based diet, as the presence of meat in these foods increases their appeal to consumers by adding a "pleasure" factor. Indeed, sensory perceptions play a major role in influencing the intention to purchase hybrid foods in relation to consumer attitudes (Banovic et al., 2022; Grasso & Jaworska, 2020). One major audience for hybrids is the flexitarian consumer, i.e. omnivores who are actively trying to reduce (but not eliminate) their meat intake for health or sustainability reasons. Public discourse on climate change, animal welfare, and healthy diets in recent years has primed a growing segment of consumers to seek compromises that reduce meat consumption without "giving up" meat entirely. Hybrid foods, positioned as a middle ground between conventional meat and plant-based substitutes, are naturally appealing to these flexitarians. This segment has expanded rapidly in many Western countries; for instance, by 2021 an estimated 30 % of Europeans identified as flexitarian (rising to ~40 % if semi-vegetarians are included) (GTFR, 2025). And the trend is particularly pronounced among younger generations: around one in five Gen Z young adults in the UK now follow a flexitarian diet, roughly double the share in 2019 (Ryder et al., 2023).

Another important element shaping consumer perception is the *psychological* disposition of the consumer toward food innovation. Food neophobia, the reluctance to eat novel or unfamiliar foods, can limit the initial adoption of hybrid foods, particularly among individuals and groups less exposed to dietary change. Empirical evidence shows that food neophobia significantly reduces willingness to pay for hybrid foods; for example, over 80 % of Chinese consumers surveyed expressed willingness to purchase a new hybrid beef-rice product, yet those with higher neophobia scores were far less likely to pay for it (Chen et al., 2025). Conversely, food neophilia, an openness and curiosity for trying new foods, tends to facilitate experimentation and early uptake. Personality traits such as openness to new experiences are associated with a higher rate of experimenting with novel foods (Keller & Siegrist, 2015).

From an ethical and environmental standpoint, hybrid foods occupy an interesting middle ground in consumer psychology. Many consumers perceive these products as a *compromise* that aligns their moral values with their desire for sensory satisfaction. Research on meat-eating behavior supports this: most dedicated meat-eaters find the idea of total abstinence daunting, but are far more receptive to smaller, incremental dietary changes that don't radically alter their habits (Prusaczyk et al., 2021). In fact, some authors argue that meat based hybrid foods could significantly "facilitate consumer acceptance" of plant-centric diets precisely because the inclusion of meat in these products makes them more acceptable to mainstream eaters (Asioli et al., 2023).

Altogether, the key to the success of hybrid food lies in aligning sensory excellence, transparent communication, cultural sensitivity, and credible sustainability narratives. When hybrids can truly deliver an enjoyable eating experience *and* convincingly tell the story of their benefits, they stand a strong chance of gaining consumer trust.

5. Environmental and economic impact of hybrid foods

Life cycle assessments (LCAs) consistently show that hybrid food products, which combine conventional and alternative ingredients, can substantially reduce environmental burdens compared to their purely animal-based counterparts. For example, a LCA found that a 50:50 beef-plant burger used about 45 % less water and emitted roughly 50 % fewer greenhouse gases than an all-beef patty, while replacing 30 % of beef with lentils cut emissions by approximately one-third (Dominguez-Lacueva et al., 2025). Land use also declines with partial meat substitution, since far less feed crop is required; one analysis reported a plant-based burger needed only about 150 g of crops versus approximately 1 kg for a beef burger (Dominguez-Lacueva et al., 2025). From an economic perspective, hybrid formulations capitalize on the lower cost of plant proteins, producing a given amount of beef protein can be an order of magnitude more expensive than the equivalent plant protein (Depra et al., 2022). These cost savings, combined with rising consumer demand for sustainable protein, translate into strong market potential: global sales of plant-based meat alternatives reached $5.6 billion in 2021 (with 8 % annual growth) and are projected to expand to approximately $34 billion by 2027 (Munialo et al., 2025). Overall, by simultaneously shrinking the environmental footprint (e.g. lower GHG emissions, water and land use) and improving cost-efficiency, hybrid food systems exemplify a strategy that aligns sustainability with economic feasibility in the future food supply (Kaplan & McClements, 2025).

From a social and policy perspective, hybrid foods can play a strategic role in national and international sustainability agendas. They align with several United Nations Sustainable Development Goals (SDGs), including responsible consumption and production (SDG 12), climate action (SDG 13), and zero hunger (SDG 2). By facilitating a gradual transition toward lower-impact diets, hybrid foods provide a politically and culturally acceptable pathway that complements broader dietary reforms. Governments and institutions can further support their development through incentives for research, eco-innovation funding, and labeling frameworks that communicate verified sustainability benefits to consumers.

6. Future trends and policy alignment

Hybrid foods are emerging as a promising strategy to enhance the sustainability and nutritional profile of diets while maintaining consumer appeal (Kaplan & McClements, 2025). This approach aligns closely with evolving policy frameworks at international and regional levels. For instance, the FAO/WHO *Guiding Principles for Sustainable Healthy Diets* call for dietary patterns that are both healthy and have low environmental impacts, while also being culturally acceptable and accessible to all. In Europe, the EU's Farm to Fork strategy under the Green Deal likewise envisions a transition to food systems that "ensure food security, nutrition and public health" and have a neutral or positive environmental impact. To support such goals, regulatory bodies are adapting: novel protein ingredients and hybrid products must undergo rigorous safety and efficacy assessments; in the EU, many fall under the *Novel Food* regulation, requiring European Food Safety Authority evaluation. Recent approvals, such as the EFSA's authorization of water lentils as a new food source, illustrate how science-based regulation is enabling innovative protein foods in line with public health standards.

Looking ahead, research and development in hybrid foods are focused on optimizing formulations, production methods, and consumer acceptance in line with policy objectives. Advances in food technology (including artificial intelligence-driven ingredient optimization and novel processing techniques such as 3D printing) are being applied to create hybrid foods with improved sensory qualities and balanced nutritional profiles, all while minimizing life-cycle environmental impacts. Such design considerations address nutritional adequacy concerns and align with the "protein transition" away from predominantly animal-based diets (Aiking & De Boer, 2020). Therefore, aligning hybrid food innovation with international and national policy frameworks involves not only technical and nutritional advancements, but also socio-economic strategies to guide consumer preferences.

7. Conclusion

Taken together, the evidence presented in this chapter underscores that hybrid food design offers a multifaceted approach to enhancing the nutritional profile and technological functionality of animal source-based foods, while concurrently advancing sustainability and public health goals. By judiciously combining diverse raw materials (from plant, fungal, or novel protein sources) with conventional raw materials of animal source (mainly meat and in a lesser extent milk), hybrid foods can achieve a superior nutrient balance (e.g. higher fiber, essential micronutrients, and favorable fatty acid profiles) without sacrificing high-quality protein content. At the same time, replacing a fraction of meat with more sustainable ingredients dramatically reduces the environmental footprint, thereby contributing to climate change mitigation. These products emerge as strategic instruments in the ongoing protein transition, providing a pragmatic compromise that resonates with flexitarian consumers and maintains sensory appeal while shifting diets toward more plant-rich paradigms. Realizing the full potential of hybrid foods will require continued technological innovation to optimize processing and product quality, efforts to tailor nutritional solutions to specific population needs (such as personalized products for different age groups or health conditions), and supportive policy frameworks (e.g. clear labeling and incentives) to foster consumer trust and industry adoption. Ultimately, hybrid foods represent not just an incremental product reformulation, but a forward-looking paradigm capable of reconciling consumer expectations with urgent sustainability and health imperatives in the global food system.

References

Ağagündüz, G., Ayakdaş, G., Katırcıoğlu, B., & Fatih Ozogul, F. (2025). Advances in non-thermal food processing: A comprehensive approach to nutrient retention, food quality, and safety. *Sustainable Food Technology, 3*(5), 1284–1308. https://doi.org/10.1039/D5FB00136F.

Ahmad, M. I., Li, Y., Pan, J., Liu, F., Dai, H., Fu, Y., ... Zhang, H. (2024). Collagen and gelatin: Structure, properties, and applications in food industry. *International Journal of Biological Macromolecules, 254*, 128037. https://doi.org/10.1016/j.ijbiomac.2023.128037.

Aiking, H., & De Boer, J. (2020). The next protein transition. *Trends in Food Science & Technology, 105*, 515–522. https://doi.org/10.1016/j.tifs.2018.07.008.

Ali, A., Devarajan, S., Manickavasagan, A., & Ata, A. (2022). Antinutritional factors and biological constraints in the utilization of plant protein foods. In A. Manickavasagan, L.-T. Lim, & A. Ali (Eds.). *Plant protein foods* (pp. 407–438). Springer International Publishing. https://doi.org/10.1007/978-3-030-91206-2_14.

Appiah, J., & Peggy, Y.-H. (2012). The use of blood and derived products as food additives. In Y. El-Samragy (Ed.). *Food additive*. InTech. https://doi.org/10.5772/32374.

Argel, N. S., Lorenzo, G., Domínguez, R., Fraqueza, M. J., Fernández-López, J., Sosa, M. E., ... Andrés, S. C. (2022). Hybrid meat products: Incorporation of white bean flour in lean pork burgers. *Applied Sciences, 12*(15), 7571. https://doi.org/10.3390/app12157571.

Ashraf, S. A. (2025). Food fortification as a sustainable global strategy to mitigate micronutrient deficiencies and improve public health. *Discover Food, 5*(1), 201. https://doi.org/10.1007/s44187-025-00512-5.

Asioli, D., Banovic, M., Barone, A. M., Grasso, S., & Nayga, R. M. (2023). European consumers' valuation for hybrid meat: Does information matter? *Applied Economic Perspectives and Policy, 45*(1), 44–62. https://doi.org/10.1002/aepp.13283.

Aung, S. H., Abeyrathne, E. D. N. S., Ali, M., Ahn, D. U., Choi, Y.-S., & Nam, K.-C. (2023). Comparison of functional properties of blood plasma collected from black goat and hanwoo cattle. *Food Science of Animal Resources, 43*(1), 46–60. https://doi.org/10.5851/kosfa.2022.e57.

Banovic, M., Barone, A. M., Asioli, D., & Grasso, S. (2022). Enabling sustainable plant-forward transition: European consumer attitudes and intention to buy hybrid products. *Food Quality and Preference, 96*, 104440. https://doi.org/10.1016/j.foodqual.2021.104440.

Bauer, J., Biolo, G., Cederholm, T., Cesari, M., Cruz-Jentoft, A. J., Morley, J. E., ... Boirie, Y. (2013). Evidence-based recommendations for optimal dietary protein intake in older people: A position paper from the PROT-AGE study group. *Journal of the American Medical Directors Association, 14*(8), 542–559. https://doi.org/10.1016/j.jamda.2013.05.021.

Benković, M., Jurinjak Tušek, A., Sokač Cvetnić, T., Jurina, T., Valinger, D., & Gajdoš Kljusurić, J. (2023). An overview of ingredients used for plant-based meat analogue production and their influence on structural and textural properties of the final product. *Gels, 9*(12), 921. https://doi.org/10.3390/gels9120921.

Bhat, Z. F., Morton, J. D., Bekhit, A. E.-D. A., Kumar, S., & Bhat, H. F. (2021). Emerging processing technologies for improved digestibility of muscle proteins. *Trends in Food Science & Technology, 110*, 226–239. https://doi.org/10.1016/j.tifs.2021.02.010.

Bhat, Z. F., Morton, J. D., Kumar, S., Bhat, H. F., Aadil, R. M., & Bekhit, A. E.-D. A. (2022). Ultrasonication as an emerging technology for processing of animal derived foods: A focus on in vitro protein digestibility. *Trends in Food Science & Technology, 124*, 309–322. https://doi.org/10.1016/j.tifs.2022.04.012.

Blutinger, J. D., Omo, E. L., Bernat, P., & Lipson, H. (2026). Multi-wavelength laser texturization with 3D-printed foods. *Journal of Food Engineering, 406*, 112798. https://doi.org/10.1016/j.jfoodeng.2025.112798.

Boukid, F., Baune, M.-C., Terjung, N., Francis, A., & Smetana, S. (2024). The 'meathybrid' concept: Bridging the gap between texture, taste, sustainability and nutrition. *International Journal of Food Science and Technology, 59*(11), 8645–8655. https://doi.org/10.1111/ijfs.17425.

Broucke, K., Van Poucke, C., Duquenne, B., De Witte, B., Baune, M.-C., Lammers, V., ... Van Royen, G. (2022). Ability of (extruded) pea protein products to partially replace pork meat in emulsified cooked sausages. *Innovative Food Science & Emerging Technologies, 78*, 102992. https://doi.org/10.1016/j.ifset.2022.102992.

Chen, Y. P., Feng, X., Blank, I., & Liu, Y. (2022). Strategies to improve meat-like properties of meat analogs meeting consumers' expectations. *Biomaterials, 287*, 121648. https://doi.org/10.1016/j.biomaterials.2022.121648.

Chen, S., Wang, D., Wang, J., & Li, J. (2025). Consumer willingness to pay for hybrid food: The role of food neophobia and information framing. *Nutrients, 17*(14), 2326. https://doi.org/10.3390/nu17142326.

Cheseto, X., Baleba, S. B. S., Tanga, C. M., Kelemu, S., & Torto, B. (2020). Chemistry and sensory characterization of a bakery product prepared with oils from African edible insects. *Foods, 9*(6), 800. https://doi.org/10.3390/foods9060800.

Conant, R.T. (2010). Challenges and opportunities for carbon sequestration in grassland systems: A technical report on grassland management and climate mitigation. FAO.

Consalez, F., Ahern, M., Andersen, P., & Kjellevold, M. (2022). The effect of the meat factor in animal-source foods on micronutrient absorption: A scoping review. *Advances in Nutrition, 13*(6), 2305–2315. https://doi.org/10.1093/advances/nmac089.

Cruz-Jentoft, A. J., Bahat, G., Bauer, J., Boirie, Y., Bruyère, O., Cederholm, T., ... Schols, J. (2019). Sarcopenia: Revised European consensus on definition and diagnosis. *Age and Ageing, 48*(1), 16–31. https://doi.org/10.1093/ageing/afy169.

Deprá, M. C., Dias, R. R., Sartori, R. B., De Menezes, C. R., Zepka, L. Q., & Jacob-Lopes, E. (2022). Nexus on animal proteins and the climate change: The plant-based proteins are part of the solution? *Food and Bioproducts Processing, 133*, 119–131. https://doi.org/10.1016/j.fbp.2022.03.006.

Domić, J., Van Loon, L. J. C., Siebelink, E., Borgonjen-van Den Berg, K. J., De Groot, L. C. P. G. M., & Grootswagers, P. (2025). The amino acid composition of commercially available vegan meat and dairy analogues. *British Journal of Nutrition, 134*(5), 377–389. https://doi.org/10.1017/S000711452510408X.

Dominguez-Lacueva, P., Beriain, M. J., & M. Aldaya, M. (2025). A comparative analysis of the water and carbon footprints of hybrid, plant-based, and animal-based burgers. *Renewable Agriculture and Food Systems, 40*, e18. https://doi.org/10.1017/S1742170525100112.

Dubin, S., McKee, K., & Battish, S. (1994). Essential amino acid reference profile affects the evaluation of enteral feeding products. *Journal of the American Dietetic Association, 94*(8), 884–887. https://doi.org/10.1016/0002-8223(94)92369-8.

FAO. (2018). The future of food and agriculture: Alternative pathways to 2050.Rome. 224 pp. Licence: CC BY-NC-SA 3.0 IGO. https://openknowledge.fao.org/server/api/core/bitstreams/e51e0cf0-4ece-428c-8227-ff6c51b06b16/content.

Gargi, A., & Sengun, I. Y. (2021). Marination liquids enriched with probiotics and their inactivation effects against food-borne pathogens inoculated on meat. *Meat Science, 182*, 108624. https://doi.org/10.1016/j.meatsci.2021.108624.

Gerber, P. J., & Food and Agriculture Organization of the United Nations (Eds.). (2013). *Tackling climate change through livestock: A global assessment of emissions and mitigation opportunities*. Food and Agriculture Organization of the United Nations.

Gomes Martins, V. M., Milano, P., Rodrigues Pollonio, M. A., Dos Santos, M., De Oliveira, A. P., Savay-da-Silva, L. K., ... De Souza Paglarini, C. (2024). Adding cricket (*Gryllus assimilis*) flour in hybrid beef patties: Physicochemical, technological and sensory challenges. *International Journal of Food Science & Technology, 59*(1), 450–459. https://doi.org/10.1111/ijfs.16826.

Gómez-Guillén, M. C., Giménez, B., López-Caballero, M. E., & Montero, M. P. (2011). Functional and bioactive properties of collagen and gelatin from alternative sources: A review. *Food Hydrocolloids, 25*(8), 1813–1827. https://doi.org/10.1016/j.foodhyd.2011.02.007.

Grasso, S. (2024). Opportunities and challenges of hybrid meat products: A viewpoint article. *International Journal of Food Science and Technology, 59*(11), 8693–8696. https://doi.org/10.1111/ijfs.17421.

Grasso, S., & Jaworska, S. (2020). Part meat and part plant: Are hybrid meat products fad or future? *Foods, 9*(12), 1888. https://doi.org/10.3390/foods9121888.

Grompone, M. A. (2011). Sunflower oil. In F. D. Gunstone (Ed.). *Vegetable oils in food technology* (pp. 137–167)(1re ed). Wiley. https://doi.org/10.1002/9781444339925.ch5.

GTFR. (2025). Europe plant-based meat market size and share analysis. Growth trends and forecast report 2025–2033. https://www.researchandmarkets.com/reports/6085642/europe-plant-based-meat-market-size-share.

Hammam, A. R. A., Martínez-Monteagudo, S. I., & Metzger, L. E. (2021). Progress in micellar casein concentrate: Production and applications. *Comprehensive Reviews in Food Science and Food Safety, 20*(5), 4426–4449. https://doi.org/10.1111/1541-4337.12795.

Han, X., Li, B., Puolanne, E., & Heinonen, M. (2023). Hybrid sausages using pork and cricket flour: Texture and oxidative storage stability. *Foods, 12*(6), 1262. https://doi.org/10.3390/foods12061262.

Hertzler, S. R., Lieblein-Boff, J. C., Weiler, M., & Allgeier, C. (2020). Plant proteins: Assessing their nutritional quality and effects on health and physical function. *Nutrients, 12*(12), 3704. https://doi.org/10.3390/nu12123704.

Hsiao, J.-T., Chang, T.-W., Chen, L.-Y., & Sheu, F. (2025). Evaluation and reinforcement of the physicochemical properties of hybrid meatball kung-wan partially replaced with rice protein. *Meat Science, 227*, 109853. https://doi.org/10.1016/j.meatsci.2025.109853.

Jakobsen, I. T., Onwezen, M. C., Wang, Y., Zhong, F., Byrne, D. V., & Andersen, B. V. (2026). Cross-cultural consumer acceptance of sustainable protein-rich foods; legumes, plant-based meat analogues and hybrids, and cell-based foods. *Food Quality and Preference, 135*, 105714. https://doi.org/10.1016/j.foodqual.2025.105714.

Jayathilakan, K., Sultana, K., Radhakrishna, K., & Bawa, A. S. (2012). Utilization of byproducts and waste materials from meat, poultry and fish processing industries: A review. *Journal of Food Science and Technology, 49*(3), 278–293. https://doi.org/10.1007/s13197-011-0290-7.

Kaplan, D. L., & McClements, D. J. (2025). Hybrid alternative protein-based foods: Designing a healthier and more sustainable food supply. *Frontiers in Science, 3*, 1599300. https://doi.org/10.3389/fsci.2025.1599300.

Kaur, H., Tian, B., Zheng, H., & Kumar, L. (2026). From texture to digestion: Understanding multi-component interactions in designing starch-based low-glycemic index soft foods. *Trends in Food Science & Technology, 169*, 105520. https://doi.org/10.1016/j.tifs.2025.105520.

Keller, C., & Siegrist, M. (2015). Does personality influence eating styles and food choices? Direct and indirect effects. *Appetite, 84*, 128–138. https://doi.org/10.1016/j.appet.2014.10.003.

Kim, J., Lee, H. E., Kim, Y., Yang, J., Lee, S.-J., & Jung, Y. H. (2021). Development of a post-processing method to reduce the unique off-flavor of Allomyrina dichotoma: Yeast fermentation. *LWT, 150*, 111940. https://doi.org/10.1016/j.lwt.2021.111940.

Kim, H.-W., Setyabrata, D., Lee, Y. J., Jones, O. G., & Kim, Y. H. B. (2016). Pre-treated mealworm larvae and silkworm pupae as a novel protein ingredient in emulsion sausages. *Innovative Food Science & Emerging Technologies, 38*, 116–123. https://doi.org/10.1016/j.ifset.2016.09.023.

Kumar, Y., Yadav, D. N., Ahmad, T., & Narsaiah, K. (2015). Recent trends in the use of natural antioxidants for meat and meat products. *Comprehensive Reviews in Food Science and Food Safety, 14*(6), 796–812. https://doi.org/10.1111/1541-4337.12156.

Lang, M. (2020). Consumer acceptance of blending plant-based ingredients into traditional meat-based foods: Evidence from the meat-mushroom blend. *Food Quality and Preference, 79*, 103758. https://doi.org/10.1016/j.foodqual.2019.103758.

Latoch, A., Stasiak, D. M., & Siczek, P. (2024). Edible offal as a valuable source of nutrients in the diet—A review. *Nutrients, 16*(11), 1609. https://doi.org/10.3390/nu16111609.

Le, T., Suttikhana, I., & Ashaolu, T. J. (2024). Exploring the potential applications of biopeptides from seaweeds, insects, and food wastes: Mechanisms, utilisation, and industry implications. *International Journal of Food Science & Technology, 59*(3), 1260–1267. https://doi.org/10.1111/ijfs.16920.

Le Tan, H. (2025). 3D food printing technologies for functional foods: Applications and antioxidant integration. *Food and Humanity, 5*. https://doi.org/10.1016/j.foohum.2025.100694.

Le Tohic, C., O'Sullivan, J. J., Drapala, K. P., Chartrin, V., Chan, T., Morrison, A. P., ... Kelly, A. L. (2018). Effect of 3D printing on the structure and textural properties of processed cheese. *Journal of Food Engineering, 220*, 56–64. https://doi.org/10.1016/j.jfoodeng.2017.02.003.

Leonard, W., Zhang, P., Ying, D., & Fang, Z. (2023). Surmounting the off-flavor challenge in plant-based foods. *Critical Reviews in Food Science and Nutrition, 63*(30), 10585–10606. https://doi.org/10.1080/10408398.2022.2078275.

Li, J., Wang, X., Chang, C., Gu, L., Su, Y., Yang, Y., ... Han, Q. (2024). Chicken egg white gels: Fabrication, modification, and applications in foods and oral nutraceutical delivery. *Foods, 13*(12), 1834. https://doi.org/10.3390/foods13121834.

Lima Nascimento, L. G., Odelli, D., Fernandes De Carvalho, A., Martins, E., Delaplace, G., Peres De Sá Peixoto Júnior, P., ... Casanova, F. (2023). Combination of milk and plant proteins to develop novel food systems: What are the limits? *Foods, 12*(12), 2385. https://doi.org/10.3390/foods12122385.

Llauger, M., Claret, A., Bou, R., López-Mas, L., & Guerrero, L. (2021). Consumer attitudes toward consumption of meat products containing offal and offal extracts. *Foods, 10*(7), 1454. https://doi.org/10.3390/foods10071454.

Logan, A., Cochet-Broch, M., Reynaud, L., Bouchama, M., Greaves, M., Puvanenthiran, A., & Tobin, A. (2025). Physicochemical and perceived textural attributes of pea and skim milk protein hybrid yoghurts compared to commercial thickened fluids. *Applied Food Research, 5*(2), 101085. https://doi.org/10.1016/j.afres.2025.101085.

Lupton, D., & Turner, B. (2018). Food of the future? Consumer responses to the idea of 3D-printed meat and insect-based foods. *Food and Foodways, 26*(4), 269–289. https://doi.org/10.1080/07409710.2018.1531213.

Ma, K. K., Greis, M., Lu, J., Nolden, A. A., McClements, D. J., & Kinchla, A. J. (2022). Functional performance of plant proteins. *Foods, 11*(4), 594. https://doi.org/10.3390/foods11040594.

Marcìa-Fuentes, J. A., Aleman, R. S., Areche, F. O., Flores, D. C., Roman, A. V., Martín-Vertedor, D., & Montero-Fernández, I. (2026). Functional foods: A review of foods ingredient and their health benefits. *Food and Humanity, 6*, 100953. https://doi.org/10.1016/j.foohum.2025.100953.

Maseko, K. H., Regnier, T., Bartels, P., & Meiring, B. (2025). Mushroom mycelia as sustainable alternative proteins for the production of hybrid cell-cultured meat: A review. *Journal of Food Science, 90*(2), e70060. https://doi.org/10.1111/1750-3841.70060.

McCarthy, N. A., Kennedy, D., Hogan, S. A., Kelly, P. M., Thapa, K., Murphy, K. M., & Fenelon, M. A. (2016). Emulsification properties of pea protein isolate using homogenization, microfluidization and ultrasonication. *Food Research International, 89*, 415–421. https://doi.org/10.1016/j.foodres.2016.07.024.

McClements, D. J., & Grossmann, L. (2021). The science of plant-based foods: Constructing next-generation meat, fish, milk, and egg analogs. *Comprehensive Reviews in Food Science and Food Safety, 20*(4), 4049–4100. https://doi.org/10.1111/1541-4337.12771.

Messina, V., Guest, N. S., Duncan, A. M., Mangels, A. R., Norris, J., Ruscigno, M., ... Messina, M. (2025). Dietary guidance on plant-based meat alternatives for individuals wanting to increase plant protein intake. *Frontiers in Nutrition, 12*, 1641234. https://doi.org/10.3389/fnut.2025.1641234.

Meurillon, M., & Engel, E. (2016). Mitigation strategies to reduce the impact of heterocyclic aromatic amines in proteinaceous foods. *Trends in Food Science & Technology, 50*, 70–84. https://doi.org/10.1016/j.tifs.2016.01.007.

Migliore, G., Di Gesaro, M., Borsellino, V., Asciuto, A., & Schimmenti, E. (2015). Understanding consumer demand for sustainable beef production in rural communities. *Quality - Access to Success, 16*(147), 75–79.

Morley, J. E. (2012). Anorexia of aging: A true geriatric syndrome. *The Journal of Nutrition, Health and Aging, 16*(5), 422–425. https://doi.org/10.1007/s12603-012-0061-9.

Munialo, C. D., Baeghbali, V., & Acharya, P. (2025). Plant-based alternatives to meat products. *Foods, 14*(8), 1396. https://doi.org/10.3390/foods14081396.

Nagatsuka, R., & Koriyama, T. (2026). Heat-treated Mucuna pruriens enhances matrix structure and nutritional quality in hybrid beef patties. *LWT, 240*, 119021. https://doi.org/10.1016/j.lwt.2026.119021.

Nie, Y., Xiong, Y. L., & Jiang, J. (2025). Texture, microstructure, and in vitro digestion of hybrid meat gel-type sausages formulated with functionalized pea protein. *Food Hydrocolloids, 167*, 111422. https://doi.org/10.1016/j.foodhyd.2025.111422.

Noreen, S., Hashmi, B., Aja, P. M., & Atoki, A. V. (2025). Health benefits of fish and fish by-products—A nutritional and functional perspective. *Frontiers in Nutrition, 12*, 1564315. https://doi.org/10.3389/fnut.2025.1564315.

O'Sullivan, J. J., Park, M., Beevers, J., Greenwood, R. W., & Norton, I. T. (2017). Applications of ultrasound for the functional modification of proteins and nanoemulsion formation: A review. *Food Hydrocolloids, 71*, 299–310. https://doi.org/10.1016/j.foodhyd.2016.12.037.

Okaiyeto, S. A., Liu, D., Zhang, C., Bai, J.-W., Chen, C., Sharma, P., ... Xiao, H.-W. (2025). Anti-nutrients of plant-based food: Physicochemical properties, effects on health and degradation techniques- a comprehensive review. *Journal of Future Foods*, S2772566925003088. https://doi.org/10.1016/j.jfutfo.2025.06.022.

Park, G., Oh, S., Park, S., Kim, Yun-a, Park, Y., Kim, Y., ... Choi, J. (2023). Physicochemical characteristics and storage stability of hybrid beef patty using shiitake mushroom (Lentinus edodes). *Journal of Food Quality, 2023*, 1–10. https://doi.org/10.1155/2023/7239709.

Parlasca, M. C., & Qaim, M. (2022). Meat consumption and sustainability. *Annual Review of Resource Economics, 14*(1), 17–41. https://doi.org/10.1146/annurev-resource-111820-032340.

Pérez-Montes, A., Rangel-Vargas, E., Lorenzo, J. M., Romero, L., & Santos, E. M. (2021). Edible mushrooms as a novel trend in the development of healthier meat products. *Current Opinion in Food Science, 37*, 118–124. https://doi.org/10.1016/j.cofs.2020.10.004.

Poore, J., & Nemecek, T. (2018). Reducing food's environmental impacts through producers and consumers. *Science (New York, N. Y.), 360*(6392), 987–992. https://doi.org/10.1126/science.aaq0216.

Portanguen, S., Meurillon, M., Duconseille, A., Drancourt, N., Peyron, M. A., Sicard, J., ... Mirade, P.-S. (2026). Towards coupling 3D printing and the glycation reaction to texturize food for people with oral frailty: A review. *Food Research International, 228*, 118419. https://doi.org/10.1016/j.foodres.2026.118419.

Portanguen, S., Meyer, V., Meurillon, M., Guittard, M., Schiell, C., Araszkiewiez, J., ... Mirade, P.-S. (2025). 3D food printing: At the intersection of social acceptance, participatory research, and specific nutritional benefits. *Innovative food packaging and processing technologies*. Elsevier 555–591. https://doi.org/10.1016/B978-0-323-91742-1.00003-9.

Portanguen, S., Tournayre, P., Gibert, P., Leonardi, S., Astruc, T., & Mirade, P.-S. (2022). Development of a 3D printer for the manufacture of functional food protein gels. *Foods, 11*(3), 458. https://doi.org/10.3390/foods11030458.

Portanguen, S., Tournayre, P., Sicard, J., Astruc, T., & Mirade, P.-S. (2019). Toward the design of functional foods and biobased products by 3D printing: A review. *Trends in Food Science & Technology, 86*, 188–198. https://doi.org/10.1016/j.tifs.2019.02.023.

Portanguen, S., Tournayre, P., Sicard, J., Astruc, T., & Mirade, P.-S. (2022). 3D food printing: Genesis, trends and prospects. *Future foods*. Elsevier 627–644. https://doi.org/10.1016/B978-0-323-91001-9.00008-6.

Prusaczyk, E., Earle, M., & Hodson, G. (2021). A brief nudge or education intervention delivered online can increase willingness to order a beef-mushroom burger. *Food Quality and Preference, 87*, 104045. https://doi.org/10.1016/j.foodqual.2020.104045.

Roozen, I., & Raedts, M. (2023). What determines omnivores' meat consumption and their willingness to reduce the amount of meat they eat? *Nutrition and Health, 29*(2), 347–355. https://doi.org/10.1177/02601060221080255.

Ross, M. M., Collins, A. M., McCarthy, M. B., & Kelly, A. L. (2022). Overcoming barriers to consumer acceptance of 3D-printed foods in the food service sector. *Food Quality and Preference, 100*, 104615. https://doi.org/10.1016/j.foodqual.2022.104615.

Ryder, C., Jaworska, S., & Grasso, S. (2023). Hybrid meat products and co-creation: What do consumers say, feel and think? *Frontiers in Nutrition, 10*, 1106079. https://doi.org/10.3389/fnut.2023.1106079.

Saffarionpour, S. (2024). Off-flavors in pulses and grain legumes and processing approaches for controlling flavor-plant protein interaction: application prospects in plant-based alternative foods. *Food and Bioprocess Technology, 17*, 1141–1182. https://doi.org/10.1007/s11947-023-03148-4.

Salim, R., Nehvi, I. B., Mir, R. A., Tyagi, A., Ali, S., & Bhat, O. M. (2023). A review on anti-nutritional factors: Unraveling the natural gateways to human health. *Frontiers in Nutrition, 10*, 1215873. https://doi.org/10.3389/fnut.2023.1215873.

Sarker, M. A. H., Pillidge, C., Afshari, R., Chandrapala, J., & Gill, H. (2026). Physicochemical properties of hybrid cheddar-type curd gels made from bovine milk and plant protein isolates. *Food Hydrocolloids, 171*, 111799. https://doi.org/10.1016/j.foodhyd.2025.111799.

Schiell, C., Portanguen, S., Diakite, I., D'Orlando, A., Mirade, P.-S., & Astruc, T. (2026). Investigation into heat-induced lentil globulin–porcine albumin protein interactions. *Food Hydrocolloids, 171*, 111811. https://doi.org/10.1016/j.foodhyd.2025.111811.

Schiell, C., Portanguen, S., Scislowski, V., Astruc, T., & Mirade, P.-S. (2023). Investigation into the physicochemical and textural properties of an iron-rich 3D-printed hybrid food. *Foods, 12*(7), 1375. https://doi.org/10.3390/foods12071375.

Singh, A., Boonchoosri, R., Saikumar, A., Maqsood, S., & Benjakul, S. (2025). Rheology, structure, and printability of surimi-based hybrid inks containing Bambara groundnut, pea, and soy protein isolates for the development of salmon fillet analogs. *Food Hydrocolloids, 169*, 111558. https://doi.org/10.1016/j.foodhyd.2025.111558.

Sipponen, M. H., Mäkinen, O. E., Rommi, K., Heiniö, R.-L., Holopainen-Mantila, U., Hokkanen, S., ... Nordlund, E. (2018). Biochemical and sensory characteristics of the cricket and mealworm fractions from supercritical carbon dioxide extraction and air classification. *European Food Research and Technology, 244*(1), 19–29. https://doi.org/10.1007/s00217-017-2931-1.

Stanišić, N., Delić, N., Cekić, B., Stojiljković, N., Gogić, M., Samolovac, L., & Stajić, S. (2025). Current processing technologies and challenges in hybrid meat production. *Processes, 13*(9), 2853. https://doi.org/10.3390/pr13092853.

Stožinić, M., Ačkar, Đ., Šojić, B., Sedlar, T., Popović, L., Pajin, B., ... Lončarević, I. (2025). Incorporation of pork meat and blood plasma proteins into a Cocoa cream matrix: Characterization, Comparison Of Functional Properties, And In Vitro Simulated Digestion. *Foods, 14*(14), 2547. https://doi.org/10.3390/foods14142547.

Suurs, P., & Barbut, S. (2020). Collagen use for co-extruded sausage casings – A review. *Trends in Food Science & Technology, 102*, 91–101. https://doi.org/10.1016/j.tifs.2020.06.011.

Theron, L. (2026). https://BioRender.com/prs4wvk.

Theron, L. (2026). https://BioRender.com/pxy26ba.

Tkaczewska, J., Mungure, T., & Warner, R. (2025). Is it still meat? The effects of replacing meat with alternative ingredients on the nutritional and functional properties of hybrid products: A review. *Critical Reviews in Food Science and Nutrition, 65*(28), 5752–5771. https://doi.org/10.1080/10408398.2024.2430750.

Toldrà, M., Lynch, S. A., Couture, R., & Álvarez, C. (2019). Blood proteins as functional ingredients. *Sustainable meat production and processing*. Elsevier 85–101. https://doi.org/10.1016/B978-0-12-814874-7.00005-5.

Tzompa-Sosa, D. A., Dewettinck, K., Gellynck, X., & Schouteten, J. J. (2021). Replacing vegetable oil by insect oil in food products: Effect of deodorization on the sensory evaluation. *Food Research International, 141*, 110140. https://doi.org/10.1016/j.foodres.2021.110140.

United Nations. (2022). Department of economic and social affairs. *world population prospects*. New York: UN.

Vanqa, N., Mshayisa, V. V., & Basitere, M. (2022). Proximate, Physicochemical, techno-functional and antioxidant properties of three edible insect (Gonimbrasia belina, Hermetia illucens and Macrotermes subhylanus) Flours. *Foods, 11*(7), 976. https://doi.org/10.3390/foods11070976.

Vlahova-Vangelova, D., Balev, D., & Kolev, N. (2023). Cricket powder (Acheta domesticus) as a lean pork meat replacer in cooked sausages. https://doi.org/10.17170/KOBRA-202210056951.

Wang, Y., Tuccillo, F., Lampi, A., Knaapila, A., Pulkkinen, M., Kariluoto, S., ... Katina, K. (2022). Flavor challenges in extruded plant-based meat alternatives: A review. *Comprehensive Reviews in Food Science and Food Safety, 21*(3), 2898–2929. https://doi.org/10.1111/1541-4337.12964.

WHO. (2023). World Health Organization. Red and processed meat in the context of health and the environment: many shades of red and green. Information brief. Geneva: World Health Organization. http://www.jstor.org/stable/resrep56541Red and processed meat in the context of health and the environment: Many shades of red and green: Information brief.

Willett, W., Rockström, J., Loken, B., Springmann, M., Lang, T., Vermeulen, S., ... Murray, C. J. L. (2019). Food in the anthropocene: The EAT–lancet commission on healthy diets from sustainable food systems. *The Lancet, 393*(10170), 447–492. https://doi.org/10.1016/S0140-6736(18)31788-4.

Xia, W., Drositi, I., Czaja, T. P., Via, M., & Ahrné, L. (2024). Towards hybrid protein foods: Heat- and acid-induced hybrid gels formed from micellar casein and pea protein. *Food Research International, 198*, 115326. https://doi.org/10.1016/j.foodres.2024.115326.

Zakki, A., Aryanti, N., & Hadiyanto, H. (2024). Functional and structural characteristic of plant protein isolates as emulsifier by ultrasound-assisted extraction: A review. *Bioactive Carbohydrates and Dietary Fibre, 32*, 100449. https://doi.org/10.1016/j.bcdf.2024.100449.

Zhang, R., Farouk, M. M., Realini, C. E., & Thum, C. (2025). Hybridization in meat--based dual protein foods: Mechanisms, challenges, and consumer insights. *Comprehensive Reviews in Food Science and Food Safety, 24*(4), e70216. https://doi.org/10.1111/1541-4337.70216.

Zhou, H., Hu, Y., Tan, Y., Zhang, Z., & McClements, D. J. (2021). Digestibility and gastrointestinal fate of meat versus plant-based meat analogs: An in vitro comparison. *Food Chemistry, 364*, 130439. https://doi.org/10.1016/j.foodchem.2021.130439.

Zinina, O., Merenkova, S., Tazeddinova, D., Rebezov, M., Stuart, M., Okuskhanova, E., … Yessimbekov (2019). Enrichment of meat products with dietary fibers: A review [PDF]. 326.4Kb. https://doi.org/10.15159/AR.19.163.

Further reading

Hu, F. B., Otis, B. O., & McCarthy, G. (2019). Can plant-based meat alternatives be part of a healthy and sustainable diet? *JAMA: the Journal of the American Medical Association, 322*(16), 1547. https://doi.org/10.1001/jama.2019.13187.

CHAPTER FOUR

Improving nutritional properties of food products by ultrasound: Basic principles and behind mechanisms

Lujuan Xing[a,b], Lei Jin[a,b], Chao Ma[a,b], Wenxuan Wang[a,b], Xianglin Yan[c], Qian Xu[c], and Wangang Zhang[a,b,*]

[a]Jiangsu Synergetic Innovation Center of Meat Processing and Quality Control, College of Food Science and Technology, Nanjing Agricultural University, Nanjing, P.R. China
[b]State Key Laboratory of Meat Quality Control and Cultured Meat Development, MOE, College of Food Science and Technology, Nanjing Agricultural University, Nanjing, P.R. China
[c]College of Food Science and Engineering, Production & Construction Group Key Laboratory of Special Agricultural Products Further Processing in Southern Xinjiang, Tarim University, Alar, P.R. China
*Corresponding author. e-mail address: wangang.zhang@njau.edu.cn

Contents

Advances in Food and Nutrition Research, Volume 122
ISSN 1043-4526, https://doi.org/10.1016/bs.afnr.2026.04.001

Abstract

Ultrasound technology is a green technology which has been widely used in the food industry. The mechanism of ultrasound relies on acoustic cavitation, thermal and mechanical effect, which physically disrupts food matrices by breaking down cell walls, tenderizing meat fibers, and modifying the structures of starch and protein in food. These structural changes directly enhance textural qualities, such as achieving softer textures in fruits and vegetables or more tender meat, while also improving water retention for food products. Nutritionally, cell disruption during ultrasound treatment can also increase the bioaccessibility and bioavailability of essential nutrients, including vitamins, antioxidants, and minerals, by releasing them from their cellular compartments. Furthermore, ultrasound can enhance the bioavailability of proteins and lipids by altering their structures. In recent years, the applications of ultrasound include accelerating marination and tenderization in meat processing, modifying texture in dairy products, and improving the stability and nutritional profile of emulsions. The non-thermal nature of ultrasound also helps preserve heat-sensitive compounds, maintaining the original flavor and nutritional value of food. This chapter focuses on the mechanism and utilization of ultrasound in improving the nutritional quality of food, with the aim of further promoting the applications of ultrasound in the food industry.

1. Introduction

Ultrasound refers to sound waves with frequencies higher than 20,000 Hz, which are mechanical longitudinal waves propagating through elastic media. Initially, the ultrasound was applied in military sonar for underwater detection. In the mid-20th century, driven by advancements in electronics and materials science, the ultrasound began to transition from laboratory research to industrial applications. Till to now, ultrasound is regarded as an effective and promising non-thermal technology, which increasingly has been investigated in the food field. As a non-thermal treatment, ultrasound offers a promising alternative in the modification and functional enhancement of food properties. The ultrasound-induced effects are primarily driven by three interconnected physical phenomena including acoustic cavitation, thermal and mechanical effect. The formation of acoustic cavitation mainly relies on the violent collapse of microscopic bubbles in a liquid medium, which is also regarded as the cornerstone mechanism for ultrasound treatments. The implosive collapse generates extreme localized conditions, including very high temperatures and pressures alongside powerful shear forces and reactive radical species, that collectively inducing targeted physical and chemical changes in food molecules. Generally, the ultrasonic forces directly influence the intermolecular interactions and conformational arrangements within foods. By altering the molecular size, surface hydrophobicity, and flexibility of main food macromolecules, the ultrasound effectively modulates key functional properties essential for food applications.

While extensive research has documented the impacts of ultrasound on the functionality and structural denaturation of different foods, a focused synthesis detailing the specific causal pathways from induced structural changes to enhanced digestibility and nutritional properties is less prevalent. Therefore, examining the mechanisms through which ultrasonic intervention improves nutritional properties, with dedicated attention to the applications of ultrasound in different food areas, represents a significant and applied research direction. This knowledge bridges fundamental science with practical formulation strategies, enabling the development of healthier, more stable and cleaner-label food products.

2. The mechanisms of ultrasound technology

2.1 Fundamentals and classification

2.1.1 The sound waves

A sound wave is a longitudinal mechanical wave that propagates through an elastic medium (like air or water) by creating compressions and rarefactions. As defined by Lorimer and Mason (1987), compression occurs when air or other gas particles are forced together, increasing the local pressure and density. In graphical representations of sound wave, the point of maximum compression is denoted as the wave crest, while the point of maximum rarefaction is identified as the wave trough (Lüders & Pohl, 2017). The propagation of sound establishes a continuous sequence of these crests and troughs within the medium. The physical properties of sound are largely governed by its frequency, defined as the number of complete wave cycles passing a point per second along with the unit of Hertz (Hz). Based on this parameter, sound waves are categorized into infrasound, audible sound and ultrasound. Infrasound encompasses frequencies below 20 Hz, which are inaudible to the human ear. Audible sound occupies the frequency range of approximately 20 Hz to 20 kHz. As shown in Fig. 1, the ultrasound refers to sound waves with frequencies exceeding 20 kHz, extending upwards into the megahertz (MHz) range.

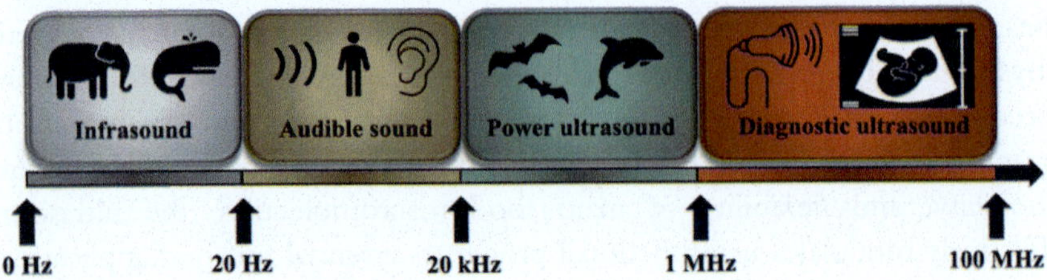

Fig. 1 Classifications of acoustic wave.

2.1.2 Principles and applications

Within the ultrasound spectrum, the subdivision is based on frequency and application. Diagnostic ultrasound typically operates at very high frequencies, generally above 1 MHz. This imaging modality operates on the principle of the pulse-echo technique, whereby emitted high-frequency sound waves are reflected by internal tissue interfaces. These returning echoes are processed to construct real-time images depicting anatomical structures and dynamic functions, including vascular flow. Consequently, it has become an indispensable non-invasive diagnostic tool in various medical fields, notably in obstetrics (Rumack & Levine, 2017). In contrast, the primary ultrasonic technology employed within the food and processing industries is power ultrasound, which occupies a lower frequency band, typically from 20 kHz to 1 MHz (Barretto et al., 2023). The effects and applications of ultrasound in food systems are critically dependent on its intensity, commonly measured in watts per square centimeter (W/cm^2). Low-intensity ultrasound (typically below 1 W/cm^2), often associated with higher frequencies, is non-destructive and serves primarily analytical and sensing functions (Zhou, Zhang, Yin, Zhang, & Yang, 2021). Its applications include the non-destructive evaluation of fruit ripeness, the assessment of compositional and qualitative properties of materials like flour, and the detection of foreign bodies (e.g., plastic or metal fragments) within food products (Chandrapala, 2015).

Conversely, high-intensity ultrasound (ranging from 10 to 1000 W/cm^2) is usually generated at lower frequencies and is inherently disruptive. This form delivers sufficient acoustic energy to induce physical and chemical changes within a medium. In food matrices, it can disrupt cellular structures, break down macromolecular assemblies, and enhance mass transfer, thereby significantly altering physicochemical attributes (Villamiel et al., 2017). Consequently, the high-intensity power ultrasound is used as a versatile processing tool to ameliorate food quality. The benefits of ultrasound include the improvement of meat tenderness, the acceleration of curing and salting processes, and the enhancement of emulsion stability and formation (Kang et al., 2016; Zhou et al., 2021; Zhang, Zhang, Zou, & Zhang, 2021). This ability, to modify texture, improve efficiency and engineer functional properties, underscores its value as an innovative technology in modern food engineering.

2.2 Mechanism of ultrasound

2.2.1 Thermal effects

When ultrasonic waves propagate through a medium, a portion of the acoustic energy is attenuated due to intrinsic absorption mechanisms within the material. This absorption arises primarily from viscoelastic and thermoviscous losses, resulting in the irreversible conversion of acoustic energy into heat. Consequently, this process induces a localized temperature rise within the medium. This heating mechanism, achieved through the absorption of acoustic energy by the medium, is known as the thermal effect of ultrasound (Bhargava et al., 2021). Compared to other heating methods, ultrasound offers a distinct pathway for temperature elevation. During propagation, high-amplitude ultrasonic waves may form sawtooth-shaped periodic shock waves, creating substantial pressure gradients at the wavefront. The vibrational energy is continuously absorbed by the medium and transformed into heat, increasing the overall temperature of the medium as well as causing localized heating near boundaries. Simultaneously, the intense high-frequency oscillations induced by ultrasound cause intermolecular friction within the medium, generating heat that can raise the temperature of both solid and fluid media. When ultrasound passes through an interface between two different media, the temperature rise tends to be more pronounced. This is due to the difference in characteristic impedance at the interface, which leads to reflection and the formation of standing waves. These standing waves enhance molecular friction, resulting in increased heat generation. The thermal action of ultrasound can be categorized into two forms of thermal effects including continuous-wave thermal and transient thermal effects. Continuous-wave thermal effects arise from the absorption of acoustic energy and internal friction losses within the medium. Under sustained exposure to continuous ultrasound, the temperature in the region of the acoustic field gradually increases. Transient thermal effects primarily refer to the instantaneous high temperatures generated during the collapse of cavitation bubbles. In food processing, thermal effects typically synergize with cavitation and mechanical effects, enhancing processing efficiency and improving food properties. However, undesirable thermal-induced reactions in food processing, such as the Maillard reaction or lipid oxidation, may be accelerated, potentially altering sensory attributes and generating compounds that affect nutritional value. Therefore, precise thermal management during ultrasound is paramount. Optimizing parameters, such as treatment time, pulse intervals, and cooling systems, allows operators to harness the beneficial mechanical and cavitation effects of ultrasound while minimizing detrimental thermal damage, thereby striking a crucial balance between achieving technological objectives and preserving the intrinsic nutritional profile of food.

2.2.2 Mechanical effects

The application of ultrasonic energy to a medium induces high-speed and minute vibrations of its particles, leading to variations in mechanical parameters such as particle velocity, acceleration, sound pressure and sound intensity. This phenomenon is referred to as the mechanical effect of ultrasound. As a form of mechanical energy propagation related to wave processes, ultrasound generates linear oscillatory effects. During its propagation through a medium, although the displacement amplitude of the particles is very small, the particle acceleration induced by ultrasound can be extremely high. For instance, a 20 kHz ultrasound with an intensity of 1 W/cm^2 propagating in water generates a sound pressure amplitude of 173 kPa. The pressure amplitude oscillates between positive and negative 173 kPa with 20,000 times per second, with the maximum particle acceleration reaching 1440 km/s^2 which is approximately 1500 times the acceleration due to gravity. The intense and rapid mechanical motion constitutes the mechanical vibration effect of power ultrasound. When the medium is not uniform, differences in acoustic impedance between layers cause reflection of the propagating waves and the formation of standing waves. The antinodes and nodes of these standing waves create localized variations in pressure, tension, and acceleration. Furthermore, since particles (e.g., biological molecules) in different media have different masses, the vibrational velocities induced by pressure also change. The resulting relative motion between particles and the associated pressure variations represents another cause of ultrasonic mechanical effects. The mechanical effects of ultrasound enhance food processing through efficient emulsification, homogenization, cell disruption for extraction, and microbial inactivation, improving product quality and process sustainability.

2.2.3 Cavitation effect

When ultrasonic waves propagate through liquid media, microscopic bubbles are generated and undergo periodic expansion and contraction under the alternating rarefaction and compression phases of the ultrasonic wave. As illustrated in Fig. 2. when these bubbles reach an unstable state, they collapse violently and rupture, instantaneously generating high temperatures (approximately 5000 K) and pressures (approximately 100 MPa), thereby giving rise to cavitation effects. The cavitation effect leads to the generation of intense shock waves in the liquid surrounding the bubbles, creating localized points of extremely high temperature and pressure. At the moment of cavitation bubble collapse, a transient temperature as high as 5000 K and a pressure of approximately 50 MPa can be produced within an extremely confined space around the bubble, with a temperature cooling rate reaching up to 10^9 K/s. Ultrasonic cavitation is typically categorized into two types including stable cavitation and transient cavitation. Under ultrasonic excitation, these bubbles can grow when the sum of the acoustic pressure and the static pressure is relatively small and shrink when it becomes larger, causing the bubbles to vibrate or pulsate. At lower ultrasonic intensities, this vibration is not violent and usually does not produce destructive forces, which is known as stable cavitation. However, even in stable cavitation, the presence of acoustic streaming can increase stress around the bubbles, potentially altering certain biological functions. When the acoustic intensity exceeds a specific threshold, the bubble vibration becomes extremely violent. When the resultant force of acoustic and static

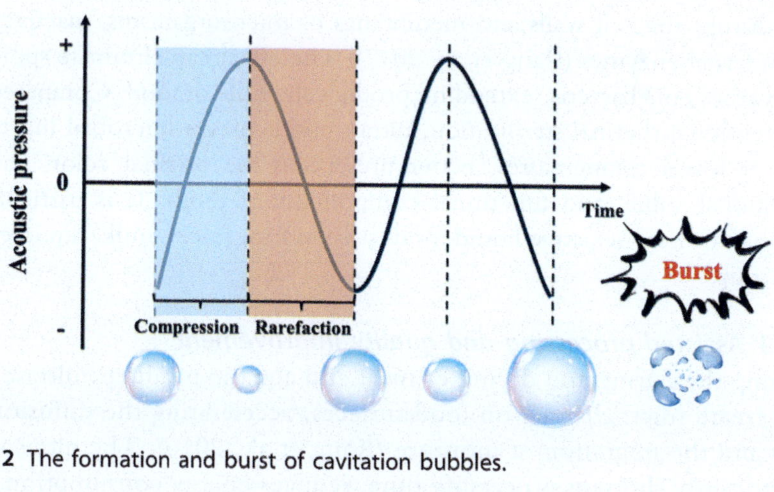

Fig. 2 The formation and burst of cavitation bubbles.

pressure approaches zero, the bubble diameter increases rapidly. Subsequently, as the acoustic pressure changes, the bubble collapses violently under a large resultant pressure, eventually rupturing into many smaller bubbles. This process generates intense shock waves along with localized high temperatures and pressures, a phenomenon referred to as transient cavitation.

2.3 The applications of ultrasound on foods

2.3.1 Pretreatment

Ultrasound is widely employed in pretreatment processes within food processing, such as cell disruption and the extraction of bioactive compounds. Ultrasonic treatment facilitates the release of intracellular bioactive substances, oils, and flavor components, thereby enhancing extraction efficiency and reducing processing time. Simultaneously, the microjets and shear forces generated by cavitation effects can effectively remove contaminants from food surfaces, serving cleaning and decontamination purposes (Ashokkumar, 2015). This is commonly applied in the cleaning of fruits, vegetables, processing equipment, and packaging materials. Furthermore, cavitation promotes reactant mixing and accelerates mass transfer, making it be suitable for homogenizing and emulsifying products such as fruit juices and sauces, thereby improving product stability.

2.3.2 Sterilization

The mechanical effects and free radicals produced by ultrasonic cavitation can disrupt the cell walls and membranes of microorganisms, inactivating the internal enzymes (Kang et al., 2017). The effect can eliminate spoilage and pathogenic bacteria, extending product shelf life of food. Compared to conventional thermal sterilization, ultrasound achieves microbial inactivation at lower temperatures, better preserving the original color, flavor, nutritional value, and functional components of foods. It is particularly suitable for heat-sensitive liquid foods such as fruit juices, milk, liquid eggs, and soups.

2.3.3 Assisted processing and quality improvement

During the curing and drying of meat and aquatic products, ultrasound can create microchannel on food surfaces, accelerating the diffusion of salts and the migration of moisture (Kang et al., 2016). The ultrasound significantly shortens processing time, reduces energy consumption and improves product quality (Zhang, Zhang, Zhou, Wang, & Zhang, 2021). During fermentation processes, low-intensity ultrasound stimulates microbial metabolism, increasing microbial growth rates and substrate utilization. This accelerates the fermentation of foods like yogurt and alcoholic beverages and enhances the yield of beneficial metabolites. During freezing processes, ultrasound promotes ice nucleation while inhibiting crystal growth, leading to the formation and uniform distribution of fine ice crystals. This minimizes mechanical damage to cellular structures caused by ice crystals, helping frozen products better retain their texture and reduce drip loss upon thawing of meat, seafood, and ice cream (Zheng & Sun, 2006).

2.3.4 Analytical detection

The propagation characteristics of ultrasound, including sound velocity, attenuation coefficient, and acoustic impedance, are closely related to the composition, structure, phase state and physical properties of foods. By analyzing variations in acoustic parameters, it is used to monitor in real-time the composition, textural changes, and internal defects of foods, providing precise monitoring and quality assurance during production (Haeggström & Luukkala, 2001). The primary advantages of ultrasound based detection include minimal sample preparation, rapid analysis, real-time monitoring capability, and the preservation of sample integrity. Low-intensity and high-frequency ultrasound (>1 MHz) is effectively employed to determine key quality parameters. By precisely measuring the speed of ultrasound and attenuation coefficient, the fat content in meat, milk and dairy products, as well as solid-fat content in oils and fats could be realized (Cho & Irudayaraj, 2003). Furthermore, ultrasound velocity is directly correlated with solute concentration, enabling the non-invasive assessment of sugar content in beverages, salt concentration in brines and overall soluble solid content in fruit juices. Beyond composition, it could also be used in assessing the viscoelastic properties of gels, doughs and emulsions, providing insights into texture and stability. The analysis of backscattered signals can reveal information about particle size distribution in colloidal systems like milk or infant formula and detect phase separation in emulsified products. In solid foods, such as cheeses and fruits, ultrasound can evaluate structural homogeneity, detect internal voids or cracks and even predict firmness and ripeness by correlating acoustic parameters with mechanical properties (Meier et al., 2015).

3. Effect of ultrasound on the nutritional components in food

3.1 Ultrasound and proteins

The impact of ultrasound on proteins is closely related to the state of the food itself. Under ultrasound treatment, the mechanical shear forces, microjets, and turbulent effects generated by cavitation can disrupt hydrogen bonds that maintain secondary structures, leading to a shift in the conformation distribution of protein secondary structures (Zhao et al., 2022). Moderate ultrasonic treatment can promote the unfolding of higher-order protein structures, exposing more enzymatic cleavage sites, thereby significantly improving the in vitro digestibility and bioaccessibility along with enhancing their nutritional value. For example, ultrasound treatment has been reported to decrease the contents of α-helices, β-turns and random coils in mussel sarcoplasmic proteins, while increasing the proportion of β-sheets (Zou et al., 2020; Zhang, Xing, Kang, Zhou, Wang, et al., 2021). However, excessive ultrasonication causes the reaggregation of already dispersed proteins, thereby exerting adverse effects on their secondary structures. Furthermore, not all proteins exhibit significant changes in secondary structure after ultrasound treatment. For instance, the secondary structure of ovalbumin showed no significant change after ultrasound treatment (Xiong et al., 2016). This indicates that the effect of ultrasound on protein secondary structure is jointly governed by protein type, aggregation state, degree of denaturation, as well as ultrasonic power and treatment duration.

The effects of ultrasound on the tertiary structure of proteins are mainly attributed to the disruption of non-covalent interactions induced by mechanical shear forces, micro-jetting and transient high temperature and high pressure. Changes in tryptophan residues, sulfhydryl content, disulfide bonds and surface hydrophobicity are commonly regarded as key indicators for characterizing alterations in protein tertiary structure (Zhang et al., 2020). Under ultrasound treatment, the internal structure of protein molecules becomes loosened or partially unfolded, resulting in the exposure of hydrophobic groups and side chains that are originally buried within the molecular interior, thereby modifying the protein tertiary structure (Zou et al., 2018). For instance, ultrasound treatment can significantly enhance the fluorescence intensity of chicken myofibrillar proteins by exposing more tryptophan residues to polar environments (Li et al., 2020). Simultaneously, sulfhydryl groups encapsulated within protein molecules are exposed to the surface under ultrasonic action, accompanied by the breaking of disulfide bonds, which increases the free sulfhydryl content of myofibrillar proteins (Zhang, Zhang, & Xing, 2021). However, excessive ultrasound treatment can cause severe disruption of the tertiary structure, resulting in protein reaggregation and structural rearrangement. For example, free radicals, such as H• and •OH generated by excessive ultrasound treatment, can promote the further oxidation of free sulfhydryl groups into disulfide bonds, causing previously relaxed protein structures to re-rigidize (Alavi, Chen, & Emam-Djomeh, 2021). Moreover, protein reaggregation causes hydrophobic regions to be re-embedded within the protein molecules, reducing surface hydrophobicity. Following ultrasonic treatment, the solubility of chickpea protein could be increased to 63.18 %. Under treatment conditions of 400 W, chickpea protein emulsion demonstrated optimal emulsifying characteristics. The treatment of 10 min markedly increased water and oil holding capacities of chickpea proteins (Habib et al., 2025). Fourier-transform infrared spectroscopy revealed protein unfolding and altered secondary structure, promoting molecular flexibility and interfacial activity. In addition, the ultrasonic treatment significantly enhanced the in vitro digestibility of chickpea protein isolates (CPIs) over extended treatment periods. Compared to the native CPI (81.61 %), it increased substantially to 85.02 % at 20 min and further to 86.29 % at 30 min. This improvement can be attributed to the protein structural modifications induced by ultrasound. Simultaneously, ultrasonic treatment can effectively degrade the structures of macromolecular proteins, such as β-lactoglobulin in milk and tropomyosin in shrimp, thereby reducing the allergenicity of these foods. However, excessive ultrasonic treatment may also lead to oxidation or aggregation of amino acid side-chain groups in proteins, consequently diminishing their nutritional quality.

3.2 Ultrasound and lipids

Oils and fats collectively refer to lipids, which are esters of glycerol and fatty acids. At room temperature, the glycerides containing predominantly unsaturated fatty acids exist in a liquid state, also known as fatty oils. Conversely, glycerides rich in saturated fatty acids remain solid under ambient conditions and are referred to as fats. As essential components of human life and ubiquitous substances in the biological world, oils and fats are widely present across the plant and animal kingdoms. The conventional extraction methods primarily include mechanical pressing, solvent extraction or a combination of both. However, these traditional techniques often suffer from drawbacks such as slow extraction rates, low oil yield, lengthy processing times and significant solvent consumption. In contrast, the application of ultrasound to intensify the solvent extraction process can improve oil quality, substantially reduce extraction time and decrease the required amount of extraction solvent. Regarding lipid extraction and functional preservation, ultrasound can significantly enhance extraction efficiency by disrupting cell walls and oil-body structures, while maintaining the stability of heat-sensitive nutrients such as polyunsaturated fatty acids. Research utilizing rice bran as the raw material demonstrated the efficacy of ultrasound-assisted ethanol extraction for rice bran oil. Results indicated that ultrasonic treatment at 40 kHz for 20 min achieved an oil extraction rate of 14.2 %, equivalent to 91.6 % of the yield obtained using n-hexane extraction. As reported in the study of Nogueira et al. (2018), ultrasound treatment (5 min) achieved a high lipid yield of 24.6 % from the *Chaetoceros calcitrans*, which was statistically equivalent to both microwave processing (24.2 % in 40 s) and conventional chemical disruption. The ultrasound treatment altered the fatty acid profile, increasing saturated and polyunsaturated fractions while decreasing monounsaturated molecules. Furthermore, ultrasound-assisted ethanol extraction showed positive effects on enhancing the content of beneficial compounds such as β-carotene and γ-oryzanol in rice bran oil, while also contributing to the improvement of its antioxidant activity. Simultaneously, ultrasound serves as an effective means for preparing stable emulsions. Its cavitation effect can break lipid droplets into micron or even nano-scale particles, thereby forming a uniform and stable emulsified system. However, intense cavitation may generate a large number of free radicals and induce localized overheating, which accelerates lipid oxidation. This process promotes the formation of harmful substances such as hydroperoxides and volatile aldehydes and ketones, ultimately reducing product stability and leading to the deterioration of the nutritional quality of the food.

3.3 Ultrasound and carbohydrates

Polysaccharides represent a significant class of nutrients found in food raw materials. They are widely distributed not only within animal and plant organisms but also across various food ingredients. Common polysaccharides include starch, inulin, intracellular plant polysaccharides and cellulose. The impact of ultrasound on carbohydrates is mainly reflected in the modification of their physicochemical structures and nutritional functions. Ultrasonic treatment would cause surface erosion, internal cracking and molecular chain breakage of starch-based polysaccharides, thereby enhancing their solubility and swelling capacity (Bera et al., 2015). It makes starch be more susceptible to hydrolysis, increasing its digestion rate and glycemic index. For dietary fiber polysaccharides such as pectin and cellulose, ultrasonic treatment effectively disrupts plant cell wall structures, reduces the polymerization of cellulose macromolecules and enhances their water-holding capacity and solubility, thus improving their dispersion stability in food systems. On the other hand, the intense cavitation effect generates free radicals that attack and oxidize reducing sugars, leading to the excessive breakdown of soluble dietary fibers. It weakens the water retention ability and impairs their functions in adsorbing fats and cholesterol. In food systems, the thermal effects and free radicals induced by ultrasonic treatment would accelerate the Maillard reaction. This process consumes reduced sugars and essential amino acids in the food system, further reducing the nutritional value of proteins and potentially generating harmful substances such as advanced glycation end products.

3.4 Ultrasound and vitamins

Compared to other nutrients, vitamins are generally less stable and susceptible to degradation by various factors such as heat, light, oxygen and environmental pH. Ultrasound, as a non-thermal processing technology, offers a distinct advantage by effectively inactivating microorganisms and enzymes at lower temperatures. The ultrasound treatment can circumvent the thermal degradation associated with traditional heat-based sterilization methods such as pasteurization and ultra-high temperature processing (UHT). Consequently, it can significantly contribute to preserving the stability and concentration of heat-sensitive vitamins (VC and VB) in fruit juices, berry purees and vegetable derivatives. For instance, it results in higher post-treatment levels and slower degradation during storage in orange juice compared to thermal processing and more effectively preserves vitamin C in watercress and tomato extract (Cruz et al., 2008; Tiwari et al., 2009). Beyond nutrient retention, ultrasonic treatment exerts minimal adverse effects on overall juice quality which has been documented in various juices including orange, guava, blackberry and strawberry (Tiwari et al., 2008; Cheng et al., 2007; Tiwari et al., 2009). Color, a key quality parameter, is also well-maintained. Studies on orange juice indicate that while both treatments induced similar chromaticity shifts (a^* and b^* values), ultrasonication yielded products with significantly higher lightness (L^* value), contributing to a more desirable visual profile and potentially greater consumer acceptance (Adekunte et al., 2010). Thus, the ultrasound would help maintain their nutritional quality. However, the application of ultrasound is a double-edged sword regarding vitamin retention. The intense cavitation bubbles generated during sonication collapse violently, producing localized high temperatures and pressures, along with reactive free radicals like hydroxyl radicals (•OH). While beneficial for microbial inactivation, these phenomena can accelerate oxidation reactions. This is particularly detrimental to oxidation-prone vitamins, including vitamin C and fat-soluble vitamin A, D and E. Excessive ultrasonic intensity or prolonged treatment times can lead to the oxidative degradation of these compounds, diminishing their nutritional value. Therefore, optimizing process parameters is crucial to maximizing the protective benefits of ultrasound while minimizing its potential degradative effects. In contrast to vitamins, mineral elements exhibit considerably stronger chemical stability. They exist primarily in ionic or inorganic compound forms within food matrices. Ultrasound treatment has a minimal direct impact on their atomic structure or valence state. Any potential effect is typically indirect, possibly influencing the solubility or bioavailability of minerals by disrupting the cellular or macromolecular structures that bind them. Overall, minerals are largely unaffected by sonication and their nutritional content remains stable.

3.5 Ultrasound and bioactive compounds

Ultrasound has been proven to be highly effective in the extraction of valuable bioactive compounds from plant based materials. The implosion of bubbles near cell walls generates powerful micro-jets and shear forces that cause physical rupture and erosion of plant tissues. This breakdown significantly increases the surface area for mass transfer and weakens the binding forces between target compounds (such as polyphenols, flavonoids, terpenoids, and pigments) and the cellular matrix including polysaccharides and proteins (Godswill et al., 2025). As a result, ultrasound-assisted extraction typically achieves higher yields in shorter times at lower temperatures compared to conventional methods. More importantly, the reduced thermal exposure helps preserve the biological activity of these thermolabile compounds. Nevertheless, similar to the challenge with vitamins, the reactive chemical environment created during sonication poses a risk to the integrity of certain extracted bioactives. The free radicals generated by cavitation, combined with localized heating, can induce oxidative degradation reactions. This is especially critical for highly sensitive molecules like anthocyanins and other natural pigments, whose color and antioxidant properties are closely linked to their chemical structure. Uncontrolled oxidation can lead to their decolorization, degradation and the loss of functionality.

4. Ultrasound and nutritional quality of food matrices

4.1 Meat and seafood

In the processing of meat, poultry and seafood products, ultrasound has emerged as a versatile pre-treatment and auxiliary technology, which is widely explored in brining, tenderization, cooking, sterilization and drying processes (Zhang et al., 2025). The incorporation of this technique into these processing stages harnesses the distinctive physical phenomena arising from acoustic cavitation and mechanical wave propagation, thereby presenting considerable prospects for augmenting both operational efficiency and end-product attributes. In terms of nutritional quality, ultrasound treatment positively influences the bioavailability of major nutrients in meat. The powerful mechanical effects, primarily from cavitation-induced micro-jets and shear forces, effectively disrupt the structure of myofibrillar proteins and connective tissues (Wang et al., 2022; Zou, Zhang, Kang, & Zhou, 2018). This structural breakdown facilitates the release of intracellular contents and partially hydrolyzes proteins, which then leading to improved protein digestibility. Furthermore, by disrupting the muscle matrix and altering the binding state of minerals, ultrasound can enhance the bioaccessibility and subsequent bioavailability of essential minerals such as calcium, iron and zinc, making them be more readily absorbable upon consumption. In addition, the ultrasound treatment significantly can modify the physical and sensory attributes of meat products. The disruption of myofiber and sarcomere structure directly contributes to meat tenderization. For instance, research on marinated beef has demonstrated that ultrasound-assisted brining promoted the degradation of key cytoskeletal proteins, including desmin and troponin-T (Kang et al., 2017). This degradation weakened the structural integrity of muscle cells, affecting the mobility and distribution of water within the meat. Specifically, it could increase the proportion of immobilized water, thereby improving water-holding capacity (WHC) and tenderness, which were critical quality parameters for products such as marinated or stewed beef. Regarding meat color, a crucial visual quality indicator, ultrasound exerted a nuanced influence with general increases in the lightness (L* value) of meat. Under shorter treatment durations, ultrasound could also enhance the redness (a* value), likely by influencing myoglobin states or surface properties. However, prolonged ultrasonic exposure may lead to a decrease in a* value, potentially due to protein denaturation or oxidation, highlighting the importance of optimized treatment time (Kang, Gao, Ge, Zhou, & Zhang, 2017). During thermal processing, the improved WHC resulting from ultrasonic pretreatment translated into reduced purge and cooking loss. The retention of inherent juices preserved the succulence and juiciness of meat products, leading to enhanced sensory quality and flavor perception.

For comminuted products like sausages and luncheon meats, ultrasound pretreatment of raw meat or application during emulsion preparation can improve functional properties (Wang, Zhang, & Zhou, 2019; Zhou, Zhang, Lorenzo, & Zhang, 2021). It promotes the extraction and solubilization of myofibrillar proteins, which are essential for forming stable emulsions and robust gel networks upon heating. Consequently, it leads to superior emulsion stability, enhanced gel strength and a more uniform texture and sliceability in the final product (Zhou et al., 2021; Zhou et al., 2025). Thawing constitutes a critical processing stage frequently associated with significant quality deterioration. Although ambient-temperature rapid thawing accelerates processing time, it may induce substantial drip loss and structural damage to muscle fibers through ice crystal recrystallization mechanisms. Combining ultrasound with low-temperature thawing presents an advantageous solution. The acoustic energy promotes more uniform and efficient heat transfer throughout the product, significantly accelerating the thawing rate while minimizing the negative physical and biochemical changes associated with slow or high-temperature thawing, thereby better preserving product quality (Bhat et al., 2022). While beneficial, the application of ultrasound requires careful parameter optimization such as time and temperature, that collectively determining the outcome. Inappropriate settings can lead to adverse effects. Most notably, the violent collapse of cavitation bubbles generates highly reactive free radicals (e.g., hydroxyl radicals), which would initiate and catalyze lipid oxidation, which is particularly detrimental to products rich in polyunsaturated fatty acids (common in fish and poultry). This oxidation not only degrades valuable lipids and fat-soluble vitamins (B, E), reducing nutritional value, but also leads to the development of off-flavors (rancidity) and potentially harmful compounds (Zhang, Zhang, Zhou, & Zhang, 2021). In summary, ultrasound serves as a powerful and non-thermal tool in modern meat and aquatic product processing. Under the optimized parameters, it can enhance nutritional bioavailability, improve key textural and sensory qualities like tenderness and water-holding capacity, accelerate processes like thawing and brining, and improve the functional properties of emulsion-based products.

4.2 Dairy products

Ultrasound technology has established as a versatile and innovative tool in modern dairy processing which could be widely applied in homogenization, microbial inactivation, fermentation, extraction of bioactive components, and cheese ripening. A primary application is the homogenization of milk and dairy emulsions. The intense shear forces and micro-jets produced during acoustic cavitation effectively disrupt and fragment large fat globules into a population of submicron-sized droplets. This physical breakdown mimics and often surpasses the effect of high-pressure homogenizers but can be achieved at lower pressures and temperatures. The resulting finer and more uniform fat emulsion significantly enhances product stability by reducing creaming or fat separation during storage. Furthermore, this micro-fragmentation contributes to a creamier mouthfeel and improved sensory attributes in products like milk, cream and yogurt drinks. As for microbial and enzymatic inactivation, ultrasound treatment can disrupt microbial cell membranes and inactivate key spoilage enzymes like lipase and peroxidase through mechanical and free-radical mechanisms. Compared to conventional high-temperature short-time pasteurization, ultrasound-assisted processing achieves comparable microbial safety while markedly improving the retention of heat-labile nutrients in milk. A higher preservation rates for vitamins B1, B2 and B12 was investigated, thereby maintaining the nutritional profile of milk closer to its raw state (Chandrapala, 2016). In fermented dairy products like yogurt and kefir, the sonication of milk before inoculation can enhance the growth and metabolic activity of starter cultures (e.g., *Lactobacillus* and *Streptococcus strains*). This is attributed to the partial release of nutrients from disrupted milk components and the modification of the fermentation medium. The result is an accelerated acid production rate, effectively shortening the total fermentation time. Moreover, ultrasound-assisted extraction is highly effective in releasing and isolating valuable bioactive compounds from the dairy matrix, such as milk-derived bioactive peptides and milk fat globule membrane (MFGM) proteins, which have been recognized as nutritional and functional benefits.

Bovine milk allergy is linked to proteins like β-lactoglobulin (β-Lg). Ultrasound processing presents a physical method to modify protein structure and potentially reduce allergenicity of milk products. The cavitation energy can induce partial unfolding or aggregation of β-Lg, altering its conformational epitopes. Generally, the digestion products of sonicated β-Lg exhibit increased antioxidant activity. Besides, these products can reduce the binding capacity to immunoglobulin G (IgG) and immunoglobulin E (IgE), correlating with the lower immune reactivity (Wu et al., 2017). Emerging evidence also suggests that consuming ultrasonically modified β-Lg may positively modulate gut microbiota, for instance, by decreasing the relative abundance of *Bacteroidetes* and increasing *Firmicutes*, thereby potentially supporting gut barrier function. In cheese manufacturing, ultrasound can be applied to accelerate the ripening process. The treatment could promote the breakdown of casein into peptides and amino acids, along with the potential effect of increasing substrate accessibility (Ylmaz et al., 2025). Based on this, the ultrasound could lead to a faster development of characteristic flavor and texture, which reducing the aging period of cheese. For frozen dairy desserts like ice cream, ultrasonic treatment was shown to control the crystallization of lactose (Mortazavi & Tabatabaie, 2008). By promoting the formation of numerous, tiny lactose crystals instead of large ones, ultrasound prevented the sandy or gritty texture defect, ensuring a consistently smooth and pleasant mouthfeel in the final product. Generally, ultrasound is a multifunctional physical processing tool capable of simultaneously achieving microbial safety, enhancing nutritional retention, modifying functional properties, improving textural and sensory quality, and accelerating production processes.

4.3 Fruit and vegetable products

Ultrasonic technology has emerged as a significant innovation during the processing of fresh fruits, vegetables and their products, with primary applications in microbial inactivation, drying and the extraction of functional components. A key benefit of ultrasound in fruit and vegetable processing is to keep the heat-sensitive nutrients. Compared to the thermal treatment, the ultrasound could decrease the losses of compounds such as vitamin C and various polyphenols, which are crucial for antioxidant activity and overall nutritional value. In addition, the mechanical effects generated by acoustic cavitation could disrupt plant cell walls and subcellular structures. This physical effect would also facilitate the release of intracellular bioactive compounds while minimizing their exposure to prolonged high temperatures. Furthermore, by breaking down the physical matrix that entraps nutrients, ultrasound can enhance the bioaccessibility and subsequent bioavailability of vitamins and minerals during human digestion. For instance, according to the study of Sun et al. (2011), the ultrasound parameters significantly impacted the extraction yield of β-carotene from mandarin peel. Yield markedly increased at particle sizes > 0.28 mm versus conventional extraction. This increase was attributed not to de novo synthesis but to the improved extraction efficiency from the cellular matrix and the dissociation from protein-carotenoid complexes, making them be more available to be extracted. In the production of concentrated juices, purees and jams, the ultrasound serves as an efficient homogenizer to improve the final quality of fruit products. The ultrasound was investigated to break down suspended pulp particles, pectin polymers and other solids into finer sizes, creating a more uniform and stable colloidal system (Maran & Priya, 2015). This action prevented phase separation, improved mouthfeel (providing a smoother and creamier texture) and enhanced overall product stability during storage. The ultrasound technology offers a multifunctional tool for the fruit and vegetable processing industry. It enhances the extraction and bioavailability of prized bioactive compounds and elevates the sensory and physical quality of a wide range of processed products.

4.4 Cereal and grain products

Ultrasonic technology has become an increasingly valuable tool in the cereal processing industry, finding diverse applications that span from primary unit operations to the targeted modification of ingredients. Minerals such as iron, zinc and calcium in grains can be bound within the cell wall matrix (e.g., to phytate or dietary fiber), limiting their release during digestion. The intense shear forces and micro-jets generated by cavitation effectively disrupt these plant cell walls and break down the surrounding matrix. This physical disintegration enhances the bioaccessibility of the mineral which meaning a greater fraction is released from the food matrix into the digestive tract, thereby increasing their potential for absorption. Ultrasound dramatically improves the efficiency of extracting valuable lipophilic and hydrophilic bioactive compounds from grains. For instance, in brown rice, conventional extraction methods for non-phenolic antioxidants like γ-oryzanol, a potent compound linking to cholesterol reduction and antioxidant activity, can be inefficient and time-consuming. Ultrasonic-assisted extraction could significantly increase both the yield and the extraction rate of γ-oryzanol

(Loypimai et al., 2020). Moreover, ultrasound based extraction exhibited enhanced antioxidant activity, potentially due to the cavitation-induced generation of more bioactive molecular forms or the co-extraction of synergistic compounds. Besides, the ultrasound can positively alter the structure of cereal proteins (glutenins, zein and oryzenin), thereby improving their digestibility. Under optimized conditions (30 min, 400 W), the ultrasonic treatment demonstrated notable increases in the in vitro digestibility of major cereal proteins. Wheat protein digestibility was elevated from 79.5 % to 89.7 %, with similar improvements being observed for rice, corn and buckwheat proteins. The mechanism involved the disruption of stabilizing molecular forces within the protein, including hydrophobic interactions, hydrogen bonds and disulfide bridges (Wang et al., 2023).

Ultrasound treatment influences the structural organization of starch granules. During cooling and storage, the modified starch molecules can realign and recrystallize into forms that are more resistant to enzymatic digestion in the small intestine, thus classified as resistant starch (RS). This increase in RS content enhances the dietary fiber functionality of the grain product, contributing to improved colonic health and a lower glycemic response. By controlling ultrasound parameters (intensity, time and temperature), it can also precisely modify the pasting and gelling properties of native starch. Treatment can alter gelatinization temperature, peak viscosity and the setback characteristics of starch pastes. This allows for the engineering of starches with specific functionalities, such as increased stability, desired texture, as well as the controlled digestibility. The quality of bread and steamed buns made from frozen dough often suffers due to ice crystal damage to the gluten network, leading to reduced gas retention and poor final texture. During dough mixing and freezing, the ultrasound treatment could positively modify the gluten protein structure along with the improving effect on the final products (Hu et al., 2013). The mechanical effects promoted the cross-linking and formation of a more uniform and elastic gluten network. This strengthened network exhibited better extensibility and superior gas-holding capacity. After freezing and thawing, ultrasound-treated dough maintained better rheological properties, resulting in bakery products with higher specific volume, finer crumb structure and improved texture. In summary, ultrasound serves as a versatile and potent physical modification tool in cereal science and technology. Its applications move beyond mere process intensification to enable targeted nutritional enhancements.

5. Ultrasonic extraction for nutritional and bioactive components

Extraction technology holds significant importance in the food industry. Processes, such as the recovery of active compounds or other valuable substances from animal, plant and microbial sources, as well as the analysis of chemical compositions, are fundamentally reliant on efficient extraction techniques. Dedicated studies on the ultrasonic extraction and separation process are essential to uncover the underlying factors that enable its widespread application across various fields and account for its distinctive characteristics. This understanding facilitates the optimization of both the process parameters and the equipment, thereby maximizing the advantages of this technology. Ultrasonic extraction and separation is a novel processing technique primarily designed for solid-liquid extraction. It is based on a scientific and rational utilization of ultrasound, taking into consideration the existing state, polarity, solubility and other properties of the target effective components or component groups within the raw material. The core principle involves using ultrasonic energy to rapidly drive the solvent into the solid matrix, facilitating the dissolution of its organic constituents as completely as possible into the solvent (Vilkhu et al., 2008). Subsequently, appropriate separation methods are employed to isolate the individual chemical compounds from this extract, followed by refining and purification steps to obtain the desired monomeric chemical compounds. This integrated technological approach, which organically combines ultrasonic extraction with subsequent separation steps, is referred to as ultrasonic extraction and separation technology. Applying ultrasonic technology to intensify the extraction and separation process can effectively enhance efficiency, significantly reduce processing time, lower operational costs, and, in some cases, even improve product quality and yield. Regarding the mechanism, for materials lacking a cellular structure, the target components can be dissolved directly or become dispersed and suspended in the solvent during extraction. However, for materials possessing a cellular structure, such as most plant or microbial tissues, the objective is to extract intracellular active components. This necessitates the disruption of the cell walls or membranes to expose the cellular contents, thereby achieving the extraction goal. Ultrasonic energy addresses this precisely through its cavitation, mechanical and thermal effects. The intense shear forces and micro-jets generated by cavitation bubble collapse, along with the high-frequency vibrations, effectively break down cell structures, enhance solvent penetration and increase mass transfer rates, overcoming the primary barriers to efficient extraction from cellular materials.

5.1 Extraction of polysaccharides and saponins

As a type of biological response modifier, the polysaccharides exhibit crucial physiological activities. For instance, the optimal separation and extraction conditions for water-soluble polysaccharides from Chinese yam (*Dioscorea opposita*) were determined as 32.89 % water content, 94.00 °C and 44.74 min. Under these optimized conditions, the maximum extraction yield of crude polysaccharides from Chinese yam peel reached 16 % (Zhang & Wang, 2017). Pectin, a primary component of dietary fiber among the seven major human nutrients, is water-soluble and capable of forming gels under appropriate conditions with sugar and acid. Research on pectin extraction from apple pomace demonstrated that ultrasound conditions (140 W, 60 min) yielded up to 32.3 % pectin, with an average MW of 163.9 kDa and a DE of 55.2 %. Simultaneously, the extracted pectin consisted predominantly of galacturonic acid (>97 %, Minjares-Fuentes et al., 2014). This ultrasonically enhanced method improved both the efficiency and yield of pectin extraction without adversely affecting its structural integrity. Another study using pomelo peel as the raw material reported that, under identical process parameters, the ultrasound-assisted method increased the pectin extraction yield by 38 % compared to other traditional techniques, highlighting its advantages of rapid processing, high efficiency, energy savings and significant potential for broader application (Liew et al., 2016).

Higher plants contain a diverse array of saponins, which possess various bioactive properties such as antimicrobial, antipyretic, sedative and anticancer activities. Using the total saponin content in cinnamon as the evaluation metric, orthogonal experimental design identified the optimal ultrasonic extraction conditions (60 °C, power of 70 % maximum and 40 min). Under these conditions, the total saponin content reached 4.39 % (Jian et al., 2008), which was higher than conventional solvent extraction with the yield of 3.41 %. Therefore, the ultrasound-assisted solvent method proved more effective and time-efficient. Similarly, for white kidney beans, the optimal extraction conditions were established with 70 % ethanol concentration, 40 min ultrasonic treatment, 75 W ultrasonic power, and a solid-to-liquid ratio of 1:20 (g/mL). This protocol achieved a saponin yield of 6.76 % from white kidney beans. The primary advantages of ultrasound include markedly increased extraction yields, substantially reduced processing times, improved efficiency, and also the preservation of structural and functional integrity for target compounds. The underlying mechanisms are attributed to the synergistic effects of ultrasound, including cavitation-induced cell disruption, enhanced solvent penetration and intensified mass transfer, which collectively overcome the limitations of conventional extraction methods.

5.2 Extraction of proteins

Proteins serve as the fundamental material basis of life and constitute essential nutrients for the human body, making them be a critical nutritional indicator in foodstuffs. The extraction and the quantification of protein content in food are therefore of paramount importance for evaluating nutritional value, rationally developing and utilizing food resources, improving product quality and optimizing food formulations and production control. Among various protein extraction methods, ultrasonic extraction has proven to be a highly effective technique. For instance, the optimal process conditions for ultrasound-assisted extraction of collagen from lophiiformes fish skin. Compared to the ultrasonic-alkaline enzyme method (yield of 3.25 %), the ultrasonic-flavor enzyme method achieved a higher extraction yield of 3.54 % with a shorter hydrolysis time of 5 h (Wang et al., 2019) Ultrasound significantly enhanced enzymatic efficiency by improving mass transfer and substrate accessibility, demonstrating its crucial role in boosting collagen recovery while reducing processing time. In addition, the ultrasonic intervention increased the porosity of the fish skin surface structure, facilitating the dissolution of collagen. The extracted collagen was identified as type I. While ultrasonic treatment induced alterations in the secondary structure composition of the collagen, its spatial network structure, subunit composition and triple-helix structure remained largely unchanged. The previous study has underscored the significant advantages of ultrasound technology in protein extraction, primarily characterized by increased yield, reduced processing time and the potential for preserving the structural integrity of the target biomolecules. The underlying mechanism is attributed to the synergistic effects of acoustic cavitation and mechanical shear forces, which disrupt cellular and tissue matrices, enhance solvent penetration and improve mass transfer rates, thereby overcoming the limitations associated with conventional extraction methods.

5.3 Extraction of alkaloids

Alkaloids are a class of naturally occurring nitrogen-containing organic compounds characterized by complex cyclic structures, basic properties, the ability to form salts with acids, and significant physiological activities. Alkaloids are predominantly found in various families of dicotyledonous higher plants and exhibit remarkable structural diversity. Their extraction methods must be tailored accordingly, as their distinct chemical structures and the forms in which they exist within plant tissues lead to differing physicochemical properties. As for the extraction of alkaloids from potato peel waste, the total yield under the optimized ultrasound conditions (amplitude 61 μm, 17 min) reached 1102 μg/g dried peel, which was 55 % higher than the solid liquid extraction yield of 710.51 μg/g. Ultrasound treatment also significantly improved the extraction of individual alkaloidsincluding α-solanine (273 μg/g), α-chaconine (542.7 μg/g), solanidine (231 μg/g) and demissidine (55.3 μg/g). As shown in the highly significant predictive model, the robustness of the ultrasound based process was an effective and time-saving method (Hossain et al., 2014). Utilizing the Sophora flavescens as raw material, a study investigated the effects of ultrasonic time, solvent concentration, solid-to-liquid ratio and ultrasonic frequency on the extraction yield of oxymatrine (Xia et al., 2012). Through orthogonal experimental design, the optimal extraction parameters were determined as pH 3, an ultrasonic frequency of 500 kHz, a liquid-to-solid ratio of 24:1 (mL/g) and an ultrasonic extraction time of 45 min. Under these optimized conditions, the extraction rate of matrine reached 80 %, surpassing the yield obtained through conventional extraction methods. The previous studies have illustrated the efficacy of ultrasound technology in enhancing the extraction of alkaloids from plant matrices. The advantages of ultrasound include improved extraction yields, significantly reduced processing times and increased efficiency. The underlying mechanism is attributed to the disruptive effects of acoustic cavitation and the associated mechanical forces, which facilitate cell wall rupture, enhance solvent penetration into the plant material and intensify mass transfer of the target compounds into the solvent phase.

5.4 Extraction of pigments

Edible pigments, also known as colorants or coloring agents, are food additives primarily used to impart or restore color to food products. They can be classified by origin into natural edible pigments (derived from plants, animals or microorganisms) and synthetic edible pigments. However, the widespread application of natural pigments has been limited due to factors such as a scarcity of pigment resources with significant commercial value, the inherent instability of many natural colorants and traditional extraction techniques that often result in low yields and high production costs. Ultrasonic extraction technology offers a novel approach to address these challenges, promoting resource efficiency and significantly improving extraction rates for natural pigments. As reported by Garcia-Ortiz et al. (2022), the extraction of natural pigments from red maize was operated by a combined microwave-ultrasound method. Here, the medium-pigmentation maize (RM) was selected for extraction under varying ethanol (0 %, 30 %, 70 %) and citric acid (0 %, 1 %) conditions. It significantly contributed to the process efficiency by promoting cell disruption and improving solvent penetration into the maize matrix. The extracted pigments exhibited distinct red and yellow hues when ethanol (30 % and 70 %) was used as a co-solvent. *Carthamus tinctorius L.* (safflower), a widely cultivated plant in China belonging to the *Asteraceae* family, was utilized in the extraction of pigments. Employing deionized water as the solvent, the ultrasound-assisted extraction recovered safflower yellow pigment (Zhang et al., 2025) Under optimized conditions (48 °C, 21 min), ultrasonic cavitation significantly enhanced mass transfer and disrupted the floral surface structure, enabling efficient simultaneous extraction of red and yellow pigments in a single step. The process achieved high extraction efficiency with a solid-to-liquid ratio of 50.5 mg/mL and water content of 12.8 %. Generally, the ultrasound synergistically strengthened hydrogen-bond interactions between choline chloride-ethylene glycol-acetamide and target compounds. The ultrasound combination methods offer a rapid and eco-friendly alternative for bioactive pigment production, reducing both time and resource consumption. These findings demonstrate that ultrasound not only facilitates efficient pigment extraction but also effectively modifies the techno-functional properties of the post-extraction flour, offering a dual advantage for the valorization of pigmented maize in food applications.

5.5 Extraction of acidic compounds

The anions of acidulants significantly influence the flavor profile of food products. Most organic acids impart a pleasantly sharp or refreshing sourness, whereas many inorganic acids tend to have a bitter or astringent aftertaste, which can degrade food quality and result in a less palatable acidity. Vinegar stands as the most crucial acidulant, followed by a range of others, including citric acid, lactic acid, gluconic acid, malic acid, oxalic acid, tartaric acid, pyruvic acid, α-ketoglutaric acid, acetic acid, fumaric acid, and succinic acid among others. Conventional methods for extracting organic acids encompass distillation, high-speed centrifugation, ultrafiltration, microwave-assisted extraction, diatomaceous earth filtration, organic solvent extraction, direct extraction with water or alkaline solutions, ion exchange, and emulsion liquid membrane separation. However, these techniques are often associated with challenges in particular high operational costs. Consequently, there has been a continuous search for new, effective and simpler alternative methods. Ultrasound-assisted extraction has emerged in this context as a promising solution.

The ultrasonic technology was applied to extract tartaric and malic acids from grape-derived matrices (Palma & Barroso, 2002). Through a fractional factorial design evaluating seven extraction variables, ultrasound was identified as a key enabling technology that significantly enhanced mass transfer and extraction efficiency. The most influential parameters were the composition of the extracting liquid and the extraction temperature, both synergistically interacting with ultrasonic energy to improve acid release. Subsequent central composite design optimization further refined these conditions, demonstrating that ultrasound enabled effective acid recovery under moderate temperatures, a notable advantage over conventional thermal extraction methods that may degrade thermolabile compounds. The optimized ultrasound method was successfully applied to both grapes and winemaking by-products, achieving high repeatability and established recovery rates for both tartaric and malic acids.

As for the organic acids extraction from honeysuckle, the optimized ultrasound-assisted treatment was determined as the time of 10 min, a solid-to-liquid ratio of 1:20 (g/mL), and the ultrasonic intensity of 0.347 W/cm^2. Analysis via high-performance liquid chromatography quantified five major organic acids in the extract. Compared to the conventional method, the ultrasound-assisted process significantly increased the yields of chlorogenic acid, caffeic acid, 3,4-dicaffeoylquinic acid, 3,5-dicaffeoylquinic acid, and 4,5-dicaffeoylquinic acid (Xu et al., 2016). Furthermore, gas chromatography-mass spectrometry analysis revealed that acidic compounds constituted its characteristic volatile profile. Importantly, the ultrasound-assisted method significantly enhanced both the extraction efficiency and the retention rate of these characteristic aroma compounds compared to ethanol reflux extraction. Thus, the ultrasound-assisted extraction serves as an efficient and viable technique for recovering organic acids. It offers advantages such as increased yield, reduced processing time and the potential for better preservation of the target compounds' native characteristics including their volatile aromatic profile.

5.6 Extraction of polyphenols components

Phenolic compounds constitute a crucial class of plant secondary metabolites, the majority of which are biosynthesized via the shikimin acid pathway. Structurally characterized by a hydroxyl group directly attached to an aromatic ring, they predominantly exist in plants in the form of glycosides. Phenolic compounds mainly encompass two categories including structurally simple low-molecular-weight phenolics and structurally complex high-molecular-weight polymers known as polyphenols and their derivatives. Low-molecular-weight phenolics are further divided into mono-aromatic and di-aromatic types, including compounds such as cinnamic acid, chlorogenic acid, protocatechuic acid, salicylic acid, anthocyanins, flavones, flavonols, and isoflavones. Ultrasound-assisted extraction technology has been widely applied for the efficient recovery of plant polyphenols and flavonoids from grape skins, pomegranate peels, onion skins, tea leaves, coffee grounds, and nut shells. Utilizing asparagus as raw material, the efficacy of ultrasound-assisted extraction for flavonoids was conducted along with optimal extraction conditions of 50 °C, a solid-to-liquid ratio of 1:5, the ultrasonic power of 150 W, and the time of 45 min. Under the optimized parameters, the total flavonoid yield reached 4.23 %. Compared to conventional hot water extraction, where the maximum yield was 2.8 %, the ultrasonic method resulted in a significantly higher recovery (Sultanova et al., 2023). This case underscores the potential of ultrasonic technology to enhance the extraction efficiency of flavonoids, offering advantages such as increased yield and potentially reduced processing time compared to traditional methods. The extraction yield, antioxidant activity and extraction kinetics of antioxidants from dried pomegranate peel were also investigated (Patil et al., 2024). Optimal ultrasound conditions (25 min, 130 W, 50 % duty cycle, 22 kHz and liquid–solid ratio 50:1) yielded a total phenolic content of 21.52 mg GAE/mL and antioxidant capacity of 17.96 mg GAE/mL, demonstrating the ability to enhance mass transfer and cell disruption within a short timeframe. In addition, the ultrasound combined enzyme treatment (55 min, 50 °C, pH 5 % and 2 % enzyme concentration) achieved significantly superior results of total phenolic content reached 30.61 mg GAE/mL and antioxidant capacity 27.70 mg GAE/mL. Similarly, for other extracts such as those from strawberries, the degradation of phenolic compounds can occur to a certain extent under ultrasonic conditions, especially when the temperature exceeded 25 °C. Generally, the ultrasound could significantly shorten extraction time, enhance target compound yield and effectively preserve thermo-sensitive bioactive components under mild temperature conditions. Furthermore, it reduces reliance on organic solvents, offering a green and efficient strategy for valorizing food-processing byproducts and producing natural functional ingredients, thereby supporting sustainable food industry development.

Appropriate ultrasound treatment can enhance the permeability and transport capacity of cell membranes, facilitating the release of intracellular contents. Research indicates that applying ultrasound twice during the processing of soybeans increased the protein content in the resulting soymilk by 68 % and the soluble solid content by 67 %. This enhancement was likely attributable to the disruptive effect of ultrasound on the soybean seed coat and cell membrane structures during the soaking stage (Zhang, Wang, Hu, Fakayode, Ma et al., 2021). As ultrasound parameters moderately increased, the protein content in soymilk showed a corresponding rise. Conversely, excessively intense ultrasound could induce protein denaturation, leading to a subsequent decrease in protein solubility. Therefore, ultrasound parameters must be carefully controlled within a relatively mild range to effectively improve the nutritional value of soymilk. Ultrasound also shows potential in improving the nutritional value of extracted oils. Studies on ultrasound-assisted extraction of walnut kernel oil using a power of 400 W and time of 28 min demonstrated that the maximum extraction yield was 59.30 % which was 1.24-fold to that of soxhlet extraction (Liu et al., 2022). Similarly, compared to conventional methods, ultrasound-assisted extraction of sunflower seed oil could be performed at lower temperatures, which helped protect unsaturated fatty acids in the oil (Rosa et al., 2025). Some studies have reported that ultrasound irradiation may induce oxidation in sunflower seed oil, leading to the development of rancid odors and a consequent decline in oil quality. This highlights the necessity of optimizing ultrasound parameters, such as intensity, duration and temperature to maximize beneficial effects while minimizing potential degradation of sensitive nutritional components.

5.7 Ultrasound assisted protein hydrolysis

Protein hydrolysis refers to the process in which peptide bonds within protein molecules are progressively cleaved, resulting in the formation of smaller peptide fragments or free amino acids. Based on the extent of hydrolysis, protein hydrolysis can be categorized into complete hydrolysis (also referred to as exhaustive hydrolysis) and partial hydrolysis. Complete hydrolysis yields a mixture of various amino acids as the end product, while partial hydrolysis produces a mixture of peptide fragments of varying lengths along with amino acids. Compared to large-molecular-weight proteins, small-molecular-weight peptides offer advantages such as higher bioactivity, lower antigenicity and easier absorption by the human body. The influence of ultrasound on protein hydrolysis is primarily achieved through the alteration of protein structure and the modulation of protease activity. Pre-treatment of proteins with ultrasound to a certain degree can loosen their spatial conformation, thereby exposing specific enzymatic cleavage sites that are otherwise buried within the native structure. The effect of ultrasound on enzymes is complex and enzyme catalytic activity can be either activated or inactivated, which depending on the conditions and intensity of the ultrasound treatment.

Pea seeds represent a plant-based protein resource that is both abundant and cost-effective, making them be highly suitable for improving human nutrition and developing healthy functional plant-based foods. However, their low solubility and the presence of other factors can significantly reduce their bioavailability and utilization. Ultrasound can assist complex proteases in hydrolyzing pea protein, yielding hydrolysates with a high degree of hydrolysis, improved emulsifying and foaming properties, enhanced antioxidant activity, and specialized functional proteins. The enzymatic hydrolysis of plant proteins such as pea protein can markedly enhance their biological activity and solubility, improve emulsifying properties and confer good foaming capacity. Research on ultrasound-assisted papain extraction of protein from shrimp heads determined the optimal conditions with a solid-to-liquid ratio of 1:10 (g: mL), enzyme addition of 8000 U/g, papain added to shrimp head particles (0.2–0.3 mm), pH 7.5, and ultrasound treatment at 120 W for 20 min, followed by enzymatic hydrolysis at 60 °C for 3 h. According to the study of Li et al. (2025), ultrasound coupled with pH-shift assisted glycosylation effectively modified shrimp head protein. The treatment significantly increased the glycosylation degree to 48.3 % and induced favorable structural changes, including looser spatial conformation and higher thermal stability. These modifications substantially enhanced the protein solubility, emulsifying performance and oil/water adsorption capacity. Importantly, ultrasound played a key role in improving the epigallocatechin gallate loading capacity of shrimp protein. The complex exhibited excellent antioxidant activity, showcasing ultrasound as a promising technique for valorizing shrimp by-products into functional protein ingredients. The optimal extraction parameters of ultrasound-coupled extraction for obtaining bioactive peptides from spirulina (*Arthrospira platensis*) were identified as 153 °C, 221 W, 64 min and 10 MPa. Validation experiments confirmed a protein extraction rate of 74.02 % under these conditions. In contrast, a combined method employing repeated freeze-thaw cycles and ultrasonic homogenization achieved an extraction rate of 45.76 % (Wang & Zhang, 2015). These results indicate that the ultrasound-assisted subcritical water method substantially enhances the extraction efficiency of bioactive peptides from spirulina.

In addition, the ultrasound pretreatment of potato protein can improve its enzymatic hydrolysis efficiency. This pretreatment increased the content of peptides and free amino acids in the final digestive products of potato protein hydrolysate, as well as enhanced the radical scavenging capacity and ferrous ion chelating ability of the digestate. It also raised the total amount of free amino acids after intestinal digestion and promoted the overall digestibility of potato protein hydrolysate (Van et al., 2025). Current recovery processes often involve heat treatment, which can degrade its nutritional value and functional properties, hindering its application in the food industry. While enzymatic hydrolysis can mitigate this to some extent, the long hydrolysis times and low protein conversion efficiency remain limitations. Single-factor optimization experiments identified the best ultrasound pretreatment parameters for increasing the degree of hydrolysis and protein conversion (ultrasound frequency of 20/40 kHz, treatment time of 10 min, and ultrasound power density of 250 W/L). Compared to conventional enzymatic hydrolysis, ultrasound pretreatment increased the degree of hydrolysis by 18.1 % and improved the protein conversion rate by 8.6 %. The pretreatment altered the secondary structure of potato protein, exposing more hydrophobic groups, unfolding the protein structure and thereby increasing the number of available enzymatic cleavage sites (Liu et al., 2023).

The different low-power-density ultrasound was also applied in the enzymatic hydrolysis of corn gluten, where the both continuous and simultaneous dual-frequency ultrasound pretreatment significantly enhanced the relative hydrolysis efficiency. Under continuous dual-frequency conditions at 20/40 kHz, the maximum enzymatic hydrolysis efficiency and protein solubility for corn gluten meal reached 15.99 % and 61.69 %, respectively. Changes in surface hydrophobicity, secondary structure and microstructure revealed ultrasound-induced conformational alterations in the corn gluten meal. Furthermore, the molecular weight distribution of the hydrolysate was primarily concentrated within the 200–500 Da range after ultrasound treatment. Amino acid composition analysis indicated that the proportion of hydrophobic amino acids in the total amino acid content of the hydrolysate increased from 38.56 % to 40.17 % following ultrasound pretreatment. Due to these structural and enzymatic susceptibility changes, low-frequency ultrasound improved the hydrolysis efficiency of corn gluten meal (Jin et al., 2016). Overall, low-frequency ultrasound plays a significant and crucial role in enhancing traditional enzymatic hydrolysis reactions and improving hydrolysis efficiency. The current findings contribute to a better understanding of the impact of ultrasound frequency modes on protein enzymatic hydrolysis. Therefore, employing various low-power-density ultrasound frequency modes for pretreating target proteins may be considered as a practical method for improving protease-mediated reactions.

5.7.1 Protease activity

Enzymes are biomacromolecules that possess relatively stable conformations. The biological activity of an enzyme molecule depends on the maintenance of its proper conformation. When ultrasound acts on protease molecules dispersed within a medium, the energy released can induce conformational changes in the enzyme molecules. When the conformation of enzyme is altered, its biological function is consequently modified. Appropriate ultrasound intensity can render the protease conformation more favorable, thereby enhancing its catalytic activity and accelerating the rate of protein hydrolysis (Ma et al., 2011). Conversely, when ultrasound intensity is excessive, it can disrupt the structure of the protease molecule, leading to an unfavorable conformational change. This diminishes enzymatic activity or even causes inactivation, ultimately inhibiting protein hydrolysis. Currently, there are two primary approaches to applying ultrasound technology for enzyme treatment. The first is ultrasound pre-treatment, where the enzyme is initially subjected to ultrasound in a medium, dried and subsequently used to catalyze the enzymatic reaction in the reaction medium. The second involves direct ultrasound treatment of the enzyme within the reaction medium, where the sonication and the enzymatic reaction occur simultaneously.

Regardless of the method employed, appropriate ultrasound conditions can enhance the rate of the enzymatic reaction. Under lower-intensity ultrasound, the acceleration of protein hydrolysis can be attributed to the following potential mechanisms: ultrasound treatment alters the tertiary structure of the enzyme molecule, promoting a more favorable conformation that increases enzymatic activity. Ultrasound exerts a homogenizing effect on the reaction substrate, increasing the contact surface area between the enzyme and substrate, thereby facilitating the reaction. The thermal effect of ultrasound can moderately raise the temperature of the reaction system, reducing the required hydrolysis time. Ultrasonic vibrations generate and transmit substantial energy, causing medium particles to enter a state of intense vibration. This increases the frequency of encounters between the enzyme and substrate, enhancing the rate of protein hydrolysis. Under high-intensity ultrasound, enzymatic activity may decrease or become inactivated, thereby inhibiting protein hydrolysis. The underlying principles likely involve the following two aspects. High-intensity ultrasound generates transient cavitation, accompanied by powerful shockwaves or microjets, which can drive enzyme molecules towards unfavorable conformations. Free radicals produced during ultrasound cavitation may chemically interact with enzyme molecules, leading to enzyme inactivation and consequent inhibition of protein hydrolysis.

5.7.2 Preparation of hypoglycemic peptides

Diabetes mellitus, a prevalent metabolic disorder, exhibits a rising annual incidence coupled with a concerning trend toward earlier onset. Currently recognized as the third leading threat to global human health, it demands substantial and focused research efforts. Most existing therapeutic drugs carry certain side effects, underscoring the importance of investigating safe and natural product-derived agents for diabetes treatment. The combination of ultrasound and enzymatic hydrolysis of soy protein was used to prepare the hypoglycemic peptides. The safety and stability of the soybean hypoglycemic peptides produced by this method were subsequently evaluated. Results indicated that the application of energy-gathered ultrasound (EGU) significantly enhanced the enzymatic production of α-glucosidase inhibitory peptides from sweet potato protein. EGU-assisted hydrolysis using alcalase, papain and flavourzyme (APF) yielded the most active hydrolysate (yield 86.39 %), with the inhibitory activity remaining highly stable after in vitro digestion. Two novel peptides, AIWGAGGGLR and FHDPMLR, were identified from the 1–2 kDa fraction, demonstrating the effectiveness of ultrasound in generating bioactive peptides with potential for diabetes management (Hu et al., 2025). This demonstrates that ultrasound pretreatment can enhance the α-glucosidase inhibitory activity of the potato protein isolate hydrolysate. According to the study of Li et al. (2020), ultrasound-assisted extraction (UAE) was also a highly effective technique for enhancing the yield of α-glucosidase inhibitors (AGIs) from *Ceratophyllum demersum L.* (CDL). By optimizing key parameters (70 % methanol, 43 mL/g liquid-to-solid ratio, 54 min, 350 W and 40 °C), the UAE achieved a remarkable half-maximal inhibitory concentration (IC_{50}) of 0.15 mg dried material/mL. The IC_{50} value was substantially lower than that of the pharmaceutical agent acarbose (0.64 mg/mL), underscoring the potent inhibitory activity obtained through ultrasound intensification. Based on the HPLC-QTOF-MS/MS analysis, there were 80 peptides identified in the extract, supporting CDL's potential as a rich source of bioactive AGIs enabled by ultrasound extraction (Li, Tu, Wang & Zhang, 2020).

5.7.3 Preparation of antioxidant peptides

Oxidative damage caused by reactive oxygen species is considered as a significant contributor to organismal aging, cellular and tissue damage, and carcinogenesis. Therefore, exploring and discovering natural and harmless antioxidants capable of effectively scavenging free radicals holds immense value for safeguarding health and mitigating aging processes. Antioxidant peptides have attracted attention due to their excellent ability to eliminate free radicals, along with their characteristics of small molecular weight, easy absorption and wide availability of raw materials. The common method for preparing antioxidant peptides is hydrolysis. Among them, the enzymatic-driven protein hydrolysis method is the most convenient and simple preparation process. At present, ultrasound, as a protein modification method, plays a very significant role in promoting proteolysis and the separation of small molecular peptide sequences. In the rice study, the optimized ultrasound conditions (1500 W, 20 min, 40 °C) increased the yield of antioxidant peptides by 43.7 % and elevated DPPH radical scavenging activity to 74.8 %, while reducing the IC_{50} value by 17.7 % compared to untreated samples (Jia et al., 2008). The obtained peptides displayed potent radical scavenging capabilities, with IC_{50} values of 0.0678 g/L (superoxide anion), 0.9315 g/L (DPPH) and 0.1173 g/L (hydroxyl radical). According to the study of Wang et al. (2017), ultrasound pretreatment elevated the degree of hydrolysis and free sulfhydryl group content of β-conglycinin and glycinin, while shifting peptide molecular weight distribution toward lower ranges. These changes facilitated enzyme accessibility and ultimately increased the yield of highly active antioxidant peptides derived from soybean proteins. Ultrasound-assisted enzymolysis also significantly enhanced the antioxidant properties of walnut meal-derived peptides (Huo et al., 2025). Under optimized conditions (180 W, 4.3 h, 10 KU/g neutral protease), walnut peptide extract demonstrated markedly improved radical scavenging capacity compared to untreated controls. Here, the DPPH and ABTS scavenging effect increased by 234.23 % and 240.22 %, respectively, while hydroxyl radical (·OH) scavenging rose by 69.52 %. Structural analysis revealed that ultrasound not only promoted the formation of smaller peptides but also enriched peptides with hydrophobic C-terminal residues. Finally, the six potent antioxidant peptides (IFW, IIPF, IVAF, IIFY, ILAFF and IFIP) were subsequently identified from the extract, which exhibited strong binding interactions with Keap1 protein via hydrogen bonding, π-alkyl and π-π stacking. These findings confirm the critical role of ultrasound in generating high activity antioxidant peptides and support the high value utilization of walnut meal. As for the animal derived proteins, the ultrasound pretreatment substantially enhanced the yield, bioactivity and functional properties of collagen hydrolysates (CHs) derived from chicken skin. Compared with the conventional enzymatic method, ultrasound-assisted processing increased the extraction yield by 6.35–16.07 % and the degree of hydrolysis from 41.9 % to 59.26 %. Structural analyses revealed that ultrasound modified the secondary structure and created a homogeneous and porous matrix, contributing to improved solubility, emulsifying capacity and water/oil holding properties. Moreover, ultrasound-pretreated CHs exhibited significantly stronger antioxidant activity, with elevated DPPH and ABTS radical scavenging and FRAP values, as well as higher angiotensin-I-converting enzyme (ACE) inhibitory activity (Lalthanmawii et al., 2024). These results indicate that ultrasound-assisted enzymatic hydrolysis of rice residue could effectively improve the protein conversion rate, the degree of hydrolysis and the antioxidant activity of the resulting peptides.

5.7.4 Preparation of antihypertensive peptides

Hypertension is the most prevalent cardiovascular disease, often leading to complications in vital organs such as the heart, brain and kidneys, posing a serious threat to human health. ACE inhibitory peptides represent a promising bioactive component for managing hypertension, offering a potential alternative to synthetic pharmaceuticals. These peptides function by modulating the renin-angiotensin and kinin-kallikrein systems, thereby contributing to blood pressure regulation (Cao et al., 2022). While they can be derived from various food proteins, including dairy, marine, egg and plant sources, conventional enzymatic hydrolysis often faces limitations in efficiency and peptide yield. As shown in Fig. 3, the overview of ACE inhibitory peptides was generally introduced. Emerging evidence highlights the significant role of ultrasound in enhancing the production and bioactivity of ACE inhibitory peptides (Kadam et al., 2015). For instance, the pretreatment with ultrasound (22 kHz, 40 W, 210 min) was shown to increase the proteolytic breakdown rate constant of defatted wheat germ protein by 22.2 %, concurrently elevating ACE inhibitory activity. Furthermore, studies indicate that probe-type ultrasound systems outperformed bath-type systems in promoting ACE inhibitor potency. Besides, the ultrasound-assisted hydrolysis also improved the antioxidant profile of protein hydrolysates. Under the treatment of ultrasound (150 W, 25 min), the radical scavenging capacity of peanut protein was significantly enhanced (Yu et al., 2012). As for the wheat gluten hydrolysates, the ultrasound treatment of 25 kHz and 0.707 W/cm^2 exhibited strengthened iron-chelating and antioxidant activities of wheat gluten derived peptides (Zhu et al., 2022). Notably, ultrasound influenced peptide molecular weight distribution,

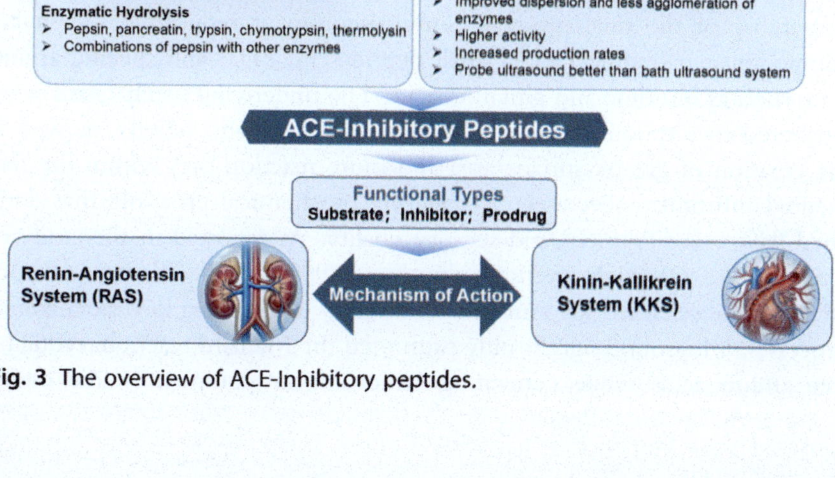

Fig. 3 The overview of ACE-Inhibitory peptides.

favoring the generation of lower-mass fragments (<1 kDa), which were closely associated with superior bioactivity. Researchers have utilized corn gluten meal as a raw material to prepare natural ACE inhibitory peptides via ultrasound-assisted protease hydrolysis. Compared to the control group without ultrasound treatment, the hydrolysate obtained after optimal ultrasound pretreatment exhibited a 74.08 % increase in ACE inhibitory activity. Furthermore, the production efficiency and protein conversion rate improved by 55.11 % and 55.06 %, respectively (Li et al., 2004). This demonstrates that the ultrasound-assisted enzymatic method was an efficient approach for preparing ACE-inhibitory peptides from corn gluten meal. Animal experiments were also conducted to evaluate the antihypertensive effect in vivo. The corn gluten meal hydrolysate exhibited a significant blood pressure-lowering effect on hypertensive rats, while showing no such effect on normotensive rats, suggesting a targeted therapeutic potential.

5.7.5 Application in soy sauce brewing

Originating in China, soy sauce boasts a long production history and distinctive national characteristics. The primary constituents of soy sauce comprise free amino acids, peptides, organic acids and volatile flavor compounds. Among these, free amino acids significantly influence the taste and aroma profile of soy sauce, serving as key determinants of its quality. Consequently, protein hydrolysis plays a vital role among the various biochemical pathways responsible for developing the unique flavor of soy sauce. Ultrasound treatment (68 kHz, 60 W/L) effectively enhanced the taste quality of soy sauce under long-term and high-salt fermentation (Gao et al., 2019). Sonication significantly increased the content of organic taste compounds by 8.4 % to 22.2 % and improved the overall harmony and palatability of the final product. Simultaneously, it reduced less desirable components, including NaCl, small peptides (≤1 kDa) and specific amino acids such as histidine and glutamic acid. The underlying mechanisms were attributed to ultrasound-induced increases in enzymatic activity, as well as the creation of greater surface area and more reaction sites within the fermented moromi, collectively accelerating biochemical processes that enriched flavor development and sensory quality. As reported in the study of Yu et al. (2024), high-intensity ultrasound (90°C, 39.48 W/cm^2, 60 min) effectively accelerated soy sauce maturation by intensifying key biochemical processes. Ultrasound significantly promoted the Maillard reaction, reducing free amino acids while generating critical flavor compounds like esters, pyrazines and ketones for an enhanced aroma. Importantly, it uniquely modulated peptide profiles, increasing the proportion of umami-imparting peptides (1–5 kDa) to 10.19 %, significantly surpassing control samples. These targeted changes, along with increased reducing sugars and solids, confirm ultrasound as a novel processing tool that improves flavor complexity and umami intensity along with shortens maturation time and reduces industrial production costs.

6. Conclusion

Ultrasound technology represents an interdisciplinary science with significant applications within the food industry. The utilization of ultrasound can lead to raw material conservation, process improvements, enhanced efficiency, superior product quality and increased yield of final products. During the process of raw material pretreatment, food manufacturing operations, quality analysis and testing, preservation and storage, as well as wastewater treatment in food processing plants, the ultrasound technology can be applied. In addition, the ultrasound effectively promotes processes such as meat tenderization and marination, texture preservation in fruits and vegetables, improved dough mixing, microbial inactivation, homogenization, sterilization, pasteurization and emulsification. As an eco-friendly green technology, it demonstrably enhances various unit operations within the food industry. Within the realm of food nutrition, beyond the previously discussed applications in bioactive compound extraction and peptide preparation, ultrasound is also employed for the encapsulation of active nutrients. With the increasing awareness of health among consumers, a multitude of functional bioactive ingredients are being incorporated into food systems. Research has established ultrasound technology as an effective means to improve encapsulation efficiency and address delivery challenges associated with these sensitive compounds. Furthermore, the application of ultrasound to food products can induce changes in certain chemical constituents, thereby influencing their nutritional value, a complex process warranting further investigation. Currently, most ultrasound applications are concentrated on optimizing extraction processes. While its broader implementation in food processing remains in a developmental stage, the inherent characteristics of ultrasound technology, high efficiency, low energy consumption, and minimal emissions align well with the food industry needs for transformation and upgrading. It possesses substantial value for further research, development and wider adoption. With the continued evolution of the food industry, the application prospects for ultrasound technology are poised to become even more expansive.

References

Adekunte, A., Tiwari, B. K., Scannell, A., Cullen, P. J., & O'donnell, C. (2010). Modelling of yeast inactivation in sonicated tomato juice. *International Journal of Food Microbiology, 137*(1), 116–120.

Alavi, F., Chen, L., & Emam-Djomeh, Z. (2021). Effect of ultrasound-assisted alkaline treatment on functional property modifications of faba bean protein. *Food Chemistry, 354*, 129494.

Ashokkumar, M. (2015). Applications of ultrasound in food and bioprocessing. *Ultrasonics Sonochemistry, 25*, 17–23.

Barretto, T. L., Sanches, M. A. R., Pateiro, M., Lorenzo, J. M., Telis-Romero, J., & da Silva Barretto, A. C. (2023). Recent advances in the application of ultrasound to meat and meat products: Physicochemical and sensory aspects. *Food Reviews International, 39*(7), 4529–4544.

Bera, S., Mondal, D., Martin, J. T., & Singh, M. (2015). Potential effect of ultrasound on carbohydrates. *Carbohydrate Research, 410*, 15–35.

Bhargava, N., Mor, R. S., Kumar, K., & Sharanagat, V. S. (2021). Advances in application of ultrasound in food processing: A review. *Ultrasonics Sonochemistry, 70*, 105293.

Bhat, Z. F., Morton, J. D., Kumar, S., Bhat, H. F., Aadil, R. M., & Bekhit, A. E. D. A. (2022). Ultrasonication as an emerging technology for processing of animal derived foods: A focus on in vitro protein digestibility. *Trends in Food Science & Technology, 124*, 309–322.

Cao, S. M., Wang, Z. X., Xing, L. J., Zhou, L., & Zhang, W. G. (2022). Bovine bone gelatin-derived peptides: Food processing characteristics and evaluation of anti-hypertensive and antihyperlipidemic activities. *Journal of Agricultural and Food Chemistry, 70*, 9877–9887.

Chandrapala, J. (2015). Low intensity ultrasound applications on food systems. *International Food Research Journal, 22*(3).

Chandrapala, J. (2016). Ultrasound processing of milk and dairy products. *Handbook of Ultrasonics and Sonochemistry*, 1287–1320.

Cheng, L. H., Soh, C. Y., Liew, S. C., & Teh, F. F. (2007). Effects of sonication and carbonation on guava juice quality. *Food Chemistry, 104*(4), 1396–1401.

Cho, B. K., & Irudayaraj, J. M. K. (2003). Foreign object and internal disorder detection in food materials using noncontact ultrasound imaging. *Journal of Food Science, 68*(3), 967–974.

Cruz, R. M., Vieira, M. C., & Silva, C. L. (2008). Effect of heat and thermosonication treatments on watercress (Nasturtium officinale) vitamin C degradation kinetics. *Innovative Food Science & Emerging Technologies, 9*(4), 483–488.

Gao, X., Zhang, J., Liu, E., Yang, M., Chen, S., Hu, F., ... Yu, X. (2019). Enhancing the taste of raw soy sauce using low intensity ultrasound treatment during moromi fermentation. *Food Chemistry, 298*, 124928.

Garcia-Ortiz, J. D., Flores-Gallegos, A. C., Ascacio-Valdés, J. A., López-Badillo, C. M., Nery-Flores, S. D., Esparza-González, S. C., & Rodríguez-Herrera, R. (2022). Microwave-ultrasound assisted extraction of red corn pigments and their effect on chemical composition and tecno-functional properties. *Food Bioscience, 50*, 102115.

Godswill, N. N., Rebezov, M., Nagdalian, A., & Shariati, M. A. (2025). Ultrasound-assisted extraction of bioactive compounds. *Application of emerging technologies and strategies to extract bioactive compounds*. Academic Press, 1–56.

Habib, M., Rani, S., Singh, S., Jan, S., Jan, K., Swer, T. L., & Bashir, K. (2025). Effects of high-frequency ultrasound on the functionality and physicochemical properties of chickpea protein isolates. *Food Biomacromolecules, 2*(3), 315–324.

Haeggström, E., & Luukkala, M. (2001). Ultrasound detection and identification of foreign bodies in food products. *Food Control, 12*(1), 37–45.

Hossain, M. B., Tiwari, B. K., Gangopadhyay, N., O'Donnell, C. P., Brunton, N. P., & Rai, D. K. (2014). Ultrasonic extraction of steroidal alkaloids from potato peel waste. *Ultrasonics Sonochemistry, 21*(4), 1470–1476.

Hu, S. Q., Liu, G., Li, L., Li, Z. X., & Hou, Y. (2013). An improvement in the immersion freezing process for frozen dough via ultrasound irradiation. *Journal of Food Engineering, 114*(1), 22–28.

Hu, Y., Sun, H. N., Zhang, M., & Mu, T. H. (2025). Production and characterization of α-glucosidase inhibitory peptides from sweet potato protein by ultrasound-assisted enzymatic hydrolysis and in vitro gastrointestinal digestion. *European Food Research and Technology, 251*(2), 257–267.

Huo, J., Cui, Z., Zhang, R., Ouyang, H., Liu, X., Wang, P., ... Li, S. (2025). Study on the effect and mechanism of ultrasonic-assisted enzymolysis on antioxidant peptide activity in walnuts. *Ultrasonics Sonochemistry, 112*, 107159.

Jia, J. Q., Ma, H. L., Qu, W. J., Ding, Q. Z., Cao, H., Luo, L., & Wang, Z. B. (2008). Ultrasonic pretreatment for preparation of antioxidant peptides from rice protein. *Transactions of the Chinese Society of Agricultural Engineering, 24*(8), 288–293.

Jian, L. I., Shun-Juan, C., Ruo-Nan, Z., & Yan, H. (2008). Study on total saponins extracting technology in cinnamon by ultrasonic-assisted solvent method. *Food Science*, 117–180.

Jin, J., Ma, H., Wang, B., Yagoub, E. G. A., Wang, K., & He, R. (2016). Effects and mechanism of dual-frequency power ultrasound on the molecular weight distribution of corn gluten meal hydrolysates. *Ultrasonics Sonochemistry, 30*, 44–51.

Kadam, S. U., Tiwari, B. K., Álvarez, C., & O'Donnell, C. P. (2015). Ultrasound applications for the extraction, identification and delivery of food proteins and bioactive peptides. *Trends Food Sci. Technol. 46*(1), 60–67.

Kang, D. C., Gao, X. Q., Ge, Q. F., Zhou, G. H., & Zhang, W. G. (2017). Effects of ultrasound on the beef structure and water distribution during curing through protein degradation and modification. *Ultrasonics Sonochemistry, 38*, 317–325.

Kang, D. C., Jiang, Y., Xing, L., Zhou, G., & Zhang, W. (2017). Inactivation of Escherichia coli O157:H7 and Bacillus cereus by power ultrasound during the curing processing in brining liquid and beef. *Food Research International, 102*, 717–727.

Kang, D. C., Zou, Y. H., Cheng, Y. P., Xing, L. J., Zhou, G. H., & Zhang, W. G. (2016). Effects of power ultrasound on oxidation and structure of beef proteins during curing processing. *Ultrasonics Sonochemistry, 33*, 47–53.

Lalthanmawii, J., Banerjee, R., Maheswarappa, N. B., Biswas, S., Belore, B., Govindaiah, P. M., & Patra, G. (2024). Ultrasound-assisted extraction for green recovery of poultry skin collagen hydrolysates with antioxidant and antihypertensive activities. *Biomass Conversion and Biorefinery*, 1–14.

Li, K., Fu, L., Zhao, Y. Y., Xue, S. W., Wang, P., Xu, X. L., & Bai, Y. H. (2020). Use of high-intensity ultrasound to improve emulsifying properties of chicken myofibrillar protein and enhance the rheological properties and stability of the emulsion. *Food Hydrocolloids, 98*, 105275.

Li, G. H., Le, G. W., Shi, Y. H., & Shrestha, S. (2004). Angiotensin I-converting enzyme inhibitory peptides derived from food proteins and their physiological and pharmacological effects. *Nutrition Research, 24*(7), 469–486.

Li, Z., Tu, Z., Wang, H., & Zhang, L. (2020). Ultrasound-assisted extraction optimization of α-glucosidase inhibitors from Ceratophyllum demersum L. and identification of phytochemical profiling by HPLC-QTOF-MS/MS. *Molecules, 25*(19), 4507.
Li, Y., Wang, J., Wu, Y., Guo, S., Zhu, L., Ou, X., ... Wang, J. J. (2025). Effects of ultrasound combined with pH-shift assisted glycosylation modification on the structural and functional properties of by-product proteins extracted from shrimp heads. *Food Chemistry: X, 31*, 103134.
Li, L., Zhang, M., & Wang, W. (2020). Ultrasound-assisted osmotic dehydration pre-treatment before pulsed fluidized bed microwave freeze-drying (PFBMFD) of Chinese yam. *Food Bioscience, 35*, 100548.
Liu, C., Ni, H. Y., Chang, Y. H., Wang, Z., Wan, N., & Cao, L. (2022). Effects of ultrasonic-microwave assisted extraction with green solvent on the chemical constituents, antioxidant and hypolipidemic activities of manchurian walnut oil. *Journal of Food Processing and Preservation*, 16603.
Lorimer, J. P., & Mason, T. J. (1987). Sonochemistry. Part 1-the physical aspects. *Chemical Society Reviews, 16*, 239–274.
Loypimai, P., Moongngarm, A., Sittisuanjik, A., & Khamanan, S. (2020). Optimization of tocols and Î³-oryzanol extraction from rice bran using ultrasound and soybean oil as a green solvent. *Food Research, 4*, 2322–2332.
Lüders, K., & Pohl, R. O. (2017). Travelling waves and acoustics. *Pohl's Introduction to Physics: Volume 1: Mechanics, Acoustics and Thermodynamics*, 291–366.
Ma, H., Huang, L., Jia, J., He, R., & Zhu, W. (2011). Effect of energy-gathered ultrasound on alcalase. *Ultrasonics Sonochemistry, 18*(1), 419–424.
Maran, J. P., & Priya, B. (2015). Ultrasound-assisted extraction of pectin from sisal waste. *Carbohydrate Polymers, 115*, 732–738.
Meier, G. O. S., Aquino, L. F. M. C., Santos, E. B., Lázaro, C. A., & Conte-Junior, C. A. (2015). Detection of raw milk adulterated with cheese whey by ultrasound method. *Journal of Advances in Dairy Research, 3*(146), 2.
Minjares-Fuentes, R., Femenia, A., Garau, M. C., Meza-Velázquez, J. A., Simal, S., & Rosselló, C. (2014). Ultrasound-assisted extraction of pectins from grape pomace using citric acid: A response surface methodology approach. *Carbohydrate Polymers, 106*, 179–189.
Mortazavi, A., & Tabatabaie, F. (2008). Study of ice cream freezing process after treatment with ultrasound. *World Applied Sciences Journal, 4*(2), 188–190.
Nogueira, D. A., da Silveira, J. M., Vidal, É. M., Ribeiro, N. T., & Burkert, C. A. V. (2018). Cell disruption of *Chaetoceros calcitrans* by microwave and ultrasound in lipid extraction. *International Journal of Chemical Engineering*, 9508723.
Palma, M., & Barroso, C. G. (2002). Ultrasound-assisted extraction and determination of tartaric and malic acids from grapes and winemaking by-products. *Analytica Chimica Acta, 458*(1), 119–130.
Patil, N., Yadav, P., & Gogate, P. R. (2024). Ultrasound assisted intensified enzymatic extraction of total phenolic compounds from pomegranate peels. *Separation and Purification Technology, 350*, 127967.
Rosa, A. C. S. D., Costa, A. J. N., Santos Júnior, Oscar, O., & Silva, C. D. (2025). Ultrasound-assisted extraction of sunflower seed oil enriched with active compounds from jambolan leaf. *Journal of the Brazilian Chemical Society, 36*(2), 20240116.
Sultanova, S., Safarov, J., Usenov, A., Samandarov, D., & Azimov, T. (2023). Ultrasonic extraction and determination of flavonoids. *AIP Conference Proceedings, 2507*(1), 050005.
Sun, Y., Liu, D., Chen, J., Ye, X., & Yu, D. (2011). Effects of different factors of ultrasound treatment on the extraction yield of the all-trans-β-carotene from citrus peels. *Ultrasonics Sonochemistry, 18*(1), 243–249.

Tiwari, B. K., O'Donnell, C. P., Patras, A., & Cullen, P. J. (2008). Anthocyanin and ascorbic acid degradation in sonicated strawberry juice. *Journal of Agricultural and Food Chemistry, 56*(21), 10071–10077.

Tiwari, B. K., O'donnell, C. P., & Cullen, P. J. (2009). Effect of sonication on retention of anthocyanins in blackberry juice. *Journal of Food Engineering, 93*(2), 166–171.

Tiwari, B. K., O'Donnell, C. P., Muthukumarappan, K., & Cullen, P. J. (2009). Ascorbic acid degradation kinetics of sonicated orange juice during storage and comparison with thermally pasteurised juice. *LWT-Food Science and Technology, 42*(3), 700–704.

Van, D. W. B., Oliveira, J., Brijs, K., & Raes, K. (2025). Impact of combined enzyme and ultrasound treatments on protein extraction from potato trimmings and soybean okara. *Food Bioscience, 63*, 105691.

Vilkhu, K., Mawson, R., Simons, L., & Bates, D. (2008). Applications and opportunities for ultrasound assisted extraction in the food industry-A review. *Innovative Food Science & Emerging Technologies, 9*(2), 161–169.

Villamiel, M., Garcia-Perez, J. V., Montilla, A., Cárcel, J. A., & Benedito, J. (2017). *Ultrasound in food processing: Recent advances.* John Wiley & Sons.

Wang, L., Li, J. J., Teng, S., Zhang, W. G., Purslow, P. P., & Zhang, R. Y. (2022). Changes in collagen properties and cathepsin activity of beef M. semitendinosus by the application of ultrasound during post-mortem aging. *Meat Science, 185*(108), 718.

Wang, Y., Liu, J., Zhang, Z., Meng, X., Yang, T., Shi, W., ... Ma, H. (2023). Insights into ultrasonication treatment on the characteristics of cereal proteins: functionality, conformational and physicochemical characteristics. *Foods, 12*(5), 971.

Wang, Y., Wang, Z., Handa, C. L., & Xu, J. (2017). Effects of ultrasound pre-treatment on the structure of β-conglycinin and glycinin and the antioxidant activity of their hydrolysates. *Food Chemistry, 218*, 165–172.

Wang, X. N., Xu, Z. S., Sun, Q. J., & Bao, J. Q. (2019). Optimization of ultrasonic assisted enzymatic extraction of collagen from fish skin of lophiiformes. *Science and Technology of Food Industry*, 175–180.

Wang, Z., & Zhang, X. (2015). Preparation of bioactive peptides derived from *Spirulina platensis* and their anti-tumor activities. *Modern Food Science and Technology, 31*(11), 25–32.

Wang, Y., Zhang, W. G., & Zhou, G. H. (2019). Effects of ultrasound assisted frying on the physiochemical properties and microstructure of fried meatballs. *International Journal of Food Science and Technology, 54*, 2915–2926.

Wu, Y., Zheng, Y., Wang, D., Liu, Z., & Deng, Y. (2017). Structural and oxidative modifications, and antioxidant activity of β-lactoglobulin treated by ultraviolet light-C and ultrasound. *International Journal of Food Properties, 20*(3), 3289–3300.

Xia, E. Q., Li, H. L., Xu, X. R., Li, F., Deng, G. F., Li, S., ... Li, H. B. (2012). Ultrasound-assisted extraction of oxymatrine from *Sophora flavescens*. *Journal of Medicinal Plants Research, 6*(10), 1991–1996.

Xiong, W., Wang, Y., Zhang, C., Wan, J., Shah, B. R., Pei, Y., ... Li, B. (2016). High intensity ultrasound modified ovalbumin: Structure, interface and gelation properties. *Ultrasonics Sonochemistry, 31*, 302–309.

Xu, W. J., Zhai, J. W., Cui, Q., Liu, J. Z., Luo, M., Fu, Y. J., & Zu, Y. G. (2016). Ultra-turrax based ultrasound-assisted extraction of five organic acids from honeysuckle (*Lonicera japonica Thunb.*) and optimization of extraction process. *Separation and Purification Technology, 166*, 73–82.

Ylmaz, F., Dademir, E., & Hayalolu, A. A. (2025). Incorporation of ultrasound-assisted treatment in cheese to accelerate ripening of kaar cheese: Changes in cheese microbiota, proteolysis, enzyme activities and volatile profiles. *Food Chemistry, 481*, 144090.

Yu, H., Jiang, L., Gao, L., Zhang, R., Zhang, Y., Yuan, S., & Li, Y. (2024). High-intensity ultrasound promoted the maturation of high-salt liquid-state soy sauce: A means of enhancing quality attributes and sensory properties. *Food Chemistry, 438*, 138045.

Yu, L., Sun, J., Liu, S., Bi, J., Zhang, C., & Yang, Q. (2012). Ultrasonic-assisted enzymolysis to improve the antioxidant activities of peanut (*Arachin conarachin L.*) antioxidant hydrolysate. *International Journal of Molecular Sciences, 13*(7), 9051–9068.

Zhang, J., Toldrá, F., Kang, D., Zhou, L., Wang, J., Zhang, W., & Hu, Y. (2025). Benefits of ultrasonic technology application in meat field and its influential mechanism: A review. *Critical Reviews in Food Science and Nutrition, 65*(28), 5726–5751.

Zhang, L., & Wang, M. (2017). Optimization of deep eutectic solvent-based ultrasound-assisted extraction of polysaccharides from Dioscorea opposita Thunb. *International Journal of Biological Macromolecules, 95*, 675–681.

Zhang, L., Wang, X., Hu, Y., Fakayode, O. A., Ma, H., & Zhou, C. (2021). Dual-frequency multi-angle ultrasonic processing technology and its real-time monitoring on physicochemical properties of raw soymilk and soybean protein. *Ultrasonics Sonochemistry, 80*, 105803.

Zhang, R., Xing, L., Kang, D., Zhou, L., Wang, L., & Zhang, W. (2021). Effects of ultrasound-assisted vacuum tumbling on the oxidation and physicochemical properties of pork myofibrillar proteins. *Ultrasonics Sonochemistry, 74*, 105582.

Zhang, J., Zhang, W., & Xing, L. (2021). Effects of ultrasound on the taste components from aqueous extract of unsmoked bacon. *Food Chemistry, 365*, 130411.

Zhang, J., Zhang, Y., Wang, Y., Xing, L., & Zhang, W. (2020). Influences of ultrasonic-assisted frying on the flavor characteristics of fried meatballs. *Innovative Food Science and Emerging Technologies, 62*, 102365.

Zhang, X., Zhang, Z., Zhao, B., Yang, X., Zhang, Y., Tang, Y., & Yue, S. (2025). Ultrasound-assisted one-step extraction of safflower red and yellow pigments by natural ternary deep eutectic solvents: Extraction optimization and mechanism, separation and biological activities of pigments. *Food Chemistry*, 147383.

Zhang, J., Zhang, W., Zhou, L., & Zhang, R. (2021). Study on the influences of ultrasound on the flavor profile of unsmoked bacon and its underlying metabolic mechanism by using HS-GC-IMS. *Ultrasonics Sonochemistry, 80*, 105807.

Zhang, R., Zhang, J., Zhou, L., Wang, L., & Zhang, W. (2021). Influence of ultrasound-assisted tumbling on NaCl transport and the quality of pork. *Ultrasonics Sonochemistry, 79*, 105759.

Zhang, J., Zhang, Y., Zou, Y., & Zhang, W. (2021). Effects of ultrasound- assisted cooking on quality characteristics of spiced beef during cold storage. *LWT-Food Science and Technology, 136*, 110359.

Zhao, Q., Xie, T., Hong, X., Zhou, Y., Fan, L., Liu, Y., & Li, J. (2022). Modification of functional properties of perilla protein isolate by high-intensity ultrasonic treatment and the stability of o/w emulsion. *Food Chemistry, 368*, 130848.

Zheng, L., & Sun, D. W. (2006). Innovative applications of power ultrasound during food freezing processes-a review. *Trends in Food Science & Technology, 17*(1), 16–23.

Zhou, L., Ali, I., Manickam, S., Goh, B., Tao, Y., Zhang, J., ... Zhang, W. (2025). Ultrasound-induced food protein-stabilized emulsions: Exploring the governing principles from the protein structural perspective. *Comprehensive Reviews in Food Science and Food Safety, 24*, 70162.

Zhou, L., Zhang, J., Lorenzo, J., & Zhang, W. (2021). Effects of ultrasound emulsification on the properties of pork myofibrillar protein-fat mixed gel. *Food Chemistry, 345*, 128751.

Zhou, L., Zhang, J., Yin, Y., Zhang, W., & Yang, Y. (2021). Effects of ultrasound assisted emulsification on the emulsifying and rheological properties of myofibrillar protein stabilized pork fat emulsions. *Foods, 10*, 1201.

Zou, Y., Zhang, W., Kang, D., & Zhou, G. (2018). Improvement of tenderness and water holding capacity of spiced beef by the application of ultrasound during cooking. *International Journal of Food Science and Technology, 53*, 828–836.

Zou, H., Zhao, N., Sun, S., Dong, X., & Yu, C. (2020). High-intensity ultrasonication treatment improved physicochemical and functional properties of mussel sarcoplasmic proteins and enhanced the stability of oil-in-water emulsion. *Colloids and Surfaces A: Physicochemical and Engineering Aspects, 589*, 124463.

CHAPTER FIVE

Ingredient innovations for food reformulation: Functional ingredients, alternative proteins, and fat, salt and sugar replacers

Nuria Prieto[a,*], Matthew G. Nosworthy[b], Óscar López-Campos[a], Rubén Agregán[c], José Manuel Lorenzo[c,d], Manuel Viuda-Martos[e], and Gema Nieto[f]

[a]Lacombe Research and Development Centre, Agriculture and Agri-Food Canada, Lacombe, Alberta, Canada
[b]Guelph Research and Development Centre, Agriculture and Agri-Food Canada, Guelph, Canada
[c]Fundación Centro Tecnolóxico da Carne, Ourense, Spain
[d]Área de Tecnoloxía dos Alimentos, Facultade de Ciencias, Universidade de Vigo, Ourense, Spain
[e]Research Group in Innovations in Food Products (IPOA), Institute for Agri-Food and Agri-Environmental Research and Innovation, Miguel Hernández University (CIAGRO-UMH), Orihuela, Alicante, Spain
[f]Department of Food Technology, Nutrition and Food Science, Veterinary Faculty, Campus de Excelencia Internacional de Ámbito Regional (CEIR), Campus Mare Nostrum (CMN), University of Murcia, Murcia, Spain
*Corresponding author. e-mail address: nuria.prietobenavides@agr.gc.ca

Contents

Abstract

The growing demand for healthier and more sustainable foods, while maintaining palatability, has driven innovation in food reformulation. This book chapter discusses

Advances in Food and Nutrition Research, Volume 122
ISSN 1043-4526, https://doi.org/10.1016/bs.afnr.2026.03.006

ingredient-based strategies focused on improving the nutritional value and eating quality of novel health-oriented food products. This chapter provides a comprehensive overview of key functional ingredients recovered from agri-food by-products, highlighting their nutritional relevance and challenges associated with their applications in food fortification, with particular emphasis on dietary fiber, polyphenols and lipid-derived bioactive compounds. The chapter further evaluates the use of alternative protein sources, including plant-, fungal- and insect-based foods, and reviews recent ingredient innovations aimed at addressing challenges related to eating experience. Additionally, novel strategies for the replacement of fat, salt and sugar with ingredients providing nutritional and health benefits are discussed. Overall, the chapter discusses how ingredient innovation and functional fortification strategies can support the development of healthier, more nutritious and sustainable, and consumer-acceptable food products.

1. Introduction

By 2050, demand for animal-based foods is projected to increase nearly 70 %, as a result of the growing world population, increasing income and urbanization (Yitbarek, 2019). This trend represents a significant threat, as conventional meat production relies on intensive and unsustainable practices that contribute to deforestation, pollution, depletion of critical water resources and biodiversity loss (Ishaq et al., 2022). Beyond environmental concerns, ethical issues related to animal welfare in intensive farming systems and the increasing prevalence of diet-related chronic diseases are also paramount. These interrelated concerns have driven a growing demand for healthier and more sustainable foods, accelerating innovation in food reformulation strategies.

In this context, the valorization of agri-food by-products as sources of functional ingredients has gained substantial attention as a sustainable reformulation strategy. Large volumes of by-products generated during food processing were historically considered waste but are now recognized as valuable sources of nutritionally relevant compounds (Galanakis, 2012; Laufenberg et al., 2003). These materials often contain high concentrations of dietary fiber, polyphenols, bioactive peptides, minerals and lipid-derived compounds, frequently exceeding levels found in the edible fractions of foods. Moreover, their incorporation into food formulations supports circular economy principles by reducing waste streams, while enhancing the nutritional and functional value of final products.

Dietary fiber is one of the most widely applied functional ingredients derived from by-products with a key role in modern food reformulation.

Insufficient dietary fiber intake, common in diets high in refined and ultra-processed foods, has been associated with increased health risks (Głabska et al., 2020; Zhang & Tsao, 2016). Fiber-rich ingredients have, therefore, been extensively investigated for food fortification due to their physiological benefits, as well as their techno-functional properties (Marczak & Mendes, 2024). Additionally, several fiber components function as prebiotics, promoting gut microbiota fermentation and the production of short-chain fatty acids that support metabolic and intestinal health (Lyu et al., 2021; Yuan et al., 2024). Nevertheless, excessive fiber incorporation can negatively affect texture and sensory attributes of the end product, highlighting the need for optimal formulation and processing techniques to improve functionality.

Alongside functional ingredients, plant-, fungal- and insect-based proteins are being explored as partial or complete substitutes for animal proteins due to their potential environmental benefits and ability to diversify protein supply (Maqsood et al., 2024; Xu et al., 2025). However, differences in amino acid composition, digestibility and protein structure compared to animal proteins introduce challenges related to nutritional value. Studies have shown that protein content and "high-protein" labelling can strongly influence consumer choice, emphasizing the importance of maintaining protein quality and quantity in reformulated foods (Antoniak et al., 2022; Kim et al., 2025). Additionally, alternative protein systems often face challenges in replicating key sensory attributes of meat, particularly texture, juiciness and flavour, which are critical determinants of consumer acceptance and repeat purchase (Opaluwa et al., 2025; Pakseresht et al., 2022). Consequently, recent research has focused on ingredient design and formulation strategies to improve the nutritional quality and eating experience of alternative protein-based foods.

Reformulation efforts have also targeted the reduction of fat, salt and added sugars due to their established links with metabolic and cardiovascular disorders when consumed in excess (Hunter et al., 2022; Witek et al., 2022). Rather than simple reduction, current approaches emphasize replacement with ingredients that provide comparable technological and sensory functionality while offering nutritional or health benefits. Carbohydrate-based fat replacers, alternative salt systems and natural sweeteners have been widely used to reduce energy, sodium and glycemic impact, while maintaining product acceptability (Nikolić et al., 2024; Tan et al., 2022; Arshad et al., 2022).

Hence, this chapter provides a comprehensive overview of ingredient innovations for food reformulation, with a focus on functional ingredients, alternative proteins, and fat, salt and sugar replacers, to support the development of healthier, more nutritious and sustainable, and consumer-acceptable food products that align with evolving dietary and environmental priorities.

2. Functional ingredients

Food by-products are materials (liquid, solid and semi-solid fractions) with significant nutritional potential, generated during the processing of raw agricultural products that are not intended for direct human consumption (Galanakis, 2012). For example, plant-based by-products such as seeds, herbs, spices, fruit peels, vegetable trimmings, pomace, cereal bran, and husks are rich sources of polyphenols, dietary fiber, minerals and vitamins. Moreover, cereal bran contains high levels of phenolic acids, insoluble fiber, and micronutrients, while fruit pomace is an important source of antioxidant compounds and soluble fiber (Ajila & Rao, 2013). In addition, animal- and marine-based by-products, including fish skin, whey, shells and bones, are rich in bioactive peptides, proteins, collagen, polyunsaturated fatty acids and minerals (Korhonen & Pihlanto, 2006).

Considering that the compositional diversity of by-products makes them suitable for a wide range of food applications, such as reformulation of bakery, meat, drinks and milk products, the valorization of food industry by-products as sources of functional ingredients and bioactive compounds has gained considerable attention in recent years due to growing consumer concerns regarding sustainability, food security and human health. This section provides a comprehensive overview of the main classes of functional ingredients recovered from agri-food by-products, their nutritional relevance, and applications in food fortification, focusing especially on the reformulation of meat products. Emphasis is placed on dietary fiber, as well as on the nutritional challenges associated with their incorporation into food systems. These residues represent a significant disposal challenge for the industry, due to high associated processing costs, inadequate infrastructure for managing large volumes of biomass and limited processing facilities. Nevertheless, such byproducts are of considerable interest because of their bioactive compounds, which confer nutraceutical value. The integration of by-products into functional foods represents a key strategy for circular economy implementation and the development of novel strategies to improve the nutritional properties of health-oriented food products.

2.1 Bioactive compounds

Bioactive compounds are naturally occurring constituents of foods derived from plant, animal, or marine sources that exert physiological effects beyond basic nutrition. Bioactive ingredients can be intrinsic to foods (e.g., dietary fiber), added through formulation (e.g., probiotics, antioxidants), or used in fortified products (e.g., prebiotics). Bioactive ingredients are typically extracted from fruits, herbs, spices, vegetables, cereals, marine organisms, or food processing by-products, while retaining their functional properties after processing. A wide range of bioactive ingredients has been identified, including antioxidants, probiotics, prebiotics, peptides, proteins, vitamins, polyunsaturated fatty acids, phytochemicals, carotenoids, minerals, dietary fiber, and phenolic compounds. Bioactive compounds can be obtained either directly from natural sources or through the valorization of food processing by-products.

Nevertheless, adding bioactive compounds from diverse sources to food product formulations remains challenging, as it necessitates precise reformulation to account for potential sensory incompatibilities that may arise in certain products, such as meat products (Oliveira Ribero et al., 2023).

2.1.1 Polyphenols and antioxidant bioactive compounds

Polyphenols constitute one of the most important and extensively studied groups of bioactive compounds recovered from fruit and vegetable by-products. Processing residues such as apple pomace, citrus peels, grape pomace, olive mill waste, and berry skins are particularly rich in phenolic compounds, including phenolic acids, flavonoids, tannins, and anthocyanins (Dai & Mumper, 2010). Structurally, polyphenols are a highly diverse class of secondary plant metabolites characterized by the presence of one or more phenolic rings, with more than 10,000 structural variants identified to date and over 500 compounds reported in commonly consumed foods (Ji et al., 2024). Among the different classes of polyphenols, flavonoids represent the most extensively investigated group, encompassing subclasses such as flavonols, flavones, catechins, isoflavones, and anthocyanins. Anthocyanins, which are glycosylated derivatives of anthocyanidins, are responsible for the red, purple, and blue pigmentation of many fruits, vegetables, grains, and flowers. These compounds have attracted particular attention due to their beneficial metabolic effects, including improved insulin secretion and the attenuation of obesity-related metabolic disorders, neurodegenerative diseases, and certain cancers (Ding et al., 2021). Other flavonoids, such as quercetin and catechin, have demonstrated the ability to modulate cellular aging pathways through their unique structural properties, conferring pronounced anti-aging and cytoprotective effects (Davinelli & Scapagnini, 2021).

In this sense, plant-derived by-products are rich sources of bioactive polyphenols with multifunctional health effects. Ginger, for example, contains high levels of phenolic compounds such as curcumin and gingerol which exhibit anti-inflammatory, antibacterial, antioxidant, and anticancer activities (Alsahli et al., 2021). Similarly, hydroxytyrosol (HXT; 4-(2-dihydroxyphenyl)ethanol) exhibits an antioxidant activity approximately ten times greater than that of green tea and twice that of coenzyme Q10 (Lee-Richards, 2014). Furthermore, the radical-scavenging capacity of HXT is comparable to that of oleuropein and catechol. Hydroxytyrosol is a phenylethanoid with well-documented antioxidant properties in vitro. It is naturally obtained from olive leaves and olive oil and is largely responsible for the characteristic intense flavour and aroma of this fruit, with oleuropein acting as its main precursor (Wang et al., 2013). In addition to its in vitro efficacy, HXT has demonstrated significant antioxidant activity in vivo, as evidenced by several studies in rat models. For instance, Merra et al. (2014) reported that HXT supplementation effectively reduced the risk of developing metabolic syndrome. From a structural perspective, hydroxytyrosol differs from tyrosol (TYR) by the presence of an additional hydroxyl group on the benzene ring. This structural feature enhances its free radical scavenging ability, thereby increasing its overall antioxidant potency and improving its effectiveness under conditions of oxidative stress.

Fruits such as citrus family generate substantial amounts of waste, including peels, pulp, seeds, and stones. Bioactive compounds have been successfully extracted from citrus, apple pomace (Rana et al., 2015) and pomegranate peels and seeds (Kaderides et al., 2015). Brewery by-products are rich in bioactive compounds such as sugars, organic acids, antioxidants and proteins. In addition, a by-product of the rice polishing process contains gamma-oryzanol, that is a phytosterol derived from rice bran with bioactive properties (Varzakas et al., 2016).

Moreover, the incorporation of agro-industrial by-products into animal nutrition as sources of natural antioxidants represents a promising strategy, as these materials minimally increase feed costs while complying with European Commission regulations (EC No. 1831/2003), which restrict the use of certain feed additives. Among such by-products, several studies have evaluated the application of rosemary (*Rosmarinus officinalis* L.) residues as natural antioxidant supplements in animal feed. Experimental diets supplemented with rosemary have been tested in pigs (Haak et al., 2006) and sheep (Moñino et al., 2008; Nieto et al., 2010). These studies demonstrated that rosemary supplementation effectively inhibited lipid oxidation in refrigerated raw meat. Moñino et al. (2008) supplemented ewe diets during gestation and lactation with 10 % or 20 % distilled rosemary leaves, achieving biologically active concentrations of rosemary polyphenols in lamb muscle without negatively affecting animal performance. The principal polyphenolic compounds identified include carnosic, rosmarinic, caffeic, ferulic, and coumaric acids, as well as carnosol, hesperidin, naringin, luteolin, apigenin, and genkwanin. Subsequent investigations demonstrated that distilled rosemary leaves supplementation extended the shelf life of raw lamb stored under retail conditions in oxygen-enriched packaging (Nieto et al., 2010).

Green leafy vegetables are nutrient-dense ingredients and important sources of bioactive compounds, including naturally occurring nitrates, phenolic acids, and flavonoids. Vegetables such as beetroot, lettuce, arugula, watercress, celery, spinach, and chard have gained increasing attention as functional ingredients in food reformulation strategies, particularly as natural alternatives to synthetic nitrates and nitrites used in dry-cured and processed meat products (Alahakoon et al., 2015).

Although excessive intake of nitrites has been associated with adverse health effects, naturally occurring nitrates from plant-based sources are typically accompanied by antioxidants, vitamins, and other micronutrients that may modulate nitrate metabolism and reduce potential risks. This intrinsic antioxidant matrix supports their application as safer and more acceptable curing agents in clean-label food products (Sindelar & Milkowski, 2012). In addition to their nitrate content, green leafy vegetables are rich in phenolic compounds, including phenolic acids and flavonoids, which contribute to antioxidant activity, color stability, and overall nutritional quality. The incorporation of these vegetables or their extracts into food formulations represents a promising strategy to improve both the eating quality and nutritional profile of novel food products (Alahakoon et al., 2015). However, the effective application of polyphenols in food systems presents several technological challenges. These compounds are highly sensitive to processing conditions such as heat, oxygen exposure, and pH variations, and their interactions with proteins and polysaccharides in complex food matrices may significantly influence both bioavailability and sensory attributes.

2.1.2 *Proteins and bioactive peptides from by-products*

Protein-rich by-products constitute an important source of bioactive peptides with diverse physiological activities. Whey proteins, oilseed meals, fish processing residues, and meat by-products have been extensively studied for their potential to release peptides with antihypertensive, antioxidant, antimicrobial, and immunomodulatory effects following enzymatic hydrolysis (Korhonen & Pihlanto, 2006).

Whey protein-derived peptides are among the most commercially successful bioactive compounds and are widely used in functional foods and sports nutrition. Similarly, fish skin and bones provide collagen and gelatin, which are associated with benefits for joint health, skin integrity, and connective tissue function. From a technological perspective, protein hydrolysates can improve emulsifying, foaming, and water-binding properties of food systems. However, bitterness and off-flavours remain significant challenges that require careful enzyme selection, hydrolysis control, and formulation strategies to ensure consumer acceptance.

2.1.3 *Lipid-based bioactive compounds and phytosterols*

Several food by-products are rich in bioactive lipids, including phytosterols, omega-3 fatty acids and conjugated linoleic acid. Fish processing residues are a major source of long-chain omega-3 fatty acids, while rice bran oil and oilseed press cakes contain significant amounts of phytosterols with cholesterol-lowering effects (Shahidi & Ambigaipalan, 2015). The incorporation of lipid-based bioactive from by-products into fortified foods presents technological challenges related to oxidative stability and sensory quality, because lipid oxidation can result in loss of nutritional value and off-flavours in the reformulated food.

2.2 Dietary fiber and prebiotic fortification

Dietary fiber is among the most extensively studied and widely applied functional ingredients obtained from food by-products. Common sources include wheat bran, rice bran, oat hulls, apple pomace, citrus peels, and sugar beet pulp, which are frequently used for fiber enrichment in a variety of food products. Chemically, dietary fiber comprises health-promoting carbohydrate polymers, predominantly non-starch polysaccharides such as cellulose, hemicellulose, and pectin, which are resistant to digestion by enzymes in the upper gastrointestinal tract (Meldrum & Yakubov, 2024). However, intensive industrial processing of staple foods, particularly refined wheat and rice products, leads to substantial losses of dietary fiber. As a result, population-level dietary fiber intake often falls below recommended values, a situation that is especially pronounced in developed countries characterized by high consumption of ultra-processed foods (Głabska et al., 2020).

From a technological perspective, dietary fiber derived from by-products contributes important functional properties to food systems, including enhanced water-holding capacity, oil-binding capacity, viscosity, and texture modification. Nevertheless, high levels of fiber incorporation may negatively affect product volume, texture, and sensory attributes, particularly in bakery and extruded products. To address these challenges, physical and biochemical treatments such as micronization, enzymatic modification, extrusion, and fermentation have been successfully applied to improve the functional performance and sensory quality of by-products-derived fiber (Galanakis, 2012).

Dietary interventions enriched with fiber have demonstrated significant potential in correcting gut microbiota dysbiosis and improving metabolic health outcomes (Srivastava, 2025). Overall, dietary fiber fortification represents a promising, non-invasive strategy for the prevention and management of a wide range of chronic diseases. Furthermore, it is important to note that several components of dietary fiber also function as prebiotics.

The incorporation of dietary fiber into fresh, cooked, or dry-cured meat products is a relatively common practice in the meat industry. Dietary fiber in meat products fulfills several functions, including: (i) increasing water-holding capacity and cooked yield, (ii) improving textural attributes such as hardness, juiciness, and chewiness, (iii) enhancing emulsifying properties and thereby increasing the stability of raw-cooked products, (iv) extending the shelf life of meat products due to the bioactive compounds associated with fiber, primarily phenolic acids and flavonoids, and (v) serving as a substitute for saturated fat in the development of meat products with reduced caloric content (Wang et al., 2024; Diao et al., 2024; He et al., 2026). On the other hand, the incorporation of dietary fiber as an ingredient in meat products could enhance their nutritional profile by providing components with prebiotic properties that may exert beneficial effects on the organism (Yuan et al., 2024).

2.2.1 Sources of dietary fiber used in meat products

Dietary fibers employed in the development of meat products may come from a wide range of plant-based materials and can be classified according to their source, techno-functional characteristics and functional properties. These fibers include cereal-based fractions, legume-derived fibers, and those obtained from fruits and vegetables, as well as emerging sources such as seaweed-derived fibers and microbial polysaccharides (Yuan et al., 2024; Ciobanu et al., 2025). In addition, emergent fiber modification techniques including enzymatic hydrolysis, ultrasonication, and extrusion processing, are being analyzed to improve the fiber functionality in meat matrix (Liu et al., 2024). Each category of fiber exhibits distinct physicochemical and techno-functional properties that contribute to specific functional roles in meat formulations.

Cereals including those from wheat, oat, barley, and rice bran, are among the most commonly utilized ingredients in the meat sector due to their content in dietary fiber. These fibers are abundant in insoluble polysaccharides, such as cellulose, hemicellulose, and lignin (Qin et al., 2025). As occurs with fruits, the use of this source of dietary fiber is generally accepted to improve several properties of meat products (Aviles et al., 2023; Adeola et al., 2024).

Vegetable-derived dietary fiber, comprising insoluble constituents (cellulose, hemicellulose, lignin) and soluble fractions (pectins, gums, β-glucans), offers multifunctional roles when incorporated into meat products (Marczak & Mendes, 2024). These fibers vary widely in content depending on species and processing, being modest on a fresh-weight basis but concentrated in peels and pomaces and in dry matter fractions. Technologically, vegetable fibers improve water-holding, emulsification, texture and fat-binding, enabling reformulation for reduced-fat and fiber-enriched meat formulations while contributing potential health benefits associated with increased dietary fiber intake (Stanišić et al., 2025).

The dietary fiber obtained from fruits and their coproducts are widely used in meat products due to the techno-functional. This fiber is composed of a mixture of soluble dietary fiber (mainly pectin and mucilage) and insoluble dietary fiber (such as cellulose and hemicellulose) in different proportions depending on the fruit type as well as the part of the fruit used such as pulp, peel, pomace, etc, (Khorasaniha et al., 2023). Flours with a high dietary fiber content, obtained from citrus fruits, apples, grape and tropical fruits, have been widely used in the development of meat products, both fresh, cooked, and dry-cured (Oliveira Ribeiro et al., 2023; Diao et al., 2024; Gracey & Tako, 2025).

2.2.2 Meat product added with dietary fiber

To develop healthier meat products, dietary fiber-rich extracts or flours with a high fiber content can be applied directly as components of the formulation to enrich meat products with dietary fiber, or they can be utilized to replace components with adverse effects on health, such as saturated fat.

2.2.2.1 Meat products added with dietary fiber from cereals

As mentioned above, cereals and their co-products are a good source of dietary fiber to improve the nutritional and techno-functional properties of fresh and cooked meat products.

In the case of fresh products, research carried out by Gomes de Campos et al. (2021) analyzed the replacement of pork meat by brewer's spent grain flour at 1.5 %, 3 %, 4.5 % and 6 % in Cuiabana-type sausage, which is classified as a fresh sausage. They found that the protein and dietary fiber content increased in formulations added with brewer's spent grain flour in a concentration-dependent manner. However, no changes were observed in textural properties nor were there any improvements in the oxidation degree. More recently, Nurymkhan et al. (2025) evaluated the effects of a mixture of cereals comprising rice (35 %), buckwheat (20 %), oats (20 %), and corn (25 %) on the chemical and techno-functional, oxidative properties of meat patties elaborated with beef and turkey. The cereal mixture was added at 5 %, 10 %, and 15 %. These authors found that patties had lower moisture, protein, and fat contents, while ash and dietary fiber levels increased with respect to control sample. The techno-functional analysis showed a dose-dependent improvement in water holding capacity and in fat retention.

Dietary fiber from cereals has also been added to cooked meat products. Thus, Jandyal et al. (2022) conducted research to analyze the effects of adding oat bran at 3 %, 6 %, and 9 % on the chemical composition and functional properties of cooked pork sausages. They reported that, in fortified sausages, cooking yield and emulsion stability both showed an upward trend with increasing oat bran content in the product. On the other hand, ash and crude dietary fiber contents were significantly higher in all treatment groups. In a similar study, Haque et al. (2024) investigated the effect of inclusion of rice bran (5 % and 10 %) on chemical composition and the sensory properties of meat nuggets. These authors found that fortified nuggets had a higher content in ash and dietary fiber than control samples. From a sensorial point of view, no differences between control and fortified nuggets were observed.

2.2.2.2 Meat products added with dietary fiber from vegetables

Vegetables and the co-products obtained from their industrialization, such as pomaces, peels, and seeds, are valuable sources of dietary fiber that can be used in the development of fresh and cooked meat products, improving both their nutritional composition and functional properties.

In this sense, Bunmee et al. (2022) examined the addition of purple eggplant flour, 2.5 %, 5.0 %, and 7.5 %, as fat replacer on nutritional composition, and product quality of beef patties. The addition of the purple eggplant flour resulted in a significant rise in moisture, ash, and total dietary fiber while protein and fat contents were reduced. The cooking yield, moisture, and fat retention of the purple eggplant flour beef patties, at all concentrations analyzed, were higher compared the control patty. In a more recent study, Karslıoğlu et al. (2024) conducted a study where dried tomato peel flour, at 1 %, 2 %, 3 %, and 4 %, was included in beef burgers. They reported that reformulated samples showed a higher ash, carbohydrate, and dietary fiber content than control. In addition, water-holding capacity, moisture, and fat retention values increased with increasing levels of tomato peel flour. Similarly, Richards et al. (2024) evaluated beef patties with carrot pomace added at 1.0 %, 3.0 %, and 4.2 %. They reported that the cooking yield and water holding capacity were improved while the chemical composition showed substantial increases in ash and dietary fiber content.

Flours with a high content of dietary fiber obtained from vegetables, also, has been added to cooked meat to enhance their nutritional properties. In this way, Cerón-Guevara et al. (2021) analyzed the effect of addition of 7.5 % and 10 % of *Agaricus bisporus* and *Pleurotus ostreatus* flours on physicochemical properties of pâté. These authors informed that the fiber content in pâté added with *A. bisporus* was 1.73 and 2.43 g/100 g whilst the fiber content in pâté added with *P. ostreatus* was 3.31 and 4.63 g/100 g. Mumyapan et al. (2022) reported that bologna sausage added with seed pumpkin flour at 2.2 % and 3 % had better emulsion stability and water holding capacity than control sample.

2.2.2.3 Meat products added with fruits and their coproducts

In recent years, the application of fruit and fruit co-products compounds in developing functional meat products has gained significant interest. In scientific literature there are several works that reported the application of fruits in meat products to improve the nutritional values as well as the techno-functional properties.

In the case of fresh products, Zhao et al. (2021) analyzed the addition of kiwi pomace powder (0.5 %, 1 %, 3 %, 5 %, and 7 %), which showed a high content of insoluble dietary fiber, and its effects on the chemical and functional properties of pork meatballs. The authors reported that the fiber content of reformulated meatballs increased from 1.04 to 5.03 g/100 g. Regarding functional properties, the inclusion of kiwi pomace powder reduced water and fat release, thereby improving emulsion stability. In other fresh meat product such as burger, de Alencar et al. (2022) analyzed the effect on chemical composition and functional properties of beef burgers added with grape skin flour at 0.5 %, 1 %, 1.5 % and 2 %. These authors reported that the dietary fiber content increased in reformulated samples with values ranged between 3.17 and 3.77 g/100 g compared with control sample (2.76 g/100 g). More recently, Sakr et al. (2023) informed that beef burger added with guava peel at 5 %, 10 % and 15 % increased the fiber content, with values ranged between 4.00 and 8.68 g/100 g, compared with control (1.68 g/100 g). In the same way, the cooking yield and water holding capacity were improved significantly with respect to control.

As regards to cooked meat product, Salazar et al. (2021) assessed a fiber-enriched Frankfurter-type sausage incorporated with green banana flours. They reported that samples added with banana peel flour had higher total dietary fiber content (5.37–10.42 g/100 g) compared with control sample (4.31 g/100 g). In addition, the samples added with banana flour showed better cooking yield than control. Vilcapoma et al. (2023) investigated the potential application of yellow dragon fruit peels, rich in insoluble dietary fiber and soluble dietary fiber, as a fat replacement in alpaca-based sausages. They found that reformulated samples had higher cooking yield than control as well a higher content in dietary fiber. More recently, Diao et al. (2024) conducted an analysis to assess the chemical composition and techno-functional of attributes of frankfurter sausages enriched with mandarin flour rich in soluble dietary fiber. These authors informed that sausages containing mandarin flour exhibited higher cooking yield compared to the control samples, indicating that this type of fiber prevented water loss during the heating process. Furthermore, the addition of mandarin flour produced a significant reduction in their fat content.

3. Plant-, fungal- and insect-based protein alternatives

It has been well established that the continued expansion of the human population presents challenges for provision of nutritious food (Hussain et al., 2025). This is occurring in conjunction with the rise of consumer interest in non-animal based protein sources for financial, ethical and heath related reasons (Marinova & Bogueva, 2022). This desire for alternative protein sources has been demonstrated not only in the vegan and vegetarian populations, but also in those who desire to reduce but not eliminate meat consumption, commonly referred to as 'flexitarians' (Marinova & Bogueva, 2022). Whatever the rationale, it is this subset of the population that have a decided interest in foods with reduced meat inclusion; however, this does not mean acceptance of reduced appearance, flavour or nutritional quality.

When reformulating foods with alternative protein sources, careful consideration must be given to both the nutritional quality of the added protein and its contribution to the final protein content of the product, as these factors can influence consumer interest and acceptance. Indeed, a study by Antoniak et al. (2022) determined that there was a positive effect of including "high protein" on the label of burger patties with consumers being more interested in labels indicating a greater protein content. As the development of hybrid food products continues and the diversity of commercially available hybrid products grows, it is important to assess how different blends impact total protein content. This section will discuss how selection of different ingredients, and their percent inclusion, alters the protein content of burger patties, sausages and frankfurters, and other hybrid meat products. Incorporation of plant-based material is presented in Table 1, fungal hybrid products in Table 2, and those incorporating insect flours in Table 3. Notably, any significant differences presented in these tables are reflective of the statistics performed within each study, included here to highlight the impact of ingredient and inclusion percentage on protein content not to provide a comparison between studies.

Beyond nutritional quality, the eating quality of new-generation protein-alternative products warrants close attention. Although hybrid and alternative-protein foods have experienced rapid market growth, many formulations still face challenges to resemble the sensory characteristics of conventional meat, particularly in terms of texture, juiciness and flavour, which are critical drivers of consumer acceptance and repeat purchase (Opaluwa et al., 2025). These challenges arise from fundamental differences in protein structure, lipid functionality and flavour chemistry between animal-derived and alternative protein sources. Consequently, considerable research efforts have been made to improve the sensory characteristics of protein-alternative products through targeted ingredient selection, structural design and formulation strategies. Hence, this section reviews recent advances in ingredient design and formulation approaches aimed at enhancing the eating quality of protein-alternative products, with a focus on innovations that address texture, juiciness, and flavour limitations in plant-, fungal- and insect-based systems.

3.1 Plant-based protein alternatives

Plant-based protein alternatives have emerged as an important category within sustainable food systems, including hybrid meat products, which partially replace meat with plant ingredients to improve nutritional quality, and plant-based meat analogues, formulated to mimic the sensory and protein characteristics of meat without animal-derived components (Petrat-Melin & Dam, 2023).

3.1.1 Nutritional quality

The use of plant flours, or concentrates/isolates, in hybrid products is relatively common with many examples being found in commercially available products. Table 1 presents the impact of incorporation of alternative ingredients on the protein burgers, sausages and a meat emulsion.

One study investigated replacing 10 % of a beef patty with either extruded faba bean protein concentrate or a 75:25 blend of faba bean protein concentrate and flaxseed extrudates (Ribes et al., 2024). In this case, incorporation of faba bean extrudate alone resulted in a significantly greater protein content than beef or the beef/fava bean/flaxseed hybrid (26.4 % vs. 22.4 %/24.2 % respectively). In another study, pork was blended with deodorized pea protein isolates alone or in combination with oat flour at inclusion rates of 25 %, 40 %, and 50 % (Flores et al., 2023). From this work, it was determined that while protein content was not significantly reduced at 25 % inclusion for either combination, the protein content at 40 % inclusion was lower than the control (18.47 % vs. 16.40 % and 15.85 %), as well as 50 % inclusion of the pea and oat blend (15.81 %). Single ingredient or dual ingredient replacement of pork is relatively straightforward as described above; however, there is interest in investigating multi-plant mixtures. Dasiewicz et al. (2024) prepared two multi-plant blends for testing in hybrid pork burgers with blend A including millet groats, dried red tomatoes, sunflower seeds, onion, and rapeseed pomace while blend B included lentils, carrots, sunflower seeds, oatmeal, onion, and camelina pomace. Although no significant differences were found in that study, the protein content ranged from 16.9 % in the control to 11.9 % with 100 % blend A and 12.2 % with 100 % blend B. Despite not being statistically different, reduced pork content in

Table 1 Effect of incorporation of plant-based ingredients on the protein content of hybrid products.

Product	Blending ingredient	Formulation (%)	Protein content		References
			Unprocessed	Processed	
Beef Burger		0	22.4^{a}	—	Ribes et al. (2024)
	Fava bean	10	26.4^{b}	—	
	Fava bean + flaxseed	10	24.2^{a}	—	
Pork Burger		0	18.47^{a}		Flores et al. (2023)
	Pea	25	16.96^{ab}	17.09^{ab}	
	Pea + oat	25	16.93^{ab}	16.57^{ab}	
	Pea	40	16.40^{b}	15.79^{bc}	
	Pea + oat	40	15.85^{bc}	15.71^{bc}	
	Pea	50	16.57^{ab}	15.39^{bc}	
	Pea + Oat	50	15.81^{bc}	13.99^{c}	
Pork Burger		0	16.9	—	Dasiewicz et al. (2024)
	Blend A	25	14.9	—	
	Blend A	50	14	—	

	Blend A	75	12.1	—	
	Blend A	100	11.9	—	
	Blend B	25	15.6	—	
	Blend B	50	15.1	—	
	Blend B	75	13.6	—	
	Blend B	100	12.2	—	
Chicken Burger		0	17.38[a]	21[a]	Chandler and McSweeney (2022)
	Yellow pea	25	15.77[b]	18.42[c]	
	Yellow pea	50	14.06[b]	15.72[d]	
	Yellow pea	75	12.81[c]	14.95e	
	Chickpea	25	18.62[a]	21.00[a]	
	Chickpea	50	18.03[a]	20.72[a]	
	Chickpea	75	18.04[a]	20.66[a]	
	Lentil	25	19.60[a]	21.66[a]	
	Lentil	50	19.56[a]	20.43[a]	
	Lentil	75	19.33[a]	20.62[a]	

(continued)

Table 1 Effect of incorporation of plant-based ingredients on the protein content of hybrid products. (*cont'd*)

Product	Blending ingredient	Formulation (%)	Protein content		References
			Unprocessed	Processed	
Pork Sausage		0	17.5	25.2	Ebert et al. (2022)
	Pumpkin seed	12.5	18.3	27.0	
	Pumpkin seed	25	18.0	26.3	
	Pumpkin seed	37.5	18.3	26.3	
	Pumpkin seed	50	19.2	27.8	
Chicken Sausage		0	14.93[a]	—	Sun et al. (2022)
	Hempseed meal	10	14.56[a]	—	
		20	14.22[a]	—	
		30	13.24[b]	—	
		40	13.73[b]	—	

Meat Emulsion		0	14.31^{a}	—	dos Santos et al. (2022)
	Soy	50	12.76^{bc}	—	
	Pea	50	12.33^{cd}	—	
	Rice	50	13.09^{b}	—	
	Sunflower	50	12.28^{cd}	—	
	Fava bean	50	11.99^{d}	—	
Beef Meatballs		0	24.09	—	Grasso et al. (2019)
	Soy	15	24.82	—	
	Soy	30	25.19	—	
	Soy + yeast	15	23.42	—	
	Soy + yeast	30	23.94	—	

Values within the same study and column (unprocessed or processed) with different superscript letters are significantly different ($P < 0.05$).

the patties was concomitant with a lower protein content. Incorporation of pulse flours into hybrid products is common, leading to one group replacing 25–75 % of chicken meat with either the flours from yellow pea, chickpea, or lentil (Chandler & McSweeney, 2022). Interestingly, while incorporating any yellow pea flour led to a significant reduction in the protein content of unprocessed patties compared to control (17.38 % vs. 12.81–15.77 %), there was no significant change in protein content when either chickpea or lentil flour was included at either 25, 50, or 75 %. A similar trend occurred post-cooking where no significant change occurred with lentil and chickpea hybrids, however inclusion of yellow pea resulted in a significantly lower protein content compared to control.

Beyond burgers, pork and chicken sausages have also been hybridized with pumpkin seed and hempseed meal, respectively (Ebert et al., 2022; Sun et al., 2022). Pumpkin seed protein concentrate at 60 % protein underwent high moisture extrusion prior to incorporation into pork sausages at 12.5–50 % and subsequent dry-curing for 21 days (Ebert et al., 2022). Although the curing process did increase the protein content by approximately 8 % in all hybrid sausages, there was no significant change in protein content with unprocessed sausages ranging from 17.5 % with 100 % pork to 19.2 % with 50 % reduced meat inclusion. A similar 2 % increase was determined after curing as well, 25.2 % in 100 % pork sausages and 27.8 % with 50 % reduced meat inclusion. While incorporation of 5 % hemp seed without meat removal has been shown to increase the protein content of a pork loaf from 20.85 % to 23.63 % (Zając et al., 2019), a study replacing up to 40 % of chicken meat in a sausage led to a significant reduction in final protein content (Sun et al., 2022). In this study, no change in protein content was determined at 10 % replacement (14.93 % vs. 14.56 %), however, protein content was significantly lower at 20 % replacement (13.22 %), 30 % (13.24 %), and 40 % (13.37 %).

There are other hybrid products of interest that incorporate plant-based proteins such as emulsions of pork and beef, as well as beef meatballs (dos Santos et al., 2022; Grasso et al., 2019). When 50 % of the lean red meat was replaced with protein concentrates of soy, pea, rice, sunflower, and fava bean in a meat emulsion, there was a significant reduction in protein content (dos Santos et al., 2022). While the closest to the protein content of the control (14.31 %), incorporation of rice concentrate only reached 13.09 % protein while fava bean was unable to reach 12 % protein (11.99 %). Replacement of 15 % and 30 % of beef mince with either texturized soy protein or texturized soy protein with additional 5 % yeast had no impact on total protein content of meatballs (Grasso et al., 2019). Although inclusion of yeast did reduce the average protein content slightly (25 % vs. 23.68 %), it was in alignment with the protein content of full beef meatballs (24.09 %).

3.1.2 Eating quality

In animal-derived meat, it is well-known that intramuscular fat positively influences sensory attributes, such as tenderness, juiciness and flavour, enhancing consumer acceptance (Hocquette et al., 2010). Recent research has focused on investigating the effects of oil addition to plant-based meat analogues. For instance, Opaluwa et al. (2025) evaluated the effect of incorporating different vegetable oils with different degrees of unsaturation (medium-chain triglyceride–MCT, coconut, olive, and grapeseed oils) into wheat gluten-based meat analogues during high-moisture extrusion and found that MCT and grapeseed oils improved tenderness and decreased firmness and chewiness without producing off-flavours, despite the highest susceptibility to oxidation of grapeseed oil. Nevertheless, the susceptibility of vegetable oils to lipid oxidation during high-temperature extrusion is a generalized concern (Ilo et al., 2014). Additionally, these oils used in liquid form have a low melting point and reduce viscosity when heated, leading to oil loss during cooking that negatively affects the texture of the product (Fasina & Colley, 2008; Maningat et al., 2021). Hence, structured lipid systems such as emulsions and oleogels have been investigated to enhance plant-based meat analogue eating quality without compromising oxidative stability or texture. In this context, Lee et al. (2023) evaluated various oil phases (MCT, coconut and peanut oils) and emulsifiers (soy protein isolate, lecithin and Tween® 80) in textured vegetable protein plant-based meat patties and observed that the oil-in-water emulsion containing MCT oil and Tween® 80 enhanced tenderness and juiciness. However, traditional emulsifiers are composed of small chemical molecules with potential adverse effects on human health (Chen et al., 2020). Alternatively, structured oils can be stabilized by solid particles (Shi et al., 2020). In this context, Keum et al. (2024) developed a Pickering emulsion (PE) in which sunflower oil was emulsioned using combinations of polysaccharides and pea protein isolate as stabilizers and the resulting PEs were incorporated into plant-based patties formulated with textured vegetable protein. These authors found that the formulation containing pea protein isolate and methylcellulose improved oil retention and texture, achieving the highest overall acceptability. These findings show that PEs stabilized by natural polysaccharide–protein complexes can be an effective replacement of vegetable oil in plant-based patties. Regarding oleogels, Hang et al. (2023) developed an oleogel by structuring soybean oil with whey protein and incorporating it into a soy protein gel model to simulate plant-based meat texture and found that the addition of the oleogel significantly enhanced the oil release during oral processing and improved fat-related texture perception for consumers. These authors suggested that this oleogel might be a potential ingredient for improving the juiciness of the plant-based meat analogues.

Despite significant progress in improving the tenderness and juiciness of plant-based meat analogues, flavour remains one of the most challenging characteristics to replicate in these food products (Pakseresht et al., 2022). Plant proteins commonly produce beany or grassy aromas, mainly due to higher levels of hexanal formed through lipid oxidation. To address this issue, recent studies have investigated the incorporation of animal-like heme proteins produced by genetically engineered yeasts to create more authentic meaty aromas and mask those undesirable flavours. In this context, Meng et al. (2025) observed that the addition of two yeast-derived heme proteins (leghemoglobin and *Sus scrofa f.domestica* myoglobin) to soy-based meat mince increased the content of aldehydes and pyrazines associated with desirable meaty flavour, while significantly reducing the content of beany flavour compounds. Additionally, Caputo et al. (2022) found that consumers preferred plant-based burgers containing leghemoglobin compared to those formulated solely with pea protein, attributing this preference to soybean leghemoglobin ability to simulate the 'bleeding' effect of cooked meat and enhance the meaty flavour of plant-based burgers. Nevertheless, the use of recombinant proteins still raises public concerns due to potential health and environmental risks that may limit their widespread adoption in foods (Fraser et al., 2018).

Another approach for improving flavour in plant-based meat analogues has been through fermentation techniques. Fermentation with lactic acid bacteria of pea protein isolates reduced the 'hay' and 'green' like aromas from plant protein (Shi et al., 2021). In plant-based matrices, Fernández-Varela et al. (2024) reported that fermenting soybean, lentil, chickpea and pea protein matrices with *Bacillus subtilis* and lactic acid bacteria strains, and their combinations, reduced the amount of hexanal, a compound associated with beany off-flavour, while producing diacetyl, a metabolite associated with pleasant dairy flavour. In contrast, Pöri et al. (2023) found that fermentating sunflower protein concentrate and pea protein isolate extrudates with *Lactobacillus helveticus* produced an undesirable sour taste in plant-based meat analogues. However, those authors observed that applying a neutral pH shift after fermentation promoted fibrillar structure formation and mitigated the sour taste of the fermented extrudates, although no significant differences in the flavour or texture were detected by a trained sensory panel when compared to the unfermented extrudates. There are several key parameters, including microbial strain selection, substrate composition, temperature, pH control, and fermentation time, that can affect the fermentation outcome. Hence, optimization of the fermentation process is key to unlock its potential for enhancing the eating quality of plant-based meat analogues.

Pea proteins are characterized for 'beany' and 'earthy' flavours, resulting in attempts to deflavour pea protein concentrates and isolates through deodorization processes (Roland et al., 2017). In a study with hybrid meat patties, Flores et al. (2023) reported that deodorized pea protein isolates blended with pork reduced the off-flavour characteristics of texturized pea protein compared to non-deodorized formulations. However, the deodorization process negatively affected textural properties due to partial solubilization and limited network formation, although the addition of oat flour partially restored hardness. These results suggest that hybrid patty formulation must be carefully optimized to balance off-flavour reduction and desirable texture.

The addition of natural antioxidants to plant-based meat analogues has also been explored as a means of improving their eating quality. In this context, Han et al. (2024) reported that incorporating 1.0 % (w/w) ethanol-extracted green tea (*Camellia sinensis*) leaf powder into textured pea protein-based meat analogues improved texture parameters, increased umami taste and reduced off-flavour intensity compared to the control.

The valorization of agricultural by-products is another promising approach to improve the eating quality of food products, aligning with circular economy principles to reduce waste and promote sustainability within the food industry. Specifically in plant-based meat analogues, Issara et al. (2024) substituted textured vegetable protein with banana peel at ratios of 25 %, 50 % and 75 % and observed that the 50 % substitution level improved texture, by enhancing oiliness and firmness. These improvements attributed to the high oil- and water-holding capacities of banana peel fiber led to the highest overall liking and consumer acceptance scores.

3.2 Fungal-based protein alternatives

Another environmentally friendly alternative protein source is filamentous fungi, which combine high nutritional value with a low environmental footprint (Wimbasara et al., 2025).

3.2.1 Nutritional quality

Mushrooms are a nutritious ingredient for development of hybrid products. Data regarding protein content of products containing mushroom as an ingredient is presented in Table 2. After replacing 1.35 % of beef with different mushrooms to reduce sodium chloride content, the protein content of a beef burger patty was significantly greater pre- and post-cooking compared to the control (Botella-Martínez et al., 2023). Interestingly, while 1.35 % portobello mushroom had the highest protein content in the unprocessed patty (16.57 %) compared to button (15.61 %) and oyster (15.72 %), there was no significant difference between mushroom types in the protein content after cooking. When 2.5–10 % of a chicken patty was replaced with oyster mushroom stalk powder, the protein content of the unprocessed burger was significantly lower than the control in all cases (Mounir et al., 2025). While the control chicken patty had a protein content of 79.36 %, the mushroom stalk hybrids ranged from 78.02 % at 2.5 % to 66.33 % at 10 % inclusion. When steamed oyster mushrooms were used to replace 5–20 % of carp meat in carp patties, the protein content was also reduced at any inclusion percentage greater than 5 % (Tokarczyk et al., 2023). While an insignificant change from 18.52 % in the control to 18.14 % at 5 % inclusion, the protein content was significantly lower at 10 % (17.32 %), 15 % (16.05 %), and 20 % (15.52 %) inclusion.

Formulations for sausages and frankfurters that include mushrooms have also been investigated. In an effort to replace pork back fat in sausage formulations, king oyster mushroom (*Pleurotus eryngii*) was incorporated into pork sausages at 14 % of total weight, equivalent to 100 % fat replacement, either as unprocessed mushrooms, or mushrooms after being boiled, deep fried, or fried (Wang et al., 2019a). In these experiments the protein content of the final sausage was greater than the control (12.75 %) after 14 % inclusion of mushrooms across all treatment groups, with no difference between treatments (average of 14.20 %). Another investigation from the same group replaced 25–100 % of lean pork meat with shiitake mushroom (*Lentinula edodes*) (Wang et al., 2019b). The protein content of all preparations with mushrooms was significantly

Table 2 Effect of incorporation of fungi-based ingredients on the protein content of hybrid products.

Product	Blending ingredient	Formulation (%)	Protein content		References
			Unprocessed	Processed	
Beef Burger		0	15.04[c]	17.44[b]	Botella-Martínez et al. (2023)
	Oyster mushroom	1.35	15.72[b]	19.55[a]	
	Button mushroom	1.35	15.61[b]	18.72[a]	
	Portobello mushroom	1.35	16.57[a]	19.36[a]	
Chicken Burger		0	—	79.36[a]	Mounir et al. (2025)
	Oyster Mushroom Stalk	2.5	—	78.02[b]	
	Oyster Mushroom Stalk	5	—	72.93[c]	
	Oyster Mushroom Stalk	7.5	—	69.9[d]	
	Oyster Mushroom Stalk	10	—	66.33[e]	
Carp Burger		0	18.52[c]	—	Tokarczyk et al. (2023)
	Oyster Mushroom	5	18.14[c]	—	
	Oyster Mushroom	10	17.32[b]	—	
	Oyster Mushroom	15	16.05[a]	—	
	Oyster Mushroom	20	15.52[a]	—	

(continued)

Table 2 Effect of incorporation of fungi-based ingredients on the protein content of hybrid products. (*cont'd*)

Product	Blending ingredient	Formulation (%)	Protein content		References
			Unprocessed	Processed	
Pork Sausage		0	12.75[a]	—	Wang et al. (2019a)
	King oyster mushroom	14	14.23[b]	—	
	Boiled king oyster mushroom	14	14.20[b]	—	
	Deep fried king oyster mushroom	14	14.19[b]	—	
	Fried king oyster mushroom	14	14.20[b]	—	
Pork Sausage		0	12.73[d]		Wang et al. (2019b)
	Shiitake Mushroom	25	9.91[c]		
		50	8.98[b]		
		75	6.23[b]		
		100	5.17[a]		

Frankfurter	Porcini mushroom	0	10.91^{b}	—	Novakovic et al. (2020)
		0.75	12.69^{a}	—	
		1.5	11.88^{a}	—	
Chicken Frankfurter	Oyster mushroom	0	18.41^{a}	—	Rosli et al. (2015)
		2	16.22^{b}	—	
		4	15.32^{c}	—	
		6	14.29^{d}	—	

Values within the same study and column (unprocessed or processed) with different superscript letters are significantly different ($P < 0.05$).

lower than that of the control (12.73 %), with a maximum of 9.91 % protein when 25 % of the lean pork was replaced to a minimum of 5.17 % when the sausage contained no pork meat. The addition of 0.75 % or 1.5 % porcini mushroom (*Boletus edulis*) decoction significantly increased the protein content of frankfurters from 10.91 % in the control to 12.69 % and 11.88 %, respectively (Novakovic et al., 2020). When oyster mushroom (*Pleurotus sajorcaju*) replaced 2–6 % of chicken meat in frankfurters, there was a significant reduction in protein content (Rosli et al., 2015). While the control sample had 18.41 % protein, replacing meat with mushrooms reduced the protein content to 16.22 % at 2 % replacement and to 14.29 % at 6 % replacement.

3.2.2 Eating quality

The interwoven mycelial network of fungi naturally forms a fibrous structure that can be molded into meat-like textures, making them promising candidates for meat analogues. However, the texture of the mycelium is too soft and weak to resemble meat fibers. Consequently, recent studies have focused on enhancing the structural integrity of fungal protein products. In this context, Wimbasara et al. (2025) investigated the effects of several non-animal binding agents (sodium tripolyphosphate, transglutaminase, pectin and modified tapioca starch) on the sensory characteristics of *Rhizopus oligosporus*–based burger patties and found that the addition of 10 % modified tapioca starch produced the most pronounced improvement in instrumental texture (increasing hardness by 105 % and chewiness by 84 % compared to the control), resulting in patties perceived by consumers as firmer and tougher. Additionally, Gu et al. (2025) incorporated additives (soy protein isolate, carrageenan and egg white protein) into *Pleurotus geesteranus*–based meat analogues and found that all the additives improved appearance (darker and more appealing coloration). Nevertheless, soy protein isolate produced the greatest improvements in textural characteristics such as hardness, springiness, cohesiveness, chewiness and resilience, resulting in the analogue most comparable to conventional pork.

An emerging trend is the development of mycoprotein hybrid meat analogues, which blend plant proteins with fungal mycelium to leverage the natural fibrous network of mycelium and its high nutritional value. Indeed, Nayak and Panda (2016) reported that fermenting soybean with *Rhizopus oligosporus* increased hardness, springiness and resilience compared to the unfermented soybean-based product. Moreover, these authors evaluated the addition of *Agaricus* mushroom powder to the meat analogue at different concentrations (3 %, 5 %, 7 %, 10 %, 20 % and 30 %) and found that a low inclusion level (3 %) of that powder produced a meat analogue with textural characteristics most closely resembling those of poultry meat.

Sustainable reuse of olive oil by-products has also been investigated in lentil-champignon mushroom hybrid meat analogues, where supplementation with a phenolic extract obtained from olive mill vegetation water enhanced product hardness, firmness and fresh vegetable and legume odours, while reducing garlic odour, gumminess and springiness (Nichita et al., 2025). This approach not only provides a natural alternative to synthetic additives for enhancing the eating quality of fungal-based food products, but also supports circular-economy valorization of olive by-products.

3.3 Insect-based protein alternatives

Using insects as food is another sustainable alternative to conventional livestock production, as insect farming requires significantly less feed, land and water, and exerts lower ecological footprint compared to the production of cattle, pigs and poultry (Abbasi & Abbasi, 2016).

3.3.1 Nutritional quality

While more recent than plant and fungal hybrid products, research has been conducted on the use of crickets, mealworms, and black soldier flies, among others, regarding incorporation of insect flours into hybrid products (Table 3). By replacing 5–10 % of lean beef with cricket flour (66 % protein content), the total protein present in the patty was significantly increased from 19.53 % in the control to 23.68 % in the unprocessed patties (Cavalheiro et al., 2024). Conversely, when flours derived from four edible insects – yellow mealworm (*Tenenbrio molitor),* white-spotted flower chafer larva *(Protaetia brevitarsis seulensis)*, Japanese rhinoceros beetle *(Allomyrina dichotoma),* and two-spotted cricket (*Gryllus bimaculatus*) were incorporated into pork patties, the protein content increased from 19 % in the control to 25 % with 20 % inclusion of the white-spotted flower chafer larva (Choi et al., 2024). Comparing between the four insects, white-spotted flower chafer larva and Japanese rhinoceros beetle had the greatest protein content (25 % and 24 %), with mealworm being significantly greater than the control (22 %) and Japanese rhinoceros beetle having similar protein content as the control (18 %).

Frankfurters, both pork and chicken, have also undergone investigation to determine the potential benefits of replacing meat with insect flours. One study replaced between 5–25 % of lean meat with protein extracted from

Table 3 Effect of incorporation of insect-based ingredients on the protein content of hybrid products.

Product	Blending ingredient	Formulation (%)	Protein content		References
			Unprocessed	Processed	
Beef burger		0	19.53[d]	—	Cavalheiro et al. (2024)
	Cricket Flour	5	21.82[c]	—	
	Cricket Flour	7.5	22.7[b]	—	
	Cricket Flour	10	23.68[a]	—	
Pork Burger		0	19[c]	—	Choi et al. (2024)
	Yellow mealworm	20	22[b]	—	
	White-spotted flower chafer	20	25[a]	—	
	Two-spotted cricket	20	24[a]	—	
	Japanese rhinoceros beetle	20	18[c]	—	
Frankfurter		0	13.06	—	Zhang et al. (2024)
	Yellow mealworm	5	12.95	—	
		10	12.59	—	
		15	12.47	—	
		20	12.5	—	
		25	12.63	—	

Pork/beef Product		0	13^{a}	—	Lemke et al. (2023)
	Yellow mealworm	10	12^{b}	—	
	Yellow mealworm	20	12^{b}	—	
	Black Soldier Fly	10	11^{c}	—	
	Black Soldier Fly	20	11^{c}	—	

Values within the same study and column (unprocessed or processed) with different superscript letters are significantly different ($P < 0.05$).

yellow mealworm (*Tenebrio molitor*) (Zhang et al., 2024). Interestingly, there was no change in protein content with 13.06 % in the control and a range of 12.47–12.95 % in the frankfurters containing mealworm protein. Products with a combination of multiple meat sources, such as a pork and beef combination, is also a point of interest in hybrid product development. When 10 % and 20 % of the pork/beef blend was replaced with either yellow mealworm (*Tenebrio molitor*) or black soldier fly larvae (*Hermetia illucens*), there was a significant reduction in protein content (Lemke et al., 2023). Between the two insect flours, replacement with mealworm flour resulted in a significantly greater protein content than that of black soldier fly larvae (approximately 12 % vs. 11 %).

3.3.2 Eating quality

Despite recent advances in the development of insect-based food products (Kozlu et al., 2024), poor taste has been identified as one of the primary factors discouraging consumers from repurchasing insect-based food products (House, 2016).

In an insect (*Tenebrio molitor*)-based model system, Wendin et al. (2019) investigated how variations in particle size, oil/water ratio and salt content influence sensory properties. These authors observed that larger particle sizes enhanced yellowness and texture attributes such as coarseness, viscosity and crispness, but reduced odour intensity, whereas higher salt levels increased saltiness and nutty flavour, a desirable characteristic of mealworm. In contrast, changes in the oil/water ratio did not affect the sensory characteristics. Moreover, the addition of rosemary extract reduced rancidity and colour changes during storage. Overall, salt enhancement and controlled particle size reduction were identified as promising strategies to improve the sensory acceptability of insect-based formulations, which could be considered in future product development.

Although plant-based meat analogues are rapidly expanding, their key limitation is the low biological quality and limited bioavailability of plant proteins, which often present incomplete amino acid profiles and lower digestibility compared to animal-derived proteins (Maqsood et al., 2024). To overcome this constraint, the incorporation of insect proteins into plant-based formulations is gaining attention, as they offer a more balanced essential amino acid profile, higher digestibility and greater nutrient density than most plant sources (Xu et al., 2025). In this context, Ardila et al. (2023) developed soy- and seitan-based burger analogues enriched with *Tenebrio molitor* flour and evaluated several binders (gellan, xanthan, guar and sodium alginate + $CaCl_2$) as well as natural ingredients (wheat flour, rice, oats, white corn and maize starch) to improve sensory characteristics. These authors found that most hydrocolloids and flours produced an undesirable mouthfeel and a strong vegetal flavour, respectively. However, sodium alginate combined with calcium chloride provided a texture very similar to beef burgers, while oat flour imparted a pleasant toasted taste, both contributing to higher sensory acceptance by consumers.

Insect fats and oils have recently been explored as food ingredients, due to their nutritional quality has been described as intermediate between vegetable oils and animal fats in terms of fatty acids (Tzompa-Sosa & Fogliano, 2017). However, their high polyunsaturated fatty acid content makes them prone to lipid oxidation, leading to undesirable odours and flavours in insect-based foods (Tzompa-Sosa et al., 2021). Consequently, recent research has focused on reducing the lipid content of insect ingredients. In this context, Ribeiro et al. (2022) evaluated defatting and drying methods in insect flours incorporated to a model of oat cereal bars and observed that defatted and microwave-dried *Tenebrio molitor* flours resulted in liking scores comparable to the control bars without insect flour, with defatted samples presenting a pleasant flavour and microwave-dried samples characterized by a cereal-like aroma – both positive drivers of overall liking. In contrast, oven-dried *Tenebrio molitor* and *Acheta domesticus* insect flours produced bars with undesirable "feed" and "rancid" odour/flavours, resulting in the lowest consumer acceptance. These off-flavours were attributed to a higher oxidation in oven-drying than microwave-drying. Additionally, Tzompa-Sosa et al. (2021) evaluated the deodorization of yellow mealworm (*Tenebrio molitor*) oil incorporated into crackers and hummus and found that deodorized insect oil successfully replaced vegetable oil without adversely affecting overall sensory experience, consumer liking or visual appearance. These findings suggest that defatting, drying and deodorization strategies can significantly enhance the sensory acceptance of insect-based ingredients and may facilitate the development of other insect-derived products with improved eating quality.

4. Fat, salt and sugar replacers with nutritional and health benefits

Fat, salt and sugars are food ingredients commonly used in the food industry, often essential for food production, depending on the type of product made. However, these ingredients are in the spotlight and severely criticized for being directly related to a number of health problems such as the risk of developing certain diseases, including cardiovascular diseases (CVDs), hypertension and type 2 diabetes. Hence, current approaches have focused on their replacement with ingredients that provide comparable technological and sensory functionality while offering nutritional or health benefits (Fig. 1).

4.1 Carbohydrates for fat replacement

Fat replacers are substances capable of totally or partially replacing the fat in food products while maintaining the same techno-functional and sensory properties. These substitutes meet two main objectives: on the one hand, they attempt to reduce the caloric content and the amount of fat used in the preparation of food products, and on the other hand, they provide physical-chemical properties similar or equal to the mass of the original food, acting as potential texturizers, stabilizers, emulsifiers, or gelling and thickening agents (Syan et al., 2024).

Complex carbohydrates have recently been used for the development of food products, including a variety of native starches, modified starches, maltodextrins, cellulose and cellulose derivatives, polydextrose, inulin, pectin, and other dietary fibers (Nikolić et al., 2024). Starch is a low-cost, multifunctional biomaterial able to mimic the properties of fats, which

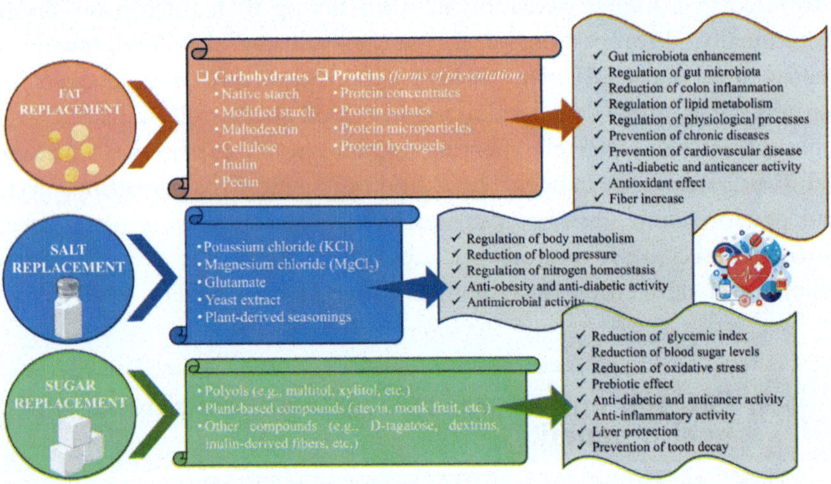

Fig. 1 Diagram showing various substitutes for fats, salt, and sugar, and the health benefits associated with their consumption.

plays crucial roles in the food industry, such as in the production of sausages, mayonnaise, and baked goods (Compart et al., 2023; Lin et al., 2024). The use of resistant starch exhibits prebiotic properties in the human gut, as it is not digested in the small intestine by amylases, passing into the colon where it undergoes fermentation by the microbiota. Starch fermentation can also alleviate inflammation in the colon and reduce the risk of chronic diseases, such as type 2 diabetes, heart disease, and colon cancer (Zhuang et al., 2024). This type of starch is being tested as a fat substitute in different products such pork batter, with promising results (Xie et al., 2024).

Maltodextrins can also serve as fat replacers, helping to reduce the risk of gastrointestinal discomfort. Despite these benefits, the glucose from digested maltodextrins is rapidly absorbed in the small intestine, possibly causing a high glycemic index (Syan et al., 2024). Although, this event may be beneficial before exercising, as suggested by a recent meta-analysis (Hartley et al., 2022). Several fatty food products were produced using this polysaccharide, including stuffed meats such as German sausages, reducing the calorie and fat content, increasing fiber, and definitely enhancing the nutritional quality (Khosravani et al., 2024).

The usefulness of cellulose materials as fat substitutes has been proven in recent research, demonstrating the potential to mimic the properties of fat. Lee and Han (2024) used a bacterial cellulose paste to develop plant-based patties, achieving rheologic behavior and an appearance closer to conventional fats when added at 3 %. Other products with added bacterial cellulose include meat products, such as Frankfurt-type sausages and Chinese-type meatballs (Oliveira Ribeiro et al., 2023). Beneficial effects of cellulose consumption have been reported in the colon, where it may be partially fermented, producing health-promoting substances, including short-chain fatty acids (e.g., propionate, acetate, and butyrate), with important functionality for the host metabolism (He et al., 2021).

Inulin is a soluble dietary fiber, capable of substituting fat in meat products, such as bologna sausages and meat burgers (Kamil et al., 2023; Paglarini et al., 2022), providing good functional properties and nutritive value. This carbohydrate can act as a probiotic, regulating the intestinal microbiota by stimulating the growth of beneficial bacteria. In addition, inulin may play a role in regulating lipid metabolism, promoting weight loss, lowering blood sugar levels, inhibiting the expression of inflammatory factors, and reducing the risk of colon cancer, among others benefits (Qin et al., 2023). Another important dietary fiber due to its ubiquity is pectin. This polysaccharide has been successfully tested in different meats products (e.g., pork sausages and beef burgers) (Círstea et al., 2023; Zhang et al., 2025a), and several healthy properties, such as anti-diabetes, anti-cancer, and hypolipidemic effects, were attributed to its intake (Zhang et al., 2025b).

4.2 Inorganic salts, glutamate, yeast extract and spices for common salt replacement

Although the consumption of sodium chloride (NaCl), commonly known as common salt by the food market, provides essential ions for the proper functioning of the body (Tan et al., 2022), excessive intake may cause high blood pressure, leading to cardiovascular problems and mortality (Hunter et al., 2022). The strategy recently tested by the food industry and academia is the substitution of common salt by other ingredients. NaCl is crucial in the development of a wide range of food products, as it is a key component for the development of the final sensory properties. Replacing NaCl with other inorganic salts, such as potassium chloride (KCl), calcium chloride ($CaCl_2$), and magnesium chloride ($MgCl_2$), is being tested in the development of meat products (Tan et al., 2022). Zou et al. (2022) found that the KCl addition could reduce serum renin, angiotensin-converting enzyme (ACE), angiotensin (Ang) II, and the levels of some indicators of dyslipidemia. On the other hand, Mg ions were found to be beneficial for osteogenesis in bone repairing or neuronal health and functions related (Kumar et al., 2024; Lau et al., 2023).

Glutamate is widely used as a replacer for NaCl, enhancing the flavour of meat and meat products, which allows reducing the amount of salt content. This non-essential amino acid is one of the most abundant in nature, representing up to 8–10 % of most dietary proteins and peptides. Only the free form of glutamate can exhibit the unique and famous "umami" taste (Kos et al., 2023). Despite the negative perception of glutamate as a food additive, which was reported to cause brain functional disruptions in the late 1960s, recent studies have found that, in the context of a normal diet and physiological conditions, glutamate was highly metabolized in the splanchnic area. Furthermore, glutamate is involved in nitrogen and energy metabolism, playing an important role in nitrogen homeostasis (Loï & Cynober, 2022). However, the excessive use of monosodium glutamate in food products has been linked to several adverse physiological effects in the reproductive, cardiovascular, and immune systems, as well as increasing the risk of obesity and metabolic diseases, among other negative effects (Kumari et al., 2023).

It has been observed that some yeast extracts improve the flavour of foods without adding odor, serving as possible salt replacers (Nurmilah et al., 2022). As flavouring ingredients, yeast extracts can provide different flavours, including meaty, kokumi, umami, and salty, as well as nutritional compounds, such as partially hydrolyzed protein with 35–40 % of free amino acids, B vitamins, and minerals at trace levels (Tomé, 2021).

Plant-derived seasonings (e.g., garlic, deadnettle family, herb blends, saffron, and hot spices) have demonstrated good acceptance when used as Na substitutes, contributing to decrease blood pressure (Taladrid et al., 2020). Garlic was found to reduce blood pressure via inhibiting angiotensin-converting enzyme (ACE) (Taladrid et al., 2020). Rosmarinic acid, naturally found in the rosemary herb, may decrease systolic, diastolic, and mean arterial blood pressure (Prasannarong et al., 2019). Anti-obesity and anti-diabetic properties have been attributed to capsaicin (Taladrid et al., 2020), a nitrogen-containing substance responsible for the spicy sensation, which also exhibits important antimicrobial properties against pathogenic fungi (e.g., *Candida* spp.) (Periferakis et al., 2023).

4.3 Polyols and plant-based compounds for sugar replacement

The addition of white or refined sugar in food formulations meets a series of technological and sensory objectives, such as providing texture and flavour, and also extending shelf life. However, the addition of refined sugar to processed food products is linked to metabolic, cardiovascular, and nervous system disorders. Of particular concern is the group of ready-to-eat foods with excessive amounts of this sugar, as well as salt and fats, making it necessary to reduce its content by developing new food formulations. This strategy represents a technological challenge for the industry, which must ensure the maintenance of food quality (Arshad et al., 2022; Gomes et al., 2023; Witek et al., 2022).

In this context, the replacement of refined sugar with other alternative natural compounds is gaining momentum due to the reduction of calories, the nutritional contribution and the presence of biocompounds. Two different types of sweeteners can be used to achieve adequate replacement of white sugars: natural sweeteners, such as polyols (e.g., maltitol and xylitol) and plant-based (e.g., stevia and monk fruit), and synthetic sweeteners (e.g., saccharin and aspartame) (Dragomir et al., 2025).

Polyols, also called polyhydric alcohols, are compounds obtained from starch, sucrose, and whey through hydrogenation and are widely used in the food industry. These white sugar substitutes have a low glycemic index derived from their low caloric value, which is 2–2.5 kcal/g, approximately half that of white sugar (4 kcal/g of sugar). Another health benefit is their incomplete absorption by the body, helping to control blood sugar levels. In addition, polyols can also act as prebiotics for the gut microbiota, improving the digestive process. However, excessive consumption may cause digestive problems, headaches, or allergic reactions (Ding & Yang, 2025; Dragomir et al., 2025; Wojtuś et al., 2022).

The low-calorie natural sweetener stevia (*Stevia rebaudiana*) is used in food product formulations as a substitute for sucrose, modifying physicochemical, nutritional, and technological properties (Schiatti-Sisó et al., 2023). This evergreen shrub of the Compositae family contains steviol glycosides, giving between 100 and 300 times more sweetness than white sugar. Stevia possesses different bioactive compounds that are related to diverse health benefits, such reducing oxidative stress and preventing and controlling type 2 diabetes, as well as other metabolic diseases (Patel & Navale, 2024). Similarly, the so-called monk fruit has been identified as a promising substitute for white sugar in the development of low-calorie functional foods and sugar-free products, especially suitable for obese and diabetic patients (Massoud & Hashem, 2023).

Other less well-known and researched compounds can be included as potential sucrose replacers in the development of low-calorie foods. This is the case of the sweetener D-tagatose, obtained from the D-galactose derived from whey by isomerization. This compound can exhibit techno-functional properties (e.g., texturizing, stabilizing, humectant, etc.) and health benefits, such as attenuating the glycemic response and preventing tooth decay (Ibrahim, 2018).

5. Conclusion

This chapter highlights ingredient innovation as a key driver in the reformulation of novel food products. The literature reviewed in this chapter demonstrates that advances in alternative ingredients, including plant-, fungal- and insect-based systems, as well as agri-food by-products, are promising for addressing the health, environmental and ethical challenges associated with conventional food production. In particular, the valorization of by-products rich in dietary fiber, polyphenols, bioactive peptides and lipid-derived compounds supports food fortification strategies that are nutritionally relevant and aligned with sustainability and circular economy principles. Despite remaining challenges related to eating quality and consumer acceptance, emerging processing technologies alongside advances in ingredient design and formulation are instrumental in reducing the sensory limitations currently associated with novel food products. Additionally, fat, salt and sugar replacers provide effective reformulation strategies to improve the nutritional and health profile of food products while preserving technological functionality and sensory quality. Further research on optimizing ingredient interactions within complex food matrices is warranted to enhance bioavailability and functionality, and to align sensory attributes with consumer expectations, thereby supporting the successful translation of these innovations into market-ready food products.

Acknowledgments

N. Prieto, M. G. Nosworthy and Ó. López-Campos thank Agriculture and Agri-food Canada for its support. R. Agregán thanks GAIN (Axencia Galega de Innovación), Xunta de Galicia, for granting an extension to his postdoctoral scholarship (grant number: IN606C−2025/002). R. Agregán and J. M. Lorenzo are members of the Competitive Reference Group "FunFood" (Functional Food), funded by GAIN (grant No. IN607A2023/01). G. Nieto thanks the Spanish Ministry of Science and Innovation project (PID2021−123628OB-C44-Innovations and development in functional bakery products using bioactive compounds and nutrients characteristic of the agri-food-co-products. Study of biological activities), which is funded by MCIN/AEI/10.13039/501100011033/ and FEDER, UE.

References

Abbasi, T., & Abbasi, S. A. (2016). Reducing the global environmental impact of livestock production: The minilivestock option. *Journal of Cleaner Production, 112*, 1754–1766. https://doi.org/10.1016/j.jclepro.2015.02.094.

Adeola, A. F., Philip, S. O., Adesola, A. A., Adegoke, B. H., & Tolulope, O. A. (2024). Response surface methodology for the optimization of process parameters during hot-air frying of chicken sausages incorporated with corn bran. *Sustainable Food Technology, 2*(3), 769–779. https://doi.org/10.1039/d3fb00204g.

Ajila, C. M., & Rao, U. J. S. P. (2013). Fruit peels: Nutritional composition and health benefits. *Food Reviews International, 29*, 1–26. https://doi.org/10.1080/87559129.2013.789004.

Alahakoon, A. U., Jayasena, D. D., Ramachandra, S., & Jo, C. (2015). Alternatives to nitrite in processed meat: Up to date. *Trends in Food Science & Technology, 45*(1), 37–49. https://doi.org/10.1016/j.tifs.2015.05.008.

Alsahli, M. A., Almatroodi, S. A., Almatroudi, A., Khan, A. A., Anwar, S., Almutary, A. G., ... Rahmani, A. H. (2021). 6-Gingerol, a major ingredient of ginger attenuates diethylnitrosamine-induced liver injury in rats through the modulation of oxidative stress and anti-inflammatory activity. *Mediators of Inflammation*, 6661937. https://doi.org/10.1155/2021/6661937.

Antoniak, M. A., Szymkowiak, A., & Pepliński, B. (2022). The source of protein or its value? Consumer perception regarding the importance of meat (-like) product attributes. *Applied Sciences, 12*(9), 4128. https://doi.org/10.3390/app12094128.

Ardila, P., Honrado, A., Marquina, P., Beltrán, J. A., & Calanche, J. B. (2023). Innovative plant-based burger enriched with Tenebrio molitor meal: Characterization and shelf-life. *Foods, 12*, 3460. https://doi.org/10.3390/foods12183460.

Arshad, S., Rehman, T., Saif, S., Rajoka, M. S. R., Ranjha, M. M. A. N., Hassoun, A., ... Aadil, R. M. (2022). Replacement of refined sugar by natural sweeteners: Focus on potential health benefits. *Heliyon, 8*(9), e10711. https://doi.org/10.1016/j.heliyon.2022.e10711.

Aviles, M. V., Naef, E. F., Abalos, R. A., Lound, L. H., Gómez, M. B., & Olivera, D. F. (2023). Use of a rice industry by-product as a meat replacer in a hybrid chicken patty: Technological and sensory impact. *International Journal of Gastronomy and Food Science, 31*, 100674. https://doi.org/10.1016/j.ijgfs.2023.100674.

Botella-Martínez, C., Muñoz-Tebar, N., Lucas-González, R., Pérez-Álvarez, J. A., Fernández-López, J., & Viuda-Martos, M. (2023). Assessment of chemical, physicochemical and sensory properties of low-sodium beef burgers formulated with flours from different mushroom types. *Foods, 12*(19), 3591. https://doi.org/10.3390/foods12193591.

Bunmee, T., Setthaya, P., Chaiwang, N., & Sansawat, T. (2022). Effect of purple eggplant flour on physicochemical, lipid oxidation, and sensory properties of low-fat beef patties. *International Journal of Food Science, 2022*, 9753201. https://doi.org/10.1155/2022/9753201.

Caputo, T., Barlocco, N., Brovarone, C. V., Parodi, M., D'Ambrosio, M., Calabrò, S., & Vegliante, F. (2022). Leghemoglobin in plant-based meat analogues: Current applications and future prospects. *Trends in Food Science & Technology, 129*, 406–419. https://doi.org/10.1016/j.tifs.2022.09.021.

Cavalheiro, C., Ruiz-Capillas, C., Herrero, A. M., Pintado, T., Avelar de Sousa, C. C., Sant'Ana Falcão Leite, J., & Costa Alves da Silva, M. (2024). Potential of cricket (Acheta domesticus) flour as a lean meat replacer in the development of beef patties. *Foods, 13*(2), 286. https://doi.org/10.3390/foods13020286.

Cerón-Guevara, M. I., Santos, E. M., Lorenzo, J. M., Pateiro, M., Bermúdez-Piedra, R., Rodríguez, J. A., ... Rangel-Vargas, E. (2021). Partial replacement of fat and salt in liver pâté by addition of *Agaricus bisporus* and *Pleurotus ostreatus* flour. *International Journal of Food Science & Technology, 56*(12), 6171–6181. https://doi.org/10.1111/ijfs.15076.

Chandler, S. L., & McSweeney, M. B. (2022). Characterizing the properties of hybrid meat burgers made with pulses and chicken. *International Journal of Gastronomy and Food Science, 27*, 100492. https://doi.org/10.1016/j.ijgfs.2022.100492.

Chen, L., Ao, F., Ge, X., & Shen, W. (2020). Food-grade pickering emulsions: Preparation, stabilization, and applications. *Molecules (Basel, Switzerland), 25*(14), 3202. https://doi.org/10.3390/molecules25143202.

Choi, N., Park, S., Park, Y., Park, G., Oh, S., Kim, Y. A., & Choi, J. (2024). Effects of edible insect powders as meat partial substitute on physicochemical properties and storage stability of pork patties. *Food Science of Animal Resources, 44*(4), 817–831. https://doi.org/10.5851/kosfa.2024.e17.

Ciobanu, M. M., Manoliu, D. R., Ciobotaru, M. C., Flocea, E. I., & Boişteanu, P. C. (2025). Dietary fibres in processed meat: A review on nutritional enhancement, technological effects, sensory implications and consumer perception. *Foods, 14*(9), 1459. https://doi.org/10.3390/foods14091459.

Cîrstea, N. (Lazăr), Nour, V., Corbu, A. R., & Codină, G. G. (2023). Efficacy of chitosan, pectin and xanthan as cold gelling agents in emulsion gels stabilized with legume proteins to be used as pork backfat replacers in beef burgers. *Gels, 9*(12), 970. https://doi.org/10.3390/gels9120970.

Compart, J., Singh, A., Fettke, J., & Apriyanto, A. (2023). Customizing starch properties: A review of starch modifications and their applications. *Polymers, 15*(16), 3491. https://doi.org/10.3390/polym15163491.

Dai, J., & Mumper, R. J. (2010). Plant phenolics: Extraction, analysis and their antioxidant and anticancer properties. *Molecules (Basel, Switzerland), 15*, 7313–7352. https://doi.org/10.3390/molecules15107313.

Dasiewicz, K., Szymanska, I., Opat, D., & Hac-Szymanczuk, E. (2024). Development and characterization of hybrid burgers made from pork and multi-ingredient plant mixtures and protected with lactic acid bacteria. *Applied Sciences, 14*(14), https://doi.org/10.3390/app14146272.

Davinelli, S., & Scapagnini, G. (2021). Interactions between dietary polyphenols and aging gut microbiota: A review. *Biofactors (Oxford, England), 48*, 274–284. https://doi.org/10.1002/biof.1785.

de Alencar, M. G., Perreira de Quadros, C., Lima Pedrosa Luna, A. L., Figueirêdo Neto, A., Matiuzzi da Costa, M., Ávila Queiroz, M. A., et al. (2022). Grape skin flour obtained from wine processing as an antioxidant in beef burgers. *Meat Science, 194*, 108963. https://doi.org/10.1016/j.meatsci.2022.108963.

Diao, X., Zhu, J., Huang, L., Li, S., Mao, X., Li, C., & Ke, W. (2024). Effect of citrus fiber on physicochemical quality of frankfurter sausages: A study on lipid oxidation and protein gel characteristics. *LWT – Food Science and Technology, 209*, 116778. https://doi.org/10.1016/j.lwt.2024.116778.

Ding, W., Liu, H., Qin, Z., Liu, M., Zheng, M., Cai, D., & Liu, J. (2021). Dietary antioxidant anthocyanins mitigate type II diabetes through improving the disorder of glycometabolism and insulin resistance. *Journal of Agricultural and Food Chemistry, 69*(45), 13350–13363. https://doi.org/10.1021/acs.jafc.1c05037.

Ding, S., & Yang, J. (2025). Polyols utilization mechanism and health benefits. In B. Zhang, & S. Dhital (Eds.). *Carbohydrate nutrition* (pp. 269–293). Elsevier. https://doi.org/10.1016/B978-0-443-15700-4.00013-4.

dos Santos, M. D., Rocha, D. A. V. F. D., Bernardinelli, O. D., Oliveira Junior, F. D., de Sousa, D. G., Sabadini, E., & Pollonio, M. A. R. (2022). Understanding the performance of plant protein concentrates as partial meat substitutes in hybrid meat emulsions. *Foods, 11*(21), 3311. https://doi.org/10.3390/foods11213311.

Dragomir, N., Grigore, D.-M., & Pogurschi, E. N. (2025). Beyond sugar: A holistic review of sweeteners and their role in modern nutrition. *Foods, 14*(18), 3182. https://doi.org/10.3390/foods14183182.

Ebert, S., Jungblut, F., Herrmann, K., Maier, B., Terjung, N., Gibis, M., & Weiss, J. (2022). Influence of wet extrudates from pumpkin seed proteins on drying, texture, and appearance of dry-cured hybrid sausages. *European Food Research and Technology, 248*(6), 1469–1484. https://doi.org/10.1007/s00217-022-03974-4.

Fasina, O. O., & Colley, Z. (2008). Viscosity and specific heat of vegetable oils as a function of temperature: 35 °C to 180 °C. *International Journal of Food Properties, 11*(4), 738–746. https://doi.org/10.1080/10942910701586273.

Fernández-Varela, R., Hansen, A. H., Svendsen, B. A., Moghadam, E. G., Bas, A., Kraĉun, S. K., ... Poulsen, V. K. (2024). Harnessing fermentation by Bacillus and lactic acid bacteria for enhanced texture, flavor, and nutritional value in plant-based matrices. *Fermentation, 10*(8), 411. https://doi.org/10.3390/fermentation10080411.

Flores, M., Hernán, A., Salvador, A., & Belloch, C. (2023). Influence of soaking and solvent extraction for deodorization of texturized pea protein isolate on the formulation and properties of hybrid meat patties. *Journal of the Science of Food and Agriculture, 103*(6), 2806–2814. https://doi.org/10.1002/jsfa.12453.

Fraser, R. Z., Shitut, M., Agrawal, P., Mendes, O., & Klapholz, S. (2018). Safety evaluation of soy leghemoglobin protein preparation derived from *Pichia pastoris*, intended for use as a flavor catalyst in plant-based meat. *International Journal of Toxicology, 37*(3), 241–262. https://doi.org/10.1177/1091581818766318.

Galanakis, C. M. (2012). Recovery of high added-value components from food wastes: Conventional, emerging technologies and commercialized applications. *Trends in Food Science & Technology, 26*, 68–87. https://doi.org/10.1016/j.tifs.2012.03.003.

Głabska, D., Guzek, D., Groele, B., & Gutkowska, K. (2020). Fruit and vegetable intake and mental health in adults: A systematic review. *Nutrients, 12*(1), 115. https://doi.org/10.3390/nu12010115.

Gomes, A., Bourbon, A. I., Peixoto, A. R., Silva, A. S., Tasso, A., Almeida, C., & Alves, V. D. (2023). Strategies for the reduction of sugar in food products. In M. Ângelo, M. P. Ribeiro Cerqueira, & L. M. Pastrana Castro (Eds.). *Food structure engineering and design for improved nutrition, health and well-being* (pp. 219–241). Elsevier. https://doi.org/10.1016/B978-0-323-85513-6.00008-6.

Gomes de Campos, K. C., Nascimento de Farias, A. K., Becker, G., Silva de Britto, G. C., Pinheiro Soares, W., Nascimento, E., et al. (2021). Quality measurements of Cuiabana-type pork sausages added with brewing by-product flours. *Meat Science, 179*, 108441. https://doi.org/10.1016/j.meatsci.2021.108441.

Gracey, P. R., & Tako, E. (2025). Apple and grape pomace: Emerging upcycled functional ingredients in processed meat products designed to increase polyphenol and fiber contents. *Sustainable Food Technology, 3*(4), 861–874. https://doi.org/10.1039/d5fb00040h.

Grasso, S., Smith, G., Bowers, S., Ajayi, O. M., & Swainson, M. (2019). Effect of texturised soy protein and yeast on the instrumental and sensory quality of hybrid beef meatballs. *Journal of Food Science and Technology, 56*(6), 3126–3135. https://doi.org/10.1007/s13197-018-3552-9.

Gu, Y., Xv, W., Zheng, Q., Ye, Z., Guo, L., Lin, J., & Zou, Y. (2025). Effects of quality-enhancing additives on texture and digestibility of *Pleurotus geesteranus* mycelia-based meat analogues. *LWT – Food Science and Technology, 214*, 118728. https://doi.org/10.1016/j.lwt.2025.118728.

Haak, L., Raes, K., Van Dyck, S., & De Smet, S. (2006). Effect of dietary rosemary and a-tocopheryl acetate on the oxidative stability of raw and cooked pork following oxidized linseed oil administration. *Meat Science, 78*(3), 239–247.

Han, J. H., Keum, D. H., Kothuri, V., Kim, Y.-J., Kwon, H. C., Kim, D. H., Jung, H. S., & Han, S. G. (2024). Enhancing emulsion, texture, rheological and sensory properties of plant-based meat analogs with green tea extracts. *Food Chemistry: X, 24*, 101807. https://doi.org/10.1016/j.fochx.2024.101807.

Hang, S., Zhao, Q., Luo, X., Tang, C., Liu, S., & Liu, X. (2023). Development of oleogel-based fat substitutes for improving texture and flavor in plant-based meat products. *Food Research International, 171*, 113008. https://doi.org/10.1016/j.foodres.2023.113008.

Haque, M. E., Mollah, M. B. R., Habib, M., Hashem, M. A., Azad, M. A. K., & Ali, M. S. (2024). Development of dietary fiber enriched chicken nugget using rice bran. *Meat Research, 4*(1), 82. https://doi.org/10.55002/mr.4.1.82.

Hartley, C., Carr, A., Bowe, S. J., Bredie, W. L. P., & Keast, R. S. J. (2022). Maltodextrin-based carbohydrate oral rinsing and exercise performance: Systematic review and meta-analysis. *Sports Medicine, 52*(8), 1833–1862. https://doi.org/10.1007/s40279-022-01658-3.

He, Q., Li, X., Liu, C., Liu, Z., Zhou, S., Li, Y., & Ma, T. (2026). Application and research progress of dietary fiber-based fat substitutes in food systems. *Food Research International, 223*(2), 117876. https://doi.org/10.1016/j.foodres.2025.117876.

He, X., Lu, W., Sun, C., Khalesi, H., Mata, A., Andaleeb, R., & Fang, Y. (2021). Cellulose and cellulose derivatives: Different colloidal states and food-related applications. *Carbohydrate Polymers, 255*, 117334. https://doi.org/10.1016/j.carbpol.2020.117334.

Hocquette, J.-F., Gondret, F., Baéza, E., Médale, F., Jurie, C., & Pethick, D. W. (2010). Intramuscular fat content in meat-producing animals: Development, genetic and nutritional control, and identification of putative markers. *Animal, 4*(2), 303–319. https://doi.org/10.1017/S1751731109991091.

House, J. (2016). Consumer acceptance of insect-based foods in the Netherlands: Academic and commercial implications. *Appetite, 107*, 47–58. https://doi.org/10.1016/j.appet.2016.07.023.

Hunter, R. W., Dhaun, N., & Bailey, M. A. (2022). The impact of excessive salt intake on human health. *Nature Reviews Nephrology, 18*(5), 321–335. https://doi.org/10.1038/s41581-021-00533-0.

Hussain, M. A., Li, L., Kalu, A., Wu, X., & Naumovski, N. (2025). Sustainable food security and nutritional challenges. *Sustainability, 17*(3), 874.

Ibrahim, O. O. (2018). D-Tagatose a new low-calorie sweetener from lactose in cheese whey as a nutraceutical value-added product. In Ö. Tokuşoğlu (Ed.). *Food by-product based functional food powders* (pp. 185–202)(1st ed.). Taylor & Francis.

Ilo, S., et al. (2014). Amylose-lipid complex formation during extrusion cooking: Effect of added lipid type and amylose level on corn-based puffed snacks. *International Journal of Food Science & Technology, 49*(2), 309–318. https://doi.org/10.1111/ijfs.12829.

Ishaq, A., Irfan, S., Sameen, A., & Khalid, N. (2022). Plant-based meat analogs: A review with reference to formulation and gastrointestinal fate. *Current Research in Food Science, 5*, 973–983. https://doi.org/10.1016/j.crfs.2022.06.001.

Issara, K., Ruangchakpet, A., Sangsawad, P., & Suwannachot, K. (2024). Utilization of banana peel flour and textured vegetable protein in plant-based patties. *Heliyon, 10*(4), e24673. https://doi.org/10.1016/j.heliyon.2024.e24673.

Jandyal, M., Malav, O. P., Mehta, N., & Wagh, R. V. (2022). Quality characteristics of functional pork sausages incorporated with oat bran powder. *Journal of Meat Science, 17*(1), 1–7. https://doi.org/10.48165/jms.2022.1701.

Ji, W., Chen, F., Chen, Z., & Jiang, H. (2024). Research advances in the bioactivity of plant polyphenols. *International Journal of Food Science & Technology, 59*(11), 8037–8044. https://doi.org/10.1111/ijfs.17494.

Kaderides, K., Goula, A. M., & Adamopoulos, K. G. (2015). A process for turning pomegranate peels into a valuable food ingredient using ultrasound-assisted extraction and encapsulation. *Innovative Food Science and Emerging Technologies, 31*, 204–215.

Kamil, M. M., Mohamed, G. F., Mohamed, G., Ali, H. S., Ragb, G. H., & Zaky, A. A. (2023). Evaluation of inulin as a fat replacer in meat burger. *Egyptian Journal of Chemistry, 67*(3), 13–23. https://doi.org/10.21608/ejchem.2023.213926.8039.

Karslıoğlu, B., Soncu, E. D., Nekoyu, B., Karakuş, E., Bekdemir, G., & Şahin, B. (2024). From waste to consumption: Tomato peel flour in hamburger patty production. *Foods, 13*(14), 2218. https://doi.org/10.3390/foods13142218.

Keum, S., Lee, H. Y., Kim, Y., & Park, J. (2024). Development of pickering emulsions stabilized by polysaccharide–protein complexes for improving oil retention and sensory properties in plant-based patties. *Food Hydrocolloids, 152*, 108129. https://doi.org/10.1016/j.foodhyd.2024.108129.

Khorasaniha, R., Olof, H., Voisin, A., Armstrong, K., Wine, E., Vasanthan, T., & Armstrong, H. (2023). Diversity of fibers in common foods: Key to advancing dietary research. *Food Hydrocolloids, 139*, 108495. https://doi.org/10.1016/j.foodhyd.2023.108495.

Khosravani, M., Nazari, S. M., & Asadi, G. H. (2024). Evaluation of chemical and sensory properties of low-fat German sausages produced by maltodextrin. *Iranian Food Science and Technology Research Journal, 20*(1), 75–83. https://doi.org/10.22067/ifstrj.2023.76773.1172.

Kim, D. E., Kalaitzandonakes, M., & Ellison, B. (2025). Added protein, added value? Consumer demand for protein labels across product types, flavors, and consumer segments. *Food Quality and Preference*, 105842. https://doi.org/10.1016/j.foodqual.2025.105842.

Korhonen, H., & Pihlanto, A. (2006). Bioactive peptides: Production and functionality. *International Dairy Journal, 16*(9), 945–960. https://doi.org/10.1016/j.idairyj.2005.10.012.

Kos, I., Bendelja Ljoljić, D., Vrdoljak, I., Glavaš, V., Kovačević, N., Pleadin, J., & Vnučec, I. (2023). Glutamate in meat processing – origin, function and novel application. *Journal of Central European Agriculture, 24*(3), 624–633. https://doi.org/10.5513/JCEA01/24.3.3940.

Kozlu, A., Ngasakul, N., Klojdová, I., & Baigts-Allende, D. K. (2024). Edible insect-processing techniques: A strategy to develop nutritional food products and novelty food analogs. *European Food Research and Technology, 250*, 1253–1267. https://doi.org/10.1007/s00217-024-04474-3.

Kumar, A., Mehan, S., Tiwari, A., Khan, Z., Gupta, G. D., Narula, A. S., & Samant, R. (2024). Magnesium (Mg2+): Essential mineral for neuronal health: From cellular biochemistry to cognitive health and behavior regulation. *Current Pharmaceutical Design, 30*(39), 3074–3107. https://doi.org/10.2174/0113816128321466240816075041.

Kumari, D., Kumar, R., Sharma, K., Thakur, K., Mahajan, D., Kumari, H., & Pankaj, P. P. (2023). Physiological effects of monosodium glutamate—a food additive—on human health: A review. *Journal of Experimental Zoology India, 26*(2), https://doi.org/10.51470/jez.2023.26.2.0000.

Lau, D. S. A., Chieng, H. E., Wang, C., Li, Z., Zhang, X., Chow, D. H. K., ... Lu, W. W. (2023). Difference of osteoporotic bone fracture healing between pure magnesium and magnesium chloride ($MgCl_2$) enriched calcium sulphate/phosphate cement (CSPC). *Materialia, 28*, 101743. https://doi.org/10.1016/j.mtla.2023.101743.

Laufenberg, G., Kunz, B., & Nystroem, M. (2003). Transformation of vegetable waste into value-added products: A review. *Bioresource Technology, 87*(2), 167–198. https://doi.org/10.1016/S0960-8524(02)00167-0.

Lee, J.-S., & Han, J. (2024). Exploring the potential of bacterial cellulose paste as a fat replacer for low-fat plant-based hamburger patties. *Food Research International, 176*, 113832. https://doi.org/10.1016/j.foodres.2023.113832.

Lee, J., Song, H., Seo, K.-H., Lee, H.-G., Kim, H., & Choi, M.-J. (2023). Physicochemical and sensory properties of plant-based meat patties using oil-in-water emulsions. *Food Bioscience, 56*, 103084. https://doi.org/10.1016/j.fbio.2023.103084.

Lee-Richards, K. (2014). The most powerful natural antioxidant discovered to date—Hydroxytyrosol. Pro-Health, Available online: http://www.prohealth.com/library/print.cfm?libid=17054.

Lemke, B., Siekmann, L., Grabowski, N. T., Plötz, M., & Krischek, C. (2023). Impact of the addition of *Tenebrio molitor* and *Hermetia illucens* on the physicochemical and sensory quality of cooked meat products. *Insects, 14*(5), 487. https://doi.org/10.3390/insects14050487.

Lin, Q., Jiang, L., Li, X., Sang, S., Ji, H., Jin, Z., & Qiu, C. (2024). Starch based fat replacers in food system: Modification, structured design, and application. *Food Bioscience, 59*, 104149. https://doi.org/10.1016/j.fbio.2024.104149.

Liu, T., Lei, H., Zhen, X., Liu, J., Xie, W., Tang, Q., et al. (2024). Advancements in modifying insoluble dietary fiber: Exploring the microstructure, physicochemical properties, biological activity, and applications in food industry—A review. *Food Chemistry, 458*, 140154. https://doi.org/10.1016/j.foodchem.2024.140154.

Loï, C., & Cynober, L. (2022). Glutamate: A safe nutrient, not just a simple additive. *Annals of Nutrition and Metabolism, 78*(3), 133–146. https://doi.org/10.1159/000522482.

Lyu, B., Wang, H., Swallah, M. S., Fu, H., Shen, Y., Guo, Z., ... Jiang, L. (2021). Structure, properties and potential bioactivities of high-purity insoluble fibre from soybean dregs (okara). *Food Chemistry, 364*, 130402. https://doi.org/10.1016/j.foodchem.2021.130402.

Maningat, C. C., Jeradechachai, T., & Buttshaw, M. R. (2021). Textured wheat and pea proteins for meat alternative applications. *Cereal Chemistry, 99*(1), 37–66. https://doi.org/10.1002/cche.10503.

Maqsood, S., Baig, M. A., Ajayi, F. F., Hamdi, M., Baba, W., Brishti, F. H., ... Maqsood, S. (2024). Recent research advances in meat analogues: A comprehensive review on processing, nutritional and functional aspects. *Food reviews international.* Advance online publication. https://doi.org/10.1080/87559129.2024.2396855.

Marczak, A., & Mendes, A. C. (2024). Dietary fibers: Shaping textural and functional properties of processed meats and plant-based meat alternatives. *Foods, 13*(12), 1952. https://doi.org/10.3390/foods13121952.

Marinova, D., & Bogueva, D. (2022). Flexitarianism. *Food in a planetary emergency.* Singapore: Springer Nature. 173–188.

Massoud, R., & Hashem, M. (2023). Monk fruit (Siraitia grosvenorii) as a natural sweetener and functional food ingredient: A review. *Critical Reviews in Food Science and Nutrition.*

Meldrum, O. W., & Yakubov, G. E. (2024). Journey of dietary fiber along the gastrointestinal tract: Role of physical interactions, mucus, and biochemical transformations. *Critical Reviews in Food Science and Nutrition, 65*, 4264–4292. https://doi.org/10.1080/10408398.2023.2241234.

Meng, Y., Xie, L., You, K., & Chen, W. (2025). Improving the aromatic profile of plant-based meat alternatives: Effects of leghemoglobin and myoglobin addition on volatiles. *Food Chemistry, 492*, 145299. https://doi.org/10.1016/j.foodchem.2025.145299.

Merra, E., Calzaretti, G., Bobba, A., Storelli, M. M., & Casalino, E. (2014). Antioxidant role of hydroxytyrosol on oxidative stress in cadmium-intoxicated rats: Different effect in spleen and testes. *Drug and Chemical Toxicology, 37*(4), 420–426.

Moñino, M. I., Martínez, C., Sotomayor, J. A., LafuenteJordán, A., & Jordán, M. J. (2008). Polyphenolic transmission to segureño lamb meat from ewe diet supplemented with the distillate from rosemary (Rosmarinus officinalis) leaves. *Journal of Agricultural and Food Chemistry, 56*, 3363–3367.

Mounir, S., Mohamed, R., Sunooj, K. V., El-Saidy, S., & Farid, E. (2025). Assessing the effects of partially substituting chicken breast meat with oyster mushroom stalk powder on the quality attributes of mushroom-chicken burgers. *Scientific Reports, 15*(1), 4361. https://doi.org/10.1038/s41598-025-86127-3.

Mumyapan, M., Aktaş, N., & GerÇekaslan, K. E. (2022). Seed pumpkin flour as a dietary fiber source in Bologna-type sausages. *Journal of Food Processing and Preservation, 46*, e16586. https://doi.org/10.1111/jfpp.16586.

Nayak, B., & Panda, B. P. (2016). Modelling and optimization of texture profile of fermented soybean using response surface methodology. *AIMS Agriculture and Food, 1*(4), 409–418. https://doi.org/10.3934/agrfood.2016.4.409.

Nichita, A., Sordini, B., Al-Olayan, E., Esposto, S., Costanzi, E., Cenci-Goga, B., ... Veneziani, G. (2025). Physicochemical, microbiological and sensory evaluation of plant-based meat analogs supplemented with phenolic extracts from olive mill by-products. *Foods, 14*(19), 3347. https://doi.org/10.3390/foods14193347.

Nieto, G., Díaz, P., Bañón, S., & Garrido, M. D. (2010). Dietary administration of ewe diets with a distillate from rosemary leaves (Rosmarinus officinalis L.): Influence on lamb meat quality. *Meat Science, 84*, 23–29.

Nikolić, I., Šoronja-Simović, D., Zahorec, J., Dokić, L., Lončarević, I., Stožinić, M., & Petrović, J. (2024). Polysaccharide-based fat replacers in the functional food products. *Processes, 12*(12), 2701. https://doi.org/10.3390/pr12122701.

Novakovic, S., Djekic, I., Klaus, A., Vunduk, J., Đorđević, V., Tomovic, V., ... Tomasevic, I. (2020). Application of porcini mushroom (*Boletus edulis*) to improve the quality of frankfurters. *Journal of Food Processing and Preservation, 44*(8), e14556. https://doi.org/10.1111/jfpp.14556.

Nurmilah, S., Cahyana, Y., Utama, G. L., & Aït-Kaddour, A. (2022). Strategies to reduce salt content and its effect on food characteristics and acceptance: A review. *Foods, 11*(19), 3120. https://doi.org/10.3390/foods11193120.

Nurymkhan, G., Kalibekkyzy, Z., Orynbekov, D., Assenova, B., Kambarova, A., Dautova, A., et al. (2025). The effect of a multi-component plant supplement on the nutritional value of meat patties. *Processes, 13*(10), 3171. https://doi.org/10.3390/pr13103171.

Oliveira Ribeiro, W., Midori Ozaki, M., dos Santos, M., Rodríguez, A. P., Soares de Castro, R. J., Harumi Sato, H., Bastianello Campagnol, P. C., & Rodrigues Pollonio, M. A. (2023). Improving the textural and nutritional properties in restructured meat loaf by adding fibers and papain designed for elderly. *Food Research International, 165*, 112539. https://doi.org/10.1016/j.foodres.2023.112539.

Opaluwa, C., De Angelis, D., Summo, C., & Karbstein, H. P. (2025). Effect of different vegetable oils on extruded plant-based meat analogs: Evaluation of oxidative degradation, textural, rheological, tribological, and sensory properties. *Food Hydrocolloids, 163*, 111038. https://doi.org/10.1016/j.foodhyd.2024.111038.

Paglarini, C. D. S., Vidal, V. A. S., Ozaki, M. M., Ribeiro, A. P. B., Bernardinelli, O. D., Câmara, A. K. F. I., ... Pollonio, M. A. R. (2022). Inulin gelled emulsion as a fat replacer and fiber carrier in healthier Bologna sausage. *Food Science and Technology International, 28*(1), 3–14. https://doi.org/10.1177/1082013220980586.

Pakseresht, A., Ahmadi Kaliji, S., & Canavari, M. (2022). Review of factors affecting consumer acceptance of cultured meat. *Appetite, 170*, 105829. https://doi.org/10.1016/j.appet.2021.105829.

Patel, S., & Navale, A. (2024). The natural sweetener stevia: an updated review on its phytochemistry, health benefits, and anti-diabetic study. *Current Diabetes Reviews, 20*(2), 26–36. https://doi.org/10.2174/1573399819666230501210803.

Periferakis, A.-T., Periferakis, A., Periferakis, K., Caruntu, A., Badarau, I. A., Savulescu-Fiedler, I., ... Caruntu, C. (2023). Antimicrobial properties of capsaicin: Available data and future research perspectives. *Nutrients, 15*(19), 4097. https://doi.org/10.3390/nu15194097.

Petrat-Melin, B., & Dam, S. (2023). Textural and consumer-aided characterisation and acceptability of a hybrid meat and plant-based burger patty. *Foods, 12*, 2246. https://doi.org/10.3390/foods12122246.

Pöri, P., Lille, M., Edelmann, M., Aisala, H., Santangelo, D., Coda, R., & Sozer, N. (2023). Technological and sensory properties of plant-based meat analogues containing fermented sunflower protein concentrate. *Future Foods, 8*, 100244. https://doi.org/10.1016/j.fufo.2023.100244.

Prasannarong, M., Saengsirisuwan, V., Surapongchai, J., Buniam, J., Chukijrungroat, N., & Rattanavichit, Y. (2019). Rosmarinic acid improves hypertension and skeletal muscle glucose transport in angiotensin II-treated rats. *BMC Complementary and Alternative Medicine, 19*(1), 165. https://doi.org/10.1186/s12906-019-2579-4.

Qin, Y.-Q., Wang, L.-Y., Yang, X.-Y., Xu, Y.-J., Fan, G., Fan, Y.-G., ... Li, X. (2023). Inulin: Properties and health benefits. *Food & Function, 14*(7), 2948–2968. https://doi.org/10.1039/D2FO01096H.

Qin, X., Wang, W., Zheng, Y., Wang, X., Chen, Z., Zhao, J., & Li, S. (2025). Effects of non-thermal processing techniques on cereal bran: Focus on changes in bran structure and polyphenol release. *Food Chemistry: X, 29*, 102875. https://doi.org/10.1016/j.fochx.2025.102875.

Rana, S., Guptad, S., Rana, A., & Bhushan, S. (2015). Functional properties, phenolic constituents and antioxidant potential of industrial apple pomace for utilization as active food ingredient. *Food Science and Human Wellness, 4*, 180–187.

Ribeiro, J. C., Santos, C., Lima, R. C., Pintado, M. E., & Cunha, L. M. (2022). Impact of defatting and drying methods on the overall liking and sensory profile of a cereal bar incorporating edible insect species. *Future Foods, 6*, 100190. https://doi.org/10.1016/j.fufo.2022.100190.

Ribes, S., Aubry, L., Kristiawan, M., Jebalia, I., Dupont, D., Guillevic, M., & Santé-Lhoutellier, V. (2024). Fava bean (*Vicia faba* L.) protein concentrate added to beef burgers improves the bioaccessibility of some free essential amino acids after in vitro oral and gastrointestinal digestion. *Food Research International, 177*, 113916. https://doi.org/10.1016/j.foodres.2023.113916.

Richards, J., Lammert, A., Madden, J., Cahn, A., Kang, I., & Amin, S. (2024). Addition of carrot pomace to enhance the physical, sensory, and functional properties of beef patties. *Foods, 13*(23), 3910. https://doi.org/10.3390/foods13233910.

Roland, W. S., Pouvreau, L., Curran, J., van de Velde, F., & de Kok, P. M. (2017). Flavor aspects of pulse ingredients. *Cereal Chemistry, 94*(1), 58–65. https://doi.org/10.1094/CCHEM-06-16-0161-FI.

Rosli, W. I., Maihiza, N., & Raushan, M. (2015). The ability of oyster mushroom in improving nutritional composition, β-glucan and textural properties of chicken frankfurter. *International Food Research Journal, 22*(1), 311–317.

Sakr, M. S., Rahma, E. H., Abd El-Hakim, H. I., & Aly-Aldin, M. M. (2023). Utilization of fruits and vegetables by-products in the manufacture of beef burger. *Menoufia Journal of Food & Dairy Science, 8*(12), 121–136.

Salazar, D., Arancibia, M., Calderón, L., López-Caballero, M. E., & Montero, M. P. (2021). Underutilized green banana (Musa acuminata AAA) flours to develop fiber-enriched frankfurter-type sausages. *Foods, 10*(5), 1142. https://doi.org/10.3390/foods10051142.

Schiatti-Sisó, I. P., Quintana, S. E., & García-Zapateiro, L. A. (2023). Stevia (Stevia rebaudiana) as a common sugar substitute and its application in food matrices: an updated review. *Journal of Food Science and Technology, 60*(5), 1483–1492. https://doi.org/10.1007/s13197-022-05396-2.

Shahidi, F., & Ambigaipalan, P. (2015). Phenolics and polyphenolics in foods, beverages and spices: Antioxidant activity and health effects. *Journal of Functional Foods, 18*, 820–897. https://doi.org/10.1016/j.jff.2015.06.018.

Shi, A., Feng, X., Wang, Q., & Adhikari, B. (2020). Pickering and high internal phase Pickering emulsions stabilized by protein-based particles: A review of synthesis, application, and prospective. *Food Hydrocolloids, 109*, 106117. https://doi.org/10.1016/j.foodhyd.2020.106117.

Shi, C., Wang, L., Zhang, H., Tang, C., & Yin, S. (2021). Fermentation as a strategy to improve the nutritional and sensory quality of plant-based meat alternatives. *Food Research International, 148*, 110602. https://doi.org/10.1016/j.foodres.2021.110602.

Sindelar, J. J., & Milkowski, A. L. (2012). Human safety controversies surrounding nitrate and nitrite in the diet. *Nitric Oxide, 26*(4), 259–266.

Srivastava, S. (2025). Functional foods and spices in the management of metabolic syndrome. In R. K. Keservani, & A. Lohani (Vol. Eds.), *Nutraceuticals in obesity management and control: 3*, (pp. 73). Apple Academic Press.

Stanišić, N., Kurćubić, V. S., Stajić, S. B., Tomasevic, I. D., & Tomasevic, I. (2025). Integration of dietary fibre for health benefits, improved structure, and nutritional value of meat products and plant-based meat alternatives. *Foods, 14*(12), 2090. https://doi.org/10.3390/foods14122090.

Sun, G., Xiong, Y., Feng, X., & Fang, Z. (2022). Effects of incorporation of hempseed meal on the quality attributes of chicken sausage. *Future Foods, 6*, 100169. https://doi.org/10.1016/j.fufo.2022.100169.

Syan, V., Kaur, J., Sharma, K., Patni, M., Rasane, P., Singh, J., & Bhadariya, V. (2024). An overview on the types, applications and health implications of fat replacers. *Journal of Food Science and Technology, 61*(1), 27–38. https://doi.org/10.1007/s13197-022-05642-7.

Taladrid, D., Laguna, L., Bartolomé, B., & Moreno-Arribas, M. V. (2020). Plant-derived seasonings as sodium salt replacers in food. *Trends in Food Science & Technology, 99*, 194–202. https://doi.org/10.1016/j.tifs.2020.03.002.

Tan, H., Tan, T., & Easa, A. M. (2022). The use of salt substitutes to replace sodium chloride in food products: A review. *International Journal of Food Science & Technology, 57*(11), 6997–7007. https://doi.org/10.1111/ijfs.16075.

Tokarczyk, G., et al. (2023). Effect of Oyster Mushroom Addition on Improving the Sensory Properties, Nutritional Value and Increasing the Antioxidant Potential of Carp Meat Burgers. *Molecules, 28*, 6975.

Tomé, D. (2021). Yeast extracts: Nutritional and flavoring food ingredients. *ACS Food Science & Technology, 1*(4), 487–494. https://doi.org/10.1021/acsfoodscitech.0c00131.

Tzompa-Sosa, D. A., Dewettinck, K., Gellynck, X., & Schouteten, J. J. (2021). Replacing vegetable oil by insect oil in food products: Effect of deodorization on the sensory evaluation. *Food Research International, 141*, 110140. https://doi.org/10.1016/j.foodres.2021.110140.

Tzompa-Sosa, D. A., & Fogliano, V. (2017). Potential of insect-derived ingredients for food applications. In V. D. C. Shields (Ed.). *Insect physiology and ecology*. InTech. https://doi.org/10.5772/67304.

Varzakas, T., Zakynthinos, G., & Verpoort, F. (2016). Plant Food Residues as a Source of Nutraceuticals and Functional Foods. *Functional foods. Foods, 5*, 1–32.

Vilcapoma, W., de Bruijn, J., Elías-Peñafiel, C., Espinoza, C., Farfán-Rodríguez, L., López, J., & Encina-Zelada, C. R. (2023). Optimization of ultrasound-assisted extraction of dietary fiber from yellow dragon fruit peels and its application in low-fat alpaca-based sausages. *Foods, 12*(15), 2945. https://doi.org/10.3390/foods12152945.

Wang, L., Guo, H., Liu, X., Jiang, G., Li, C., Li, X., & Li, Y. (2019b). Roles of *Lentinula edodes* as the pork lean meat replacer in production of the sausage. *Meat Science, 156*, 44–51. https://doi.org/10.1016/j.meatsci.2019.05.016.

Wang, L., Li, C., Ren, L., Guo, H., & Li, Y. (2019a). Production of pork sausages using *Pleaurotus eryngii* with different treatments as replacements for pork back fat. *Journal of Food Science, 84*(11), 3091–3098. https://doi.org/10.1111/1750-3841.14839.

Wang, X. X., Wang, L. Y., Li, S. M., & Zhou, A. K. (2024). Amelioration of tiger nut insoluble dietary fiber as a partial substitute for fat in meat emulsions: Techno-functional properties and in vitro protein digestibility. *Food Research International, 197*(1), 115167. https://doi.org/10.1016/j.foodres.2024.115167.

Wang, D., Williams, B. A., Ferruzzi, M. G., & D'Arcy, B. R. (2013). Microbial metabolites, but not other phenolics derived from grape seed phenolic extract, are transported through differentiated Caco-2 cell monolayers. *Food Chemistry, 138*, 1564–1573.

Wendin, K., Olsson, V., & Langton, M. (2019). Mealworms as food ingredient—Sensory investigation of a model system. *Foods, 8*(8), 319. https://doi.org/10.3390/foods8080319.

Wimbasara, D. F., Wikandari, R., Setiowati, A. D., Sar, T., & Taherzadeh, M. J. (2025). Texture improvement of filamentous fungi burger derived from *Rhizopus oligosporus*: Impact of binding agent on physical, sensory, and microstructure. *Trends in Sciences, 22*(3), 8731. https://doi.org/10.48048/tis.2025.8731.

Witek, K., Wydra, K., & Filip, M. (2022). A high-sugar diet consumption, metabolism and health impacts with a focus on the development of substance use disorder: A narrative review. *Nutrients, 14*(14), 2940. https://doi.org/10.3390/nu14142940.

Wojtuś, M., Tomaszuk, S., & Wąsik, K. (2022). Polyols - what do we know about their impact on the gut microbiome? *Journal of Education, Health and Sport, 12*(12), 146–151. https://doi.org/10.12775/JEHS.2022.12.12.023.

Xie, C., Liu, G.-H., Liang, M.-H., Li, S.-H., & Kang, Z.-L. (2024). Applying resistant starch to improve the gel and water retention of reduced-fat pork batter. *Gels, 10*(5), 347. https://doi.org/10.3390/gels10050347.

Xu, X., Feng, M., Wei, T., Pan, F., Zhao, L., & Zhao, L. (2025). Edible insects as future proteins: Nutritional value, functional properties, bioactivities, and safety perspectives. *Nutrients, 17*(19), 3165. https://doi.org/10.3390/nu17193165.

Yitbarek, M. B. (2019). Livestock and livestock product trends by 2050: Review. *International Journal of Animal Research, 4*(30).
Yuan, D., Liang, X., Kong, B., Xia, X., Cao, C., Zhang, H., Liu, Q., & Li, X. (2024). Influence of seaweed dietary fibre as a potential alternative to phosphates on the quality profiles and flavour attributes of frankfurters. *Meat Science, 213*, 109511. https://doi.org/10.1016/j.meatsci.2024.109511.
Zając, M., Guzik, P., Kulawik, P., Tkaczewska, J., Florkiewicz, A., & Migdał, W. (2019). The quality of pork loaves with the addition of hemp seeds, de-hulled hemp seeds, hemp protein and hemp flour. *LWT – Food Science and Technology, 105*, 190–199. https://doi.org/10.1016/j.lwt.2019.02.013.
Zhang, F., Li, X., Liang, X., Kong, B., Sun, F., Cao, C., & Liu, Q. (2024). Feasibility of *Tenebrio molitor* larvae protein to partially replace lean meat in the processing of hybrid frankfurters: Perspectives on quality profiles and in vitro digestibility. *Food Research International, 176*, 113846. https://doi.org/10.1016/j.foodres.2023.113846.
Zhang, H., & Tsao, R. (2016). Dietary polyphenols, oxidative stress, and antioxidant and anti-inflammatory effects. *Current Opinion in Food Science, 8*, 33–42. https://doi.org/10.1016/j.cofs.2016.02.002.
Zhang, A., Yang, B., Xu, S., Liu, K., Dong, H., Pan, J., ... He, Q. (2025a). Emulsion gels based on soy protein–pectin complexes as fat replacers: Enhancing pork sausage quality through structural and molecular insights. *International Journal of Biological Macromolecules, 321*, 146331. https://doi.org/10.1016/j.ijbiomac.2025.146331.
Zhang, B., Zhu, J., Kontogiorgos, V., & Dhital, S. (2025b). Pectin: Chemical structure and health benefits. In B. Zhang, & S. Dhital (Eds.). *Carbohydrate nutrition* (pp. 237–267). Elsevier. https://doi.org/10.1016/B978-0-443-15700-4.00008-0.
Zhao, D., Guo, C., Liu, X., & Xiao, C. (2021). Effects of insoluble dietary fiber from kiwi fruit pomace on the physicochemical properties and sensory characteristics of low-fat pork meatballs. *Journal of Food Science and Technology, 58*, 1524–1537. https://doi.org/10.1007/s13197-020-04665-2.
Zhuang, J., Zhu, J., Cheung, P. C. K., & Li, C. (2024). The physical and chemical interactions between starch and dietary fiber: Their impact on the physicochemical and nutritional properties of starch. *Trends in Food Science & Technology, 149*, 104566. https://doi.org/10.1016/j.tifs.2024.104566.
Zou, X., Xu, X., Chao, Z., Jiang, X., Zheng, L., & Jiang, B. (2022). Properties of plant-derived soluble dietary fibers for fiber-enriched foods: A comparative evaluation. *International Journal of Biological Macromolecules, 223*, 1196–1207. https://doi.org/10.1016/j.ijbiomac.2022.11.008.

CHAPTER SIX

Nutritional adequacy and novel food products

Sher Ali[a,*], Nasir Rajput[b], Tahseen Fatima Miano[a], Asadullah Marri[a], Nida Shaikh[a], and Muhammad Ammar Khan[c]

[a]Institute of Food Sciences and Technology, Sindh Agriculture University, Tandojam, Sindh, Pakistan
[b]Department of Poultry Science, Faculty of Animal Husbandry & Veterinary Sciences, Sindh Agriculture University, Tandojam, Sindh, Pakistan
[c]Department of Food Science & Technology, Faculty of Agriculture & Environment, The Islamia University of Bahawalpur, Bahawalpur, Pakistan
*Corresponding author. e-mail address: sherali@sau.edu.pk

Contents

Abstract

Adequate nutrition is an essential requirement for health, growth, prevention of nutritional diseases and normal physiological functions throughout life. Although there has been significant progress in agriculture, food systems, nutrition science, food insecurity and dietary insufficiency continue to be a significant public health challenge globally. This might be due to factors including limited dietary diversity, socioeconomic inequalities, infectious diseases, climate change and poor governance. This chapter thoroughly explains the concept of nutritional adequacy, its definition, evolution over time and the components of an adequate diet. It also explains the recommended dietary allowances, dietary diversity, nutrient density, bioavailability and commonly used approaches for nutritional assessment. It also reviews the current state of nutritional

Advances in Food and Nutrition Research, Volume 122
ISSN 1043-4526, https://doi.org/10.1016/bs.afnr.2026.06.003

status worldwide and sheds light on key factors that are associated with nutritional inadequacy in various populations. Furthermore, it demonstrates the growing role of novel food products as an intervention to achieve nutritional adequacy by producing nutrient-rich, safe and sustainable food. Special attention is given to new functional ingredients such as algae, edible insects, yeast, fungi, legumes and wild plants. It also focused on novel food processing technologies such as food fortification, alternative protein innovation, precision fermentation and multidimensional food printing. Conclusively, the scientifically formulated novel food products are a promising approach to improve the nutritional quality of the diet, overcome nutrient deficiencies and provide sustainable nutrition goals to improve health of human population.

1. Introduction

1.1 Overview

The exact coinage of the term "nutritional adequacy" is not fully clear in the history of human nutrition. It is believed that the concept of nutritional adequacy has mainly evolved from the available scholarly literature that gradually gained popularity in the field of human nutrition and dietetics. Governmental bodies further emphasize the concept of nutritional adequacy during policy discussions and establishing regulations for health and nutrition. During the early 1990's, health professionals recognized diseases (such as scurvy, rickets and beriberi etc.) that were manifested due to the deficiency of some essential nutrients. This discovery opened debate among the health professionals on idea about "adequate" or "inadequate" diet. They started to understand the concept of an adequate diet that can supply essential nutrients to prevent an individual from deficiency diseases. In 1941, a milestone initiative was taken by U.S. scientific body, the Food and Nutrition board (FNB) and Recommended Dietary Allowances (RDA) for nutrient intake was introduced. Later, these recommendations by FNB properly structured the concept of "nutritional adequacy". The term "nutritional adequacy" finally emerged as a defined concept at broader spectrum in nutrition science during 20th century. Food and Agriculture Organization (FAO) and World Health Organization (WHO) also adopted and started the formal usage of the term in dietary standards from mid-20th century (Wretlind, 1982). In primitive times, food was cherished mainly for its role in providing energy and means for survival. Empirically, through experiences and observations people understand its value in keeping an individual physically and mentally healthy. Today, an adequate food can be referred to as "food that can fulfil the nutritional requirement of an individual to maintain overall health, growth, and normal bodily activity". Adequate food provides sufficient energy and essential nutrients (protein, vitamins, and minerals) on daily basis without causing malnutrition.

Typically, nutritional adequacy refers to as a provision of appropriate amount of essential nutrients such as carbohydrate, protein, fat, vitamins and minerals etc. in a diet (Miano & Dogan, 2021). Nutritional adequacy has been defined by various global organizations and nutritionists in marginally different perspectives depending upon their conceptual focus and insight knowledge. Some international food advisory bodies crucially define nutritional adequacy as per their own standards, recommendations and guidelines. The WHO states nutritional adequacy as "an intake of nutrients that fulfills the body's need for growth and maintenance, prevents deficiency diseases, and supports optimal health all through the life". FAO jointly with WHO states that "nutritional adequacy is a nutritionally adequate diet that facilitates availability of sufficient nutrients and energy to fulfil physiological needs of individuals, supports health, growth, and well-being." In FNB viewpoint "nutritional adequacy is an intake of nutrients that meets or exceeds the Estimated Average Requirement (EAR) or RDA to maintain health in an individual". European Food Safety Authority (EFSA, 2018) also defines nutritional adequacy as "condition in which a diet delivers all vital nutrients in sufficient amounts to maintain physiological functions, growth, and overall health". Moreover, according to Wretlind (1982), nutritional adequacy may be achieved when diet provides adequate nutrients to fulfil the fundamental dietary requirements of different population groups, in accordance with EFSA, WHO and FAO standards. At this point, understanding the term "nutritional inadequacy" is equally important. Nutritional inadequacy is a converse state of nutritional adequacy which refers to as a situation in which an individual or population is not receiving adequate amount of essential nutrients in the diet. Nutritional inadequacy induces detrimental impacts upon health, and the severity of the adverse effects depends on the extent of unavailability of essential nutrients in the diet of an individual. The consequences of nutritional inadequacy on health are extremely damaging (Kiani et al., 2022). It impairs health, results stunt growth, retards cognitive functions, leads to diet-related deficiency diseases, compromises bodily functions and increases the rate of morbidity and mortality.

Ensuring nutritional adequacy is crucial for good health and wellbeing but is quite challenging to be achieved in the modern context. Guaranteeing nutritional adequacy on global scale is a big complexity since several factors influence it globally such as food availability crises, food insecurity, economic inaccessibility of food, inadequate dietary practices and poor nutritional awareness etc. Ensuring nutritional adequacy involves result oriented coordinated actions that encompass agriculture, health, education and policy making etc. Among various actions, one specific action is intervention, i.e., novel food product development. This intervention can be implemented to effectively achieve nutritional adequacy worldwide (Baugreet et al., 2017). Novel food products (NFPs) are innovative foods in modern food system which can be developed by utilizing nutritionally rich ingredients by means of different processing techniques. NFPs can be produced by emerging technologies and processes. These innovative foods can significantly meet nutritional needs of individual or population.

Nowadays, the intervention of novel food product development is strongly rooted in food systems and human nutrition. Developing suitable NFPs may ensure appropriate nutritional intake of macro and micronutrients, availability of balanced diets, prevention from malnutrition (stunting, wasting, micronutrient deficiencies etc.) and chronic diseases, food security, sustainable nutritional goals and healthy population (Ribeiro et al., 2025). The production of NFPs is a complex process which delivers newly developed foods into marketplaces after organized ideation, concept development, ingredient sourcing, recipe formulation, prototype development, nutritional and safety assessment (Onwezen et al., 2021). The development of NFPs is like an intervention that provides diversity in nutritionally rich diets. However, the development of NFPs is an extremely crucial strategy for ensuring nutritional adequacy among consumers and requires meticulous, thorough, and careful evaluation of nutritional profile, safety assessment and consumer acceptance of newly developed food (Perez-Puyana et al., 2024; Ribeiro et al., 2025). Moreover, the development of NFPs requires a multidisciplinary team that usually comprises food scientists, food technologists/processors, food engineers, nutritionists/dieticians, food safety and supply chain experts etc. These professionals contribute to novel food product development from concept to tangible marketable products. This should be kept in consideration prior NFP development that nutritional demand of every individual varies for growth, development and physiological functions. This demand should be fulfilled by technologically enhancing the nutritional adequacy of food through fortification of specific essential nutrients in recommended concentrations (Bryant & Barnett, 2020; Fanta, 2015).

Across the regulatory frameworks worldwide, novel foods are referred to as newly developed foods for human diet, which are developed from advanced/unconventional/innovative technologies, or lacking a well-established record of safe consumption within a particular region. These products typically require pre-market safety evaluations to ensure safety from health risks. Numerous global food regulatory agencies have defined the term "novel food products" as per their respective viewpoint. According to EFSA, a novel food is any food that cannot be consumed to a significant degree by individuals in the European Union (before 15 May 1997) and requires safety and nutritional evaluation prior to market authorization. In authorization process, the EFSA assesses the nutrient profile of novel food, and the bioavailability of its nutrients alongside dietary impact. The Food Standards Agency (FSA) of United Kingdom also provides a similar definition of novel foods that is given by the EFSA. As per FSA, novel food was not extensively consumed in United Kingdom or European Union (before 15 May 1997) and needs pre-market authorization before sell in Great Britain. In definitions provided by EFSA and FSA, the date 15 May 1997 is crucial since this time indicates the reference point used to determine whether a food is "traditional food" or "novel food". In the EU, any food that was commonly consumed before 15 May 1997 is considered as traditional food and therefore does not require special approval whereas foods that were rare or not consumed significantly before 15 May 1997 are categorized as novel foods (Ververis et al., 2020). There is not any strict legal definition provided by FAO for novel foods. Instead, FAO states the concept that "foods or their ingredients availed from new sources or produced by innovative technologies that traditionally not consumed by humans". They may be insect-based foods, algae-based foods or cultured foods etc. Codex Alimentarius Commission of FAO and WHO provides an informal expression for novel foods that foods are obtained through modern technologies with new characteristics and require risk assessment prior consumption.

1.2 Current global situation of nutritional adequacy

Despite advancements in the field of agriculture, food system, and nutrition science, attaining goal of nutritional adequacy remains a critical challenge worldwide. As per the available global estimations by (FAO, 2024), around 733 M people across the world experienced hunger during the year 2023. This represents one individual among eleven worldwide and indicates that global population is still far from zero hunger/Sustainable Development Goal-2. Moreover, approximately 28 % of the global population (which comprised 2.3 B people) went through moderate to severe food insecurity during the year 2024 (FAO, 2025) and lacked access to adequate diet. Based on recent United Nation's reports, global estimations exhibit insufficient progress toward ending hunger (Sustainable Development Goal-2:Target 2.1). According to State of Food Security and Nutrition, during the year 2024 around 8.2 % of the global population experienced undernourishment. Global malnutrition of all forms also remains another serious challenge under Sustainable Development Goal-2:Target 2.2. Current estimations indicate that 23.2 % children under five years are stunted, 6.6 % are wasted while 5.5 % are overweight. Moreover, micronutrient deficiencies are extremely prevalent worldwide and nearly 1/3 of women of reproductive age are anemic. Dietary diversity on the other hand remains limited across many regions globally, merely 34 % of children and around 65 % of are women meeting minimum dietary diversity. Even though some progress has been perceived in certain areas, this progress is uneven and sometimes extremely insufficient to fulfil global nutrition targets by the year 2030. Malnutrition is undoubtedly a fundamental barrier force against achieving certain Sustainable Development Goals and is responsible for reduced productivity, increased healthcare costs, and slowed economic growth across the globe.

Based on above facts it can be stated that global challenges for the provision of nutritionally adequate nutrient rich diet are increasing exponentially since global food system has nearly failed in ensuring access to nutritionally balanced diet for all people across the world. Though global food production can supply enough calories to the global population, still many regions in the world lack sufficient dietary diversity and availability of nutrients which leads to a persistent nutrient gap (Beal et al., 2017). According to Global Burden of Nutritional Deficiencies report, sub-Saharan Africa, south Asian countries, Middle east and some North African countries are facing nutritional inadequacy. Table 1 presents information about top ten countries across the globe which are facing nutritional inadequacy. These countries reflect lack of access to sufficient food, caloric intake, and vital nutrients.

Global ongoing challenge of double burden malnutrition (coexistence of under and over nutrition in same population) is also one leading cause of nutritional inadequacy. Both undernutrition (such as stunting, wasting and micronutrient deficiencies) and overnutrition (overweight and obesity) are predominant issues in low and middle-income countries (Hasan et al., 2022). Poor people frequently suffer from undernutrition in their early life and later obesity due to the consumption of cheap, low-quality sugary and fat containing diets. Generally, undernutrition was predominant in countries that lie

Table 1 Top ten countries across the globe facing severe nutritional inadequacy.

Countries	Continent	Prevalence (%)	References
Samalia	Africa	51.3	Huey et al., Yilmaz and Yilmaz
Haili	North America (Caribbean)	50.4	Huey et al., Yilmaz and Yilmaz
Madagascar	Africa	39.7	Huey et al., Yilmaz and Yilmaz
Yemen	Asia (middle east)	39.5	Solon-Biet et al., Huey et al., Yilmaz and Yilmaz
Liberia	Africa	38.4	Huey et al., Yilmaz and Yilmaz
Zimbabwe	Africa	38.1	Huey et al., Yilmaz and Yilmaz
DR Congo	Africa	37.0	Solon-Biet et al., Huey et al., Yilmaz and Yilmaz
Uganda	Africa	36.9	Huey et al., Yilmaz and Yilmaz
Zambia	Africa	35.4	Huey et al., Yilmaz and Yilmaz
Chad	Africa	35.1	Solon-Biet et al., Huey et al., Yilmaz and Yilmaz

below the poverty line; however, obesity is now also rapidly increasing among poor people of these countries. This means that low and middle-income countries are simultaneously facing both issues of under and over nutrition (Kiosia et al., 2024). The nutritional transition is considered as a main driving force behind double burden malnutrition. Global trend of switching to a healthier diet (fiber and protein) and adopting highly processed, energy dense foods rich in sugar and fats is rising. Urbanization is perhaps also increasing the trend of consumption of ready-to-eat fast foods and sugary drinks which is further enhancing the challenge of double-burden malnutrition.

As per mainstream global food agencies, the United States of America exhibits significantly higher state of nutritional adequacy. Some major nutritionally vulnerable countries worldwide are facing significant challenges for ensuring nutritional adequacy. Table 2 summarizes the nutritional adequacy

Table 2 Information regarding nutritional adequacy comparison among selected countries and United States of America the table needs to be adjusted to make the same word in one line.

Country	Prevalence of undernourishment (PoU)/hunger	Child stunting (% <5)	Child wasting (% <5)	Anaemia — women (15–49)	Avg dietary energy supply (kcal/day per capita)	% population unable to afford a healthy diet (approx.)	Age groups most affected	Major diet-related diseases/notes
Pakistan	High — large vulnerable share (several reports show widespread food insecurity; many households vulnerable to acute food insecurity). (IPCInfo)	~37–42 % (children < 5 stunted — various sources: GNR/NNS show ~37–42 %). (Global Nutrition Report)	~7–18 % (wasting estimates vary by survey; GNR reports ~7.1 %; some surveys show higher local/seasonal peaks). (Global Nutrition Report)	~40–43%+(-women of reproductive age anaemia and widespread micronutrient deficiencies reported). (GAIN)	~2500–2900 kcal/day (FAO/national balance estimates; calorie supply rose in 2019–22 but distribution & access remain unequal). (pakalumni.com)	Very high — many estimates show majority cannot afford a fully healthy diet (news/Gain/FAO-derived analyses report very large shares — country variation by method; local reports cited ~ >60–80 %). (Associated Press of Pakistan)	Infants & children (0–24 months) — stunting and micronutrient deficits; Women of reproductive age — anaemia & micronutrient gaps; Rural poor & displaced groups — acute food insecurity. (UNICEF)	Micronutrient deficiencies (iron, vitamin D, zinc), child stunting; pockets of wasting and rising overweight/obesity in adults in urban groups. Climate shocks (floods) increase acute risk. (GAIN)

India	Moderate PoU (historic improvements but many still undernourished; Global Hunger Index & FAO show significant numbers). (Global Hunger Index)	~33 % (stunting remains high regionally). (Global Hunger Index)	~18 % (wasting—one of the highest country wasting rates). (Global Hunger Index)	Anaemia in women high (varies by survey — tens of %; national programs target reduction). (Global Nutrition Report)	~2400–2700 k-cal/day (national supply varies widely by state). (FAOHome)	High — a substantial share of the poor cannot afford healthy diets (FAO/World Bank estimates show large vulnerable groups). (Our World in Data)	Young children (stunting/ wasting) and adolescent girls / women (anaemia) are priority groups. (Global Hunger Index)	Child wasting and stunting, widespread anaemia among women and girls; at the same time rising adult overweight/ obesity and diet-related NCDs in urban/affluent groups. (Global Hunger Index)
Bangladesh	Moderate/ declining PoU but food & micronutrient insecurity remains. (Global Nutrition Report)	~28–31 % (stunting; progress but still high). (Global Nutrition Report)	~8–10 % (wasting). (Global Nutrition Report)	~36–42 % (women 15–49 — anaemia remains a public health problem). (Global Nutrition Report)	~2200–2700 k-cal/day (improving supply; regional/ seasonal gaps exist). (FAOHome)	High — many poorer households struggle to afford a healthy diet (FAO/World Bank estimates). (Our World in Data)	Children under 5 (stunting/ wasting) and pregnant women (anaemia/low birthweight). (Global Nutrition Report)	Undernutrition- + high anaemia burden; progress on breastfeeding but double burden (rising overweight in some groups). (Global Nutrition Report)

(continued)

Table 2 Information regarding nutritional adequacy comparison among selected countries and United States of America the table needs to be adjusted to make the same word in one line. *(cont'd)*

Country	Prevalence of undernourishment (PoU)/ hunger	Child stunting (% <5)	Child wasting (% <5)	Anaemia — women (15–49)	Avg dietary energy supply (kcal/day per capita)	% population unable to afford a healthy diet (approx.)	Age groups most affected	Major diet-related diseases/ notes
Ethiopia	High vulnerability to food insecurity (climate shocks, conflict affect access). (Global Nutrition Report)	~36–37 % (stunting — substantial). (Global Nutrition Report)	~6–7 % (wasting — variable by region/ season). (Global Nutrition Report)	Anaemia common in women (regional estimates high — WHO data). (Global Nutrition Report)	~2000–2400 k-cal/day (lower than many regions; wide rural/seasonal variation). (FAOHome)	Very high — a large share cannot afford a healthy diet (FAO/ OWID/World Bank pointers). (Our World in Data)	Young children and rural households — stunting/ wasting; pregnant women micronutrient risks. (Global Nutrition Report)	Child stunting & micronutrient deficiencies; pockets of acute malnutrition (wasting) during shocks; emerging overweight in urban elites is low but rising slowly. (Global Nutrition Report)

United States (USA)	Low prevalence of caloric undernourishment (food availability high) but food insecurity persists in low-income groups. (FAOHome)	<2–3 % (stunting in children is rare on national average). (Global Nutrition Report)	~<2–3 % (wasting uncommon nationally). (Global Nutrition Report)	Anaemia prevalence lower than South Asia/ Africa in general but present in subgroups. (World Health Organization)	~3400 kcal/day (high) — national dietary energy supply is high, but diet quality is uneven. (FAOHome)	Low overall — healthy diet affordability much better than low-income countries, but food insecurity pockets exist (SNAP recipients, food deserts). (Our World in Data)	Low prevalence of child undernutrition overall; higher risk groups: low-income children, some minority communities, elderly on fixed incomes. (CDC)	Major problem: overweight/ obesity and NCDs (high adult obesity ~40 % in latest CDC data), diet-related diabetes, cardiovascular disease. Undernutrition is rare at population level. (CDC)

comparison among selected countries and United states of America. Global nutritional inadequacy is affecting millions of people across continents by manifesting stunting, wasting and micronutrient deficiencies etc. The severity of nutritional inadequacy varies region to region due to several factors like poverty, climate change, and impaired food system etc. Addressing nutritional adequacy in effective way requires targeted interventions as per unique context of each continent. The situation of nutritional adequacy is different in different continents. The continent-wise global nutritional inadequacy is stated here:

1.2.1 Africa

Africa is the most severely affected continent regarding nutritional inadequacy worldwide. In Africa, nutritional adequacy is greatly intricate and varies across the continent. The situation of nutritional adequacy in Africa is influenced by certain cultural, environmental, economic, and geo-political factors. Previous records on nutritional adequacy in Africa have reflected a gradual transition from diverse and balanced diet to nutritionally deficient diet. During European colonial era (pre-independence of Africa) cash crops such as cotton, cocoa, coffee, palm oil, sugarcane, tea, tobacco etc. were introduced for expanding international trade. This almost vanished cultivation of traditional food crops, devastated local dietary diversity and led to dietary deficiencies across African nations. After colonial era downfall (post-independence era), newly independent nations of Africa established food policies for improving agriculture, emphasizing the production of staple crops and thus achieved a bit food self-sufficiency. Yet African nations were not fully succeeded to attain micronutrient gaps and protein-energy malnutrition.

African countries greatly experiencing nutritional inadequacy include Somalia, Madagascar, Liberia, Sudan, Chad, Niger, Burundi, and DR Congo. According to a report published by the United Nations, more than 20 % of the population (around 307 M people) in Africa is chronically undernourished due to consistently lacking enough calories in diet This data are from which year?. Moreover, 1 B people were unable to afford nutritionally adequate diet in the year 2024. Alongside, moderate to severe food insecurity is at its highest in most African countries particularly in Sub-Saharan regions (Mali, Sudan, DR Congo etc.). These projections suggest that by the year 2030, around 60 % of all chronically undernourished individuals across the globe will be in Africa, if trends continue. Economic instability, climate change, poor agricultural systems, food inflation, regional conflicts etc. remain main contributing factors of nutritional inadequacy in

African countries (CCARDESA, 2025). The AIDS/HIV epidemic has further worsened food security and nutritional adequacy in Southern and Eastern regions of Africa (Ojo et al., 2022). Food insecurity is linked with undernutrition which weakens immunity and complicates HIV management. This does not merely increase vulnerability to progression in disease but provokes chances of co-infections like tuberculosis, hepatitis etc.

Recent developments suggest that ongoing research activities and nutrition policies by different global institutions, UN agencies, and African Union can combat hidden hunger in deprived African countries. Interventions like dietary diversity, biofortification and food fortification can subsequently suppress the burden of undernutrition in Africa.

1.2.2 Asia

In most Asian countries, low income, economic or financial constraints remain most dominant drivers of nutritional inadequacy. Unlike Africa, the low-and middle-income Asian countries have higher prevalence of micronutrient deficiencies rather than chronic undernourishment. Altogether, the situation of nutritional adequacy in different Asian countries is complex and differ widely due to distinct economic status, climatic conditions, agricultural practices, food systems, multiple cultures, and health infrastructure etc. Therefore, nutritional adequacy in many regions of Asia is influenced by a complex interplay above different factors. Asia also faces a double burden of malnutrition, both undernutrition/micronutrient deficiencies coexist with overnutrition/obesity. Asia represents a notable proportion of the world's undernourished individuals due to its large population size; however providentially prevalence rates are far lower than Africa. In Aisa, the South Asia is most affected by food insecurity. Most nutritionally deprived South Asian countries include India, Pakistan, Bangladesh, Afghanistan, and Nepal. In these countries, undernourishment is highly prevalent while stunting and wasting are frequent among children. In many South Asian Countries, hundreds of millions of individuals do not get enough dietary energy per day. According to (FAO, 2024), these countries alone have around 281 M undernourished people making 40 % of the global population. Food insecurity, poor dietary diversity, infectious diseases and poverty are leading to challenging issues in many developing countries of South Asia. Rapid growth in population is also affecting dietary requirements while low-income individuals are dealing with imbalanced diets while affluent circles are overconsuming calories. According to (UNICEF, 2024), in South Asian Countries stunting and wasting are frequent among under five-year-old children. One baby in four born with low birth weight while half of the population of adolescent girls and women are anemic. Around 72 % South Asian population is suffering from various forms of micronutrient deficiencies while micronutrients like folic acid, iron, zinc, vitamin A are commonly lacking in diet causing anemia, impaired immunity, night blindness etc. Moreover, approximately 40 % of global undernourished or hungry population live in these countries. However, it is also on records that around 6.4 % undernutrition has declined in the past recent years from the continent. This situation indicates a slight progression towards alleviation of hunger, but an absolute burden of undernourishment or hunger is still massive.

Many other Asian countries are also facing conflict-driven food insecurities and persistent issues of famine such as Yemen, Myanmar, Cambodia, Loas etc. Some Asian countries like China, Thailand, Malaysia, Japan, South Korea etc. exhibit marginal levels of undernutrition due to rooted dietary diversity in their cuisine. Gulf countries such as United Arab Emirates, Qatar, and Saudi Arabia represent very small extent of undernutrition due to strong economic status. Iran on the other hand is that Asian country which comes under moderate-to-low undernutrition category. Global organizations such as WHO, World Food Program (WFP), UNICEF, (United Nations International Children's Emergency Fund, 2025), FAO, and World Bank (WB) etc. are playing their impressive role in countering this serious predicament of food insecurity in South Asia. Experts further suggest that promoting dietary diversity via nutrient rich agriculture, subsidizing nutritious food, controlling inflation on food commodities, and strengthening distribution system of food commodities can bring nutritional adequacy up to a certain extent in South Asian countries.

1.2.3 North and South America

North America denotes better signs of nutritional adequacy while issues like obesity, overnutrition, and diet-associated chronic diseases are yet persistent. In South America, a substantial share of urban populations exhibits adequate nutritional status. Instead, many low-income communities are facing crises of undernutrition and micronutrient deficiencies. Parts of south and north America like Latin America and Caribbean at present exhibit marginal nutritional adequacy. Nevertheless, these regions exhibit reduced rates of undernutrition than Africa and Asia. A significant burden of overnutrition or obesity is an additional challenge in these regions. As per previous records, during early 20th century Latin America and Caribbean, the diets were mainly comprised of traditional foods like cereals, tubers, fruits, vegetables and seafoods. Nutritional deficiencies had locally existed among rural indigenous populations (Llorca-Jaña et al., 2025) while socioeconomic situations like poverty, social inequality and scarcity of food commodities were among the prominent driving factors of nutritional inadequacy. Rapid industrialization and urban expansion begun after the middle of 20th century, but food systems remained weak and rate of hunger was frequent particularly among children. Later, government initiated modern agriculture practices, feeding programs, health interventions, economic growth policies etc. that dramatically declined food insecurity to an extent, however the issues of climate change, micronutrient deficiencies and urban obesity remained persistent.

Based on current reports by FAO, some prominent countries of Latin America and Caribbean are facing significant challenges of food insecurity. Ongoing crises of food insecurity in Haiti is extremely massive. A big share of population is confronting acute hunger, malnutrition, and inaccessibility of food due to internal political and economic crises. Other countries such as Guatemala, Honduras, El Salvador, Nicaragua, Venezuela etc. are facing challenges of food insecurity. In Colombia, moderate or severe food insecurity can be seen in remote rural regions despite overall local advancement. In many small Caribbean nations high-priced food and import dependency are among major causes of exacerbated nutritional inadequacy (Basurko et al., 2025).

1.2.4 Australia/ Oceania

The nutritional inadequacy across Oceania differs broadly. The prevalence of undernourishment in the Australia and New Zealand is extremely low while regions like Melanesia and some Pacific islands exhibit significant nutritional inadequacies (Kitchener et al., 2026). The situation of nutritional adequacy in some areas is compromised due to remoteness, dependency on imported foods, food commodity inflation, and reduced access to diverse dietary commodities etc. Regions with nutritional inadequacy in marginally deprived regions require peripheral actions such as strengthening system of local food production, improving food distribution infrastructure etc. to reduce nutritional inadequacy.

1.2.5 Europe

Europe typically shows the lowest prevalence of under-nutrition and therefore overall rates of nutritional adequacy are substantially above the global averages. Although food insecurity issues persist among low-income communities. These communities are facing emerging challenges such as poor diet quality, deficiencies of essential micronutrients, and uneven distribution of healthy foods. Minimal actions are sufficient to achieve overall nutritional adequacy across Europe by focusing on strengthening social welfare systems, by subsidizing food commodities, and by controlling obesity and other non-communicable diseases etc.

1.3 Factors contributing to nutritional inadequacy

Nutritional inadequacy is manifested through multidimensional factors. Generally, main contributing factors linked to promoting nutritional inadequacy in an individual or group of people include global demographic expansion due to accelerating population growth rate, malnutrition and related micronutrient deficiencies, prevalence of diabetes, obesity and other non-communicable diseases, climate change issues, and famine etc. Some socioeconomic determents further indirectly contribute to nutritional inadequacy among the people such as educational deprivation, unemployment, inflation, poor health facilities, war and terror situations etc. An inadequate distribution of food supplies, economic barriers, and inefficiencies in food systems makes it harder for vulnerable groups to obtain nutritious foods, even where overall food availability seems adequate (Ribeiro et al., 2025).

Besides this, in many food-insecure countries, some widely consumed staples lack essential nutrients and significantly contribute to global nutritional inadequacy. Moreover, due to modern agricultural practices such as soil depletion, refining processes and poor food storage practices, some widely consumed staple crops (wheat, rice, etc.) have declined in concentrations of key essential micronutrients such as iron, zinc, and important vitamins (Gupta et al., 2024). Global populations rely heavily on these crops and are suffering from caloric deficits and micronutrient deficiencies (Azeem et al., 2025). Some other complex barriers limiting the availability of adequate nutrition on a global scale include unnecessary mechanical processes (milling, refining, excess washing, peeling etc.), inefficient food processing techniques (deep frying, baking, high-temperature pasteurization, UHT etc.), poor storage and preservation methods (sun drying, canning, irradiation etc.). These food processing methods are crucially responsible for causing major loss in essential nutrients (Singh et al., 2023). Principle factors contributing to nutritional inadequacy worldwide are outlined as following.

1.3.1 Low-income socioeconomic status

Low-income households often remain resource limited. Poor socioeconomic status is linked with poverty and is the most consistent factor of nutritional inadequacy worldwide. It reduces access to safe, sufficient, and nutritious food, further restricts healthcare facilities, hinders attainment of education, and constrains sanitation (Katoch, 2022). Households with poor socioeconomic status hardly afford nutritious food and therefore mainly rely on low-cost poor-quality foods. Sustenance on nutritionally poor food leads to undernutrition and micronutrient deficiencies.

1.3.2 Maternal educational and nutritional status

An educated mother is aware of idea behind healthy diets and feeding practices. Education and awareness not merely enable her to know for own nutritional needs but ensure adequate nutrition for her children and family. Educated mothers adopt breastfeeding, and follow complementary-feeding and healthcare-seeking practices. Maternal education further signifies the value to hygiene, sanitation, and healthcare. In contrast, lower maternal education is typically liked with reduced dietary diversity and nutritional inadequacy (Shahid et al., 2021). Maternal malnutrition is directly linked to low birth weight. Women of reproductive age with inadequate nutritional status are more likely to give birth of undernourished children (Katoch, 2022).

1.3.3 Suboptimal breastfeeding and inadequate complementary feeding practices

Suboptimal breastfeeding together with inappropriate complementary feeding practices are main contributors to nutritional inadequacy among infants and young children. Both suboptimal breastfeeding and inappropriate complementary feeding practices influence nutrient needs of children during growth and developmental phases. Suboptimal breastfeeding refers to discontinued breastmilk feeding (Jiang et al., 2025) and reduced exclusivity (Huang et al., 2020). As per the recommendations of WHO, suboptimal breastfeeding does not meet recommended nutritional standards for infants and young children and contributes to the burden of disease, infection, malnutrition or mortality (Roberts et al., 2013). Too early or too late commencement of complementary foods, poor dietary diversity, and low meal frequency increase the risk of undernutrition (Ashraf et al., 2024). Traditional or cultural feeding practices further strongly influence optimal feeding routines (Makwela & Mashaba, 2025). Similarly, delayed breastfeeding, early breastfeeding termination enhances nutrient deficiencies among infants (Katoch, 2022).

1.3.4 Food insecurity and hidden hunger

Food insecurity refers to the state when individuals/households are deprived of access to enough safe and nutritious food required by them for sustaining a healthy life. This includes physical and economic limitations of availing food (Cassells et al., 2024). Food insecurity is not merely meant for insufficient caloric intake, but it considerably decreases dietary diversity and lifts nutrient inadequacy (Johnson et al., 2018). It leads people to heavily rely on inexpensive micronutrient-deficit staple diets. Systematic reviews on food insecurity state that food insecurity is strongly connected to micronutrient deficiencies since inaccessibility to nutrient-dense foods inhibits healthy dietary patterns (Lopes et al., 2023). Low dietary diversity is linked with high prevalence of stunting in infants and young children. This demonstrates how poor-quality foods render human health (Gassara & Chen, 2021). Hidden hunger is also one cause of nutritional adequacy among people. It is a form of malnutrition that occurs due to the deficiency of essential micronutrients even when calorie intake is adequate. The diet without dietary diversity manifests "hidden hunger" and increases susceptibility to faltering growth and impaired cognitive development. In fact, hidden hunger develops from nutritionally inadequate diet. Staple diets high in only one specific type of cereal limits diversity. Economic limitations also lift micronutrient deficiencies. Iron deficiency anemia (IDA) remains the common form of micronutrient deficiency and in prevalent in India (Natekar et al., 2022). In Sub-Saharan Africa, deficiency in vitamin A is a leading cause of higher child mortality and night blindness (Zhao et al., 2022). Goiter and some cognitive issues are common in Himalayan regions due to iodine deficiencies (Ali et al., 2024); however, the rate of zinc deficiency is higher in South Asian countries. These micronutrient deficiencies stem through diets dominating with staple foods, showing that adequate calories cannot guarantee proper nutrition.

1.3.5 Burden of infectious disease

Infectious diseases (such as cholera, tuberculosis, AIDS, hepatitis, malaria etc.) or pandemic situations contribute to nutritional inadequacy through vicious cycles or bidirectionally. Infection reinforces malnutrition and vice versa. When infection worsens nutritional status and causes malnutrition, this weakens immunity and further increases the risk of infectious diseases (Fan et al., 2022). Recurrence of infectious diseases on the other hand leads to stunting, wasting and micronutrient deficiencies. Persistent nutritional inadequacy due to repeated onsets of infectious diseases significantly contributes to higher mortality rates among vulnerable groups. The pandemic COVID-19 started in 2019 from Wuhan, China and lasts 2023. COVID-19 was pandemic and people across the world were affected due to contagious respiratory virus (SARS-CoV-2). During this pandemic situation, a great number of undernourished individuals were devastatingly suffered from COVID-19 due to impaired immunity (UNICEF, 2022). Not even this but hospitalized COVID-19 patients with undernutrition experienced higher rates of mortality in comparison to nourished counterparts. This exhibits a direct association between nutritional status and burden of infectious disease (Wei et al., 2020). Moreover, outbreak of measles (a contagious viral infection) in the Democratic Republic of the Congo during 2019–2020 highlighted how malnutrition may contribute to an increased rate of morbidity and mortality among vulnerable groups (Sodjinou et al., 2022).

1.3.6 Poor water, sanitation, and hygiene (WASH)

Safe drinking water, adequate sanitation and hygienic conditions are all strongly linked with nutritional adequacy. Contaminated drinking water may have variety of pathogenic agents (such as norovirus, hepatitis A virus, E-coli, salmonella, candida, Giardia lamblia, hookworm etc.) that cause infections, impaired nutrient absorption, increased nutrient loss, environmental enteric dysfunction and poor growth and development (Zavala et al., 2021). Epidemiological studies suggest that contaminated drinking water and inadequate sanitation correlate with higher prevalence of stunting and undernutrition among children (Rizaldi et al., 2025). Unsanitary environments cause repeated infections which affect gut health, weaken immune system and influence nutritional status (Gizaw & Worku, 2019). Poor WASH situations contribute to diarrhea and intestinal parasitic infections (Yaya et al., 2018) while repeated diarrhea decreases nutrient absorption and amplifies cycle of malnutrition and micronutrient deficiencies particularly in developing countries where WASH practices are poorly implemented. Available scholarly data from WHO/UNICEF Joint Monitoring Programs show that African countries like Eritrea, Niger, Chad, South Sudan, and Benin follow poorer wash practices and thus the prevalence of malnutrition is common among these countries (UNICEF, 2025). However, countries like Sweden, Japan and Germany since having robust WASH systems show adequate nutritional status among their people (WHO & UNICEF, 2025).

1.3.7 Gender and sociocultural inequalities

Sociocultural and gender inequalities remarkably affect nutritional adequacy by manipulating food availability, health care facilities and social pressures. At community level, gender bias or inequality is associated with high prevalence of low birth weight, stunting, wasting, and micronutrient deficiencies among female etc. suggesting women's limited access to schooling, health facilities, and economic participation (Marphatia et al., 2016). Gender disparities lead to unequal distribution of food (often inferior food quality and food portion) and health care among female children and women. Gender inequalities therefore reinforce undernutrition and are a key determinant of nutritional inadequacy (Shafiq et al., 2021).

Sociocultural inequalities represent an unequal distribution of social, cultural, and economic resources which often influence access of nutritious food and health facilities among the ignored societies. This creates disparities and manifests inadequate dietary practices. Most often certain cultural norms and poor socioeconomic status interfere with dietary diversity and contribute a crucial role in creating nutritional inadequacy (Alamirew et al., 2023).

1.3.8 Household size and demographic expansion

Large family size and increased population are closely related to nutritional inadequacy in various ways (Katoch, 2022). Both raise the demand for nutritious food and shrink overall basic livelihood resources. These also ultimately reduce quantity and quality of food, particularly among food-insecure societies. Larger families frequently confront higher food insecurity and lower dietary diversity among family members, which are two main indicators of nutritional inadequacy. Furthermore, demographic expansion or increased population growth is strictly associated with higher rates of undernourishment particularly in developing countries. It is obvious that when the number of individuals to feed raises rapidly than financial incomes, the quantity and quality of food becomes inadequate. Consequently, both household size and macro-demographic densities contribute to nutritional inadequacy by decreasing access to sufficient and nutrient-dense diets across the poor societies.

1.3.9 Poor governance and political determinants

Good governance is a broader structural factor that lifts nutritional inadequacy through establishing strong food systems, robust agriculture or economic policies, food laws and social equalities. However, poor governance, policy and legislative determinants, and inappropriate enforcement and compliance of policies etc. strongly influence nutritional adequacy (Biadgilign et al., 2019). According to United Nation's World Food Security report, inadequate governance structures together with impaired transparency and poor institutional accountability are key root causes of malnutrition worldwide and are indicators of nutritional inadequacy (FAO, 2025). In different poorly governed countries, around 282 M people are facing acute food insecurity due to policy system fragility (WFP, 2024).

1.3.10 War and terror situations

War and terror are no doubt contributing factors of nutritional inadequacy. Both can disrupt food systems, limit food supply chain, displace populations, and enhance vulnerability towards undernutrition and related issues. These conflicts often devastate agricultural production and lead to food crises, high prices of food commodities and thus limited access to adequate nutrition. During these situations, population displacement into refugee camps further increases the risk of nutritional inadequacy as the quality of food in these shelter camps is not even poor but also insufficient. In addition, stress and trauma during conflicts further contribute to the loss of appetite, digestive issues, and nutrient deprivation etc. It is suggested that children in conflict-affected areas experience higher rates of undernutrition/hunger in comparison to children in war- and- terror free regions (Carroll et al., 2017). Review from various war-torn countries revealed that conflicts severely destroyed food and healthcare systems and contributed to malnutrition. In regions like South Sudan, Yemen, Ukraine and Gaza large populations of vulnerable groups faced displacement and hunger due to disrupts agriculture, and food supply chain (Saber, 2026).

1.3.11 Natural disasters

Floods, landslides, earthquakes, storms, droughts etc. greatly interrupt food production, disturb food supply chains, and interfere with livelihoods. Natural disasters of various types affect food security and nutritional adequacy indirectly to a much extent (Tidiane Ndour et al., 2025). Flood destroys crops through waterlogging, impacts soil fertility, destroys storage facilities and leads to food scarcity and high prices of food commodities.

Drought or water scarcity halts growth in plants and increases rate of livestock mortality due to fodder scarcity. Storm or cyclones through strong gusts of wind destroy orchards and crops in fields. Conversely, earthquakes mainly affect infrastructures, damages irrigation, storage, and transport facilities which indirectly affect food security. Biological natural disasters (locust swarm, pest invasion, and plant diseases etc.) are kind of most destructive natural disaster caused by living organisms (insect, pest, rodents, virus, bacteria, fungi etc.). These can inflict extensive damage of food crops. Desert locust is severely destructive to food crops in many African and Asian countries (Barasa et al., 2022). Locust swarms can damage hundreds of acres of crops in a single day. An incident happened during the year 2020, when locust swarm outbreak affected crops in some East African and South Asian countries and subjected food security among millions of peoples (Newsom et al., 2021). Overall natural disasters are responsible for food system destruction and reducing food availability, dietary diversity, and nutritional adequacy. Therefore, populations distressed by these disasters experience under or overnutrition due to reliance on nutritionally poor-quality emergency foods (Farahmandnia et al., 2024; Nour & Altintaş, 2025).

2. Nutritional adequacy: Concepts and criteria

All human beings of different ages require a balanced diet loaded with adequate concentration of important nutrients that can meet bodily requirements for sustaining a healthy and physically active life. Some crucial concepts in this aspect are discussed below.

2.1 Components of adequate diet

An adequate diet is a complex combination of essential nutrients and provides sufficient energy to maintain overall health, growth, development, physiological and cognitive functions. A nutritious food is a complex matrix of carbohydrates, proteins, fats, natural organic acids (such as citric and malic acid), dietary fibers (like cellulose), vitamins, minerals, and phytochemicals. Nutritional adequacy is directly associated with both quantity and quality of these dietary components such as macro and micronutrients, fiber, water, and phytochemicals. The components of an adequate diet are described here.

2.1.1 Macronutrients

Caloric sufficiency depends on eating appropriate proportion of carbohydrate, protein and fat in the diet. An adequate caloric intake fulfills the need of energy needs for different cellular activities, growth, voluntary and involuntary bodily functions. Not even this, but caloric sufficiency regulates basal metabolism. Macronutrients, such as carbohydrates and fats, are primary energy-yielding nutrients needed in sufficient amounts. Conversely, protein is considered a secondary source or last resort of energy in the body since its main role is supplying amino acids for building muscle mass, tissue repair, regulating metabolism (as enzymes and hormone) etc. Carbohydrates provide around 50–55 % of total energy intake from complex forms by releasing glucose. Fats and oils are concentrated energy sources and are also needed for constructing cell membranes, synthesizing hormone etc. (Bray & Siri-Tarino, 2016).

2.1.2 Micronutrients

Micronutrients (vitamins and minerals) are needed in trace amounts, but are essential for many metabolic functions (Shergill-Bonner, 2017). Both fat-and water-soluble vitamins are obtained from diet which support growth, immune function, vision cycle, and metabolic reactions. Minerals are important for bone health, neurotransmission, enzymatic activities, and fluid balance, etc. An inadequate micronutrient intake can lead to "hidden hunger" or micronutrient deficiencies. Hence, nutritional adequacy requires nutrient density not merely caloric intake. A satisfactory intake of micronutrients prevents deficiency-diseases and promotes metabolic activities (Table 3).

2.1.3 Dietary fiber and water

Dietary fiber and water are crucial parts of adequate diet. Dietary fibers are non-digestible forms of carbohydrates. They support process of digestion, regulate blood glucose level, prevent chronic diseases, and act as laxatives (Anderson et al., 2009). Water is vital for cellular hydration, metabolic activities, temperature regulation, and overall homeostasis (Rani et al., 2018). Both dietary fibers and water are essential constituents of functionally adequate diet. Dietary fiber is very important and its key role is largely researched in recent years. I would suggest the authors enlarge the contents of this part.

2.1.4 Phytochemicals

Phytochemicals are crucial health-promoting components that can be obtained from plant-based foods. They are secondary metabolites biosynthesized in the plants in response to several external environmental

Table 3 Sources and roles of different micronutrients in maintaining nutritional adequacy and metabolic/ physiological activities

Micronutrients	Deficiency diseases	Metabolic/ physiological activities	Dietary sources
Vitamins (fat and water-soluble)			
A (Retinol)	Xerophthalmia, night blindness and impaired immunity	Vision cycle, maintenance of epithelial cells and immunity	Carrots, spinach, sweet potatoes, liver and eggs
D (D2 Ergocalciferol, D3 =Cholecalciferol)	Osteomalacia in adults and rickets in children	bone mineralization, absorption of minerals like calcium and phosphorus	Exposure to sunlight, vitamin D fortified milk, egg yolk and fish liver oil
E (Tocopherols)	Neuropathy and hemolytic anemia	Antioxidative functions, protection of cellular membranes	Seed oils, nuts and green leafy vegetables
K (K1 =Phylloquinone, K2 =Menaquinones)	Frequent bruising and bleeding disorders	Clotting of blood and bone metabolism	Fermented foods, Green leafy vegetables and broccoli
C (Ascorbic acid)	Bleeding gums (scurvy) and delayed wound healing	Collagen biosynthesis, antioxidative action and iron absorption	Guava, citrus fruits, tomatoes and peppers
B1 (Thiamine)	Beriberi and Wernicke-Korsakoff	Energy generation through carbohydrate metabolism and nerve function	Whole cereal grains, legumes and nuts

B2 (Riboflavin)	Ariboflavinosis such as cheilitis, glossitis etc. and eye disorders	Energy metabolism, and enzyme cofactor	Green vegetables, liver, milk and eggs
B3 (Niacin)	Pellagra	Metabolism of NAD/NADP and repairing of DNA	Legumes, nuts, meat, fish and poultry
B5 (Pantothenic acid)	Numbness and fatigue	Synthesis of Coenzyme A and metabolism of fatty acids	Legumes, cereal grains, meat and eggs
B6 (Pyridoxine)	neurological disorders, anemia and dermatitis	Metabolism of amino acids and synthesis of neurotransmitters	Banana, potatoes, meat, fish and poultry
B7 (Biotin)	hair loss and skin problems	Synthesis of fatty acids and metabolism of carbohydrate	Legumes, whole cereal grains, eggs and nuts
B9 (Folic acid)	Anemia (megaloblastic) and neural tube defect in fetal stage	Synthesis of nucleic acids and red blood cells	Legumes, green leafy vegetables and liver
B12 (Cobalamin)	Anemia (pernicious) and neurological disorders	Synthesis of red blood cell and DNA, nerve functioning	Dairy products, meat, fish and poultry

(continued)

Table 3 Sources and roles of different micronutrients in maintaining nutritional adequacy and metabolic/ physiological activities (*cont'd*)

Micronutrients	Deficiency diseases	Metabolic/ physiological activities	Dietary sources
Macrominerals (needed in larger proportions)			
Calcium	Osteomalacia, rickets and osteoporosis	Formation of bone and teeth, muscle contraction and nerve function	Milk, cheese, yogurt, leafy greens
Phosphorus	bone pain and weakness	Mineralization of bone, ATP synthesis and energy metabolism	Meat, dairy, eggs, legumes, nuts
Magnesium	Arrhythmias, fatigue and muscle cramps	Action as an enzyme cofactor, muscle/nerve activity and energy metabolism	Nuts, seeds, green vegetables, whole grains
Sodium	Hyponatremia	Body fluid balancing, nerve impulsion, contraction of muscles	Salt, processed foods
Potassium	Hypokalemia	Nerve impulsion, contraction of muscles and fluid balancing	Bananas, oranges, potatoes, beans
Chloride	Metabolic alkalosis	Synthesizes gastric acid and fluid balancing	Salt, vegetables, seafood

Microminerals (needed in traces)			
Iron	IDA (iron deficiency anemia)	Oxygen transport, synthesis of hemoglobin and myoglobin	Legumes, spinach red meat and liver
Zinc	Poor wound healing, retard growth and hair loss	Action as an enzyme cofactor, boosts immunity, growth and wound healing	Nuts, seeds, whole cereal grains, meat and seafood
Iodine	Hypothyroidism, goiter and mental retardation	Synthesis of thyroid hormone and regulation of metabolism	Fortified or Iodized salt, sea and dairy foods
Selenium	Cardiomyopathy and Keshan disease	Antioxidative action and thyroid function	Brazil nuts, eggs, seafood and meat,
Copper	Bone abnormalities, anemia and neutropenia	Action as an enzyme cofactor, Iron metabolism and formation of connective tissues	Nuts, seeds, liver and shellfish
Manganese	Skeletal muscle abnormalities	Action as an enzyme cofactor, bone formation and metabolic activities	Whole cereal grains, legumes, nuts and green leafy vegetables
Fluoride	Weak bones and dental caries	Formation of tooth enamel and mineralization bone	Tea, fluoridated water and seafood
Chromium	Impaired glucose tolerance	Glucose metabolism and insulin activity	Whole cereal grains, broccoli and Meat,
Molybdenum	Metabolic disorders	Action as an enzyme cofactor	Legumes, nuts, and grains

stresses as a plants' own defense system (Bitwell et al., 2023). Fruits, vegetables, cereals, seeds, nuts and herbs are rich sources of phytochemicals. These plant-based biomolecules promote nutritional adequacy by exerting sophisticated roles in boosting antioxidant defense against oxidative stress, lowering inflammation, and protecting cellular damage etc. (Almodaifer et al., 2017). They further play vital role in improving nutrient absorption in the gut by the metabolic activities of gut microbiota.

2.2 Recommended dietary allowance (RDA)

Attaining nutritional adequacy is a complex criterionand it needs implementation of some rigorous protocols or guidelines set by combined efforts of different national and international food organizations/agencies. These guidelines are set based on scientific research in the field of nutrition and public health. Typically, nutritional adequacy can be attained by depending on several crucial guidelines and without implementing them, the essence of adequate nutritional status cannot be gained. The primary goal of setting RDA for different nutrients is to prevent diseases and promote optimal health (Agrawal et al., 2020). Nevertheless, nutritional adequacy can be better attained by consuming nutrients (macro and micronutrients) as per their respective Recommended Dietary Allowances (RDA). RDA is referred to as "average daily intake sufficient to fulfil around 97–98 % of the nutrient requirements of a healthy individual", making it a criterion for ensuring the nutritional adequacy. RDAs differ broadly based on age, gender, physiological states (e.g., pregnancy, lactation, illness etc.), and mode of lifestyle (vigorous or sedentary). It suggests planning nutritionally adequate diets and evaluates whether an individual's dietary intake fulfils nutritional requirements (Wolfe et al., 2017). In this manner, RDA is one prominent scientific benchmark that can be used for ensuring and assessing nutritional adequacy. Multiple macro and micronutrient recommendations are covered by RDA and are globally used for developing dietary guidelines. RDAs mainly focus on specific nutrient groups such as proteins from macronutrients and vitamins and minerals from micronutrients (FAO, 2004). Hence, for maintaining nutritional adequacy, it is immensely crucial for an individual to consume all nutrients in the diet as per RDA.

2.3 Food groups, dietary diversity and nutrient density

Food groups such as fruits and vegetables, cereals or grains, dairy products, meat and eggs, oils and sugar etc. are all basic food sources which provide enough essential macro and micronutrients (Moore et al., 2012). Intake of a balanced mix of all food groups or dietary diversity is necessary for acquiring nutritional adequacy since no single food group can provide all nutrients therefore combining all food groups is important to prevent deficiency diseases and support long-term healthy life (Jacobs & Tapsell, 2013). Nutritional adequacy strictly demands dietary diversity and appropriate nutrient density for wellbeing. Dietary diversity can be defined as the consumption of a variety of foods from the food groups over a given time typically in 24 h or in a week (Habte & Krawinkel, 2016). Nutrient density on the other hand can be achieved by consuming diet with high nutrients and caloric content. Since relying repeatedly on merely one or two staples (as in Asian countries people mainly rely on rice or wheat rather than other cereals) exerts negative health impacts. A diverse diet mixed of all foods groups supplies a broader range of nutrients in the body and limits the risks of nutrient deficiencies, growth faltering, compromised immune system, chronic diseases etc.

2.4 Bioavailability

Bioavailability is again a crucial mechanism for body to maintain nutritional adequacy through ensuring appropriate nutrient distribution across the body. An impaired setup of nutrient absorption and utilization leads to overall disruption of bodily function despite the presence of essential nutrients in the diet (Schönfeldt et al., 2016). Excessively processed foods destroy their key essential nutrients (vitamin C, folic acid etc.) due to elevated thermal treatments, irradiation, microwaving etc. Moreover, sometimes food's own naturally occurring components (phytates, oxalates, tannins etc.) interfere with the bioavailability of essential nutrients (Soni et al., 2022). These anti-nutrients bind with important minerals (such as iron, zinc and calcium etc.) and limit their availability for absorption and utilization. Nevertheless, for attaining nutritional adequacy it is imperative that anti-nutrient components must not be exceeded from tolerable limits.

3. Assessment of nutritional adequacy

Assessing nutritional status is one of the strongest means of evaluating nutritional adequacy in an individual or population in a real sense. Nutritional adequacy evaluation tells whether diet provides enough energy and key nutrients to maintain health and prevent deficiencies or excesses. Commonly, by comparing dietary intake to standards such as

RDA, Estimated Average Requirement (EAR), and Adequate Intake (AI) etc. as per age, sex and physiological state of an individual of the deficiency risk, possible harm, and health range can be judged. For comprehensive assessment of nutritional adequacy, certain tools and techniques commonly referred to as A, B, C and D of nutritional assessment (i.e., Anthropometric measurements, Biochemical testing, Clinical observations, and Dietary intakes) are performed. Table 4 presents some tools and techniques frequently used for assessing nutritional adequacy. However, recently (Zambrano-Villacres et al., 2026) updated the model for nutritional assessment as ABCDEFG in which E was incorporated for the Ecological–microbiota, F for Functional indicators and G for the Genomic–nutrigenomic approaches.

Anthropometric measurements include measurement of weight, height, body mass index (BMI), arm and head circumferences, skinfold measurements etc. (Frison et al., 2016). In anthropometric assessment, BMI and body circumferences are key indicators of nutritional adequacy. BMI helps in identifying the category of normal body weight, overweight or obesity class. It also predicts the risk of cardiovascular disease, type 2 diabetes and hypertension etc. Body circumferences include mid-upper arm circumference, waist circumference, waist-to-hip ratio and hip circumference identify fat reserves in different body parts while skinfolds measurements indicate body fat percentage and are commonly used in sports science and clinical nutrition. Biochemical testing includes the evaluation of nutrients and metabolite concentrations in body fluids based on laboratory examination (Zheng & Tako, 2024). In biochemical testing blood and urine samples are mainly analyzed for identifying nutrient biomarkers. Clinical observations are done by physical examination of visible signs, symptoms, and functional indicators of nutrient deficiencies. Dietary recalls include gathering information about an individual's dietary diversity scores (DDS), Dietary Reference Intakes (DRIs), nutrient adequacy ratio (NAR) and 24-hour dietary recall etc. (Valença et al., 2025). The DDS remains the most prominent indicator for nutrient adequacy measurements. This technique is used to identify dietary intake from various food groups over a reference period typically prior 24 h. The ratio of food quantity and nutrient uptake is referred to as NAR, while Mean Adequacy Ratio (MAR) is the average of NAR for a set of nutrients reflecting overall micronutrient adequacy. Moreover, 24-hour dietary recall remains most frequently used tool in which an individual recalls everything consumed in past 24 h.

Table 4 Prominent tools and techniques frequently used for assessing nutritional adequacy among individuals.

Assessment tools/techniques	Objective	Nutritional adequacy revelation	Advantages	Disadvantages
Anthropometric Assessment				
Height	Anthropometric assessment is used to measure size, proportions, composition, and mode of growth and development of human body. Identify stunting, wasting and micronutrient deficiency	Identify stunting, wasting, micronutrient deficiencies and detect situations like undernutrition, overnutrition and obesity grade	Easy, simple, cost-effective, and rapid screening tool	Specific nutrient deficiency cannot be detected. Edema condition interferes with outcomes
Weight				
Body mass index				
Head circumference				
Mi-upper arm circumference				
Waist circumference				
waist-to-hip ratio and hip circumference				
Triceps skin fold thickness				
Subscapular skin fold thickness				

(continued)

Table 4 Prominent tools and techniques frequently used for assessing nutritional adequacy among individuals. (*cont'd*)

Assessment tools/techniques	Objective	Nutritional adequacy revelation	Advantages	Disadvantages
Biochemical Assessment				
Hemoglobin	Biochemical tests are used to measure the level of nutrients and metabolic indicators in body fluids (mainly blood and urine)	Reflects individual's nutritional status at molecular or cellular level Detects micronutrient deficiencies	Sensitive	Costly, needs efficient laboratory facilities
Lipid profile				
Serum vitamin D				
Serum albumin				
Serum ferritin				
Blood glucose				
Urinary iodine				
Serum calcium				

Clinical assessment

General appearance (edema, paller, posture etc.)	To identify different nutrient deficiencies via visible physical signs	Indicates protein deficiency, iron deficiency anemia etc.	Rapid, equipment-free, inexpensive	Detect deficiencies at advanced stages
Hair (dryness, discoloration, brittleness)		Indicates deficiency of protein, zinc and fatty acid		
Skin (dryness, hyperpigmentation, dermatitis)		Indicates protein, niacin, vitamin A and C deficiency		
Eye (Bitot's spot, pale conjunctiva, night blindness)		Indicates vitamin A and iron deficiency		
Mouth (cracked lips, bleeding gums, inflamed tongue)		Indicates vitamin B-complexes and C deficiency		
Nails (brittle nails with ridges, koilonychia)		Indicates protein, and iron deficiency		

(continued)

Table 4 Prominent tools and techniques frequently used for assessing nutritional adequacy among individuals. (*cont'd*)

Assessment tools/techniques	Objective	Nutritional adequacy revelation	Advantages	Disadvantages
Dietary assessment				
Estimated Average Requirement (EAR) Recommended Dietary Allowance (RDA)	This compares dietary intake data with standard dietary requirements	For determining nutritional adequacy, nutrient deficiency risk, or even excessive intake of nutrients	Scientifically standardized questionnaire based	Accurate/correct data collection is needed
Adequate Intake (AI)				
Dietary Reference Intakes (DRIs),				
dietary diversity scores (DDS),				
Nutrient adequacy ratio (NAR)				
24-hour dietary recall				

4. Overview- novel food products

Novel food products are newly developed foods that have not been conventionally eaten by humans to a significant extent before. They are produced from newly sourced functional ingredients or nutrients by using innovative or advanced technologies. In nutrition science, they aim to develop for reducing macro and micronutrient deficiencies, attaining functional health benefits and reducing associated risks (Miazzi et al., 2025). New sources of nutrient-rich ingredients play a vital role in the development of novel foods by introducing alternative and sustainable substances in human diet. Development of novel food products is a powerful strategy to counter nutritional inadequacy. Novel foods are not merely safe for consumption but are appealing and enriched with key nutrients that are often lacked in diets. Through novel food formulations specific dietary deficiencies can be addressed.

4.1 Newly emerging sources of functional ingredients or nutrients

Most of the recent studies on novel food development suggest that many newly emerged functional ingredients or nutrients have been effectively used for developing novel food products.

4.1.1 Macro and microalgae

Macroalgae (also seaweed) are photosynthetic eukaryotic organisms grouped as red, brown, and green algae (Wang et al., 2017). They produce a wide variety of novel bioactive components or metabolites like high biological value proteins, polysaccharides, polyunsaturated fatty acids (PUFAs), and phlorotannins etc. Microalgae on the other hands are simple photosynthetic microscopic organisms that can synthetize complex molecules (Borowitzka, 2018) like carbohydrates, proteins, lipids, PUFAs, vitamins, and a variety of secondary metabolites (Sidari & Tofalo, 2019; EFSA, 2019) has approved several macro and microalgae species for novel food development such as Palmaria palmata, Laminaria spp., Ulva spp., Saccharina latissimi, Tetraselmis chuii, Euglena gracilis, Haematococcus pluvialis, Schizochytrium etc. (EFSA NDA Panel, 2014, 2020).

4.1.2 Edible insects

Edible insects such as beetles, crickets, grasshoppers, mealworms, silkworm etc. are sustainable sources of alternative proteins, fats, vitamins, iron, and zinc etc. They are being used for developing many novel foods such as protein powders, snack bars, and other fortified foods. Insects need fewer resources (feed, water, and land) for survival and this characteristic makes them an environmentally friendly choice (Scaffardi & Formici, 2022).

4.1.3 Yeast and fungi

Yeast is becoming a crucial novel food ingredient due to its unique nutritional value, versatility and sustainability (Ma et al., 2024). Saccharomyces cerevisiae remains the most common species of nutritional yeast used for food fermentation. Through precision fermentation yeast synthesizes animal-free proteins or mycoproteins. Fusarium venenatum, a yeast-like fungi, can produce sustainable high-protein foods. Fungi and edible mushroom are also exceptional sources of mycoprotein, dietary fiber, B-complex vitamins, and minerals (González et al., 2020). Mycoprotein can be widely used as meat substitutes.

4.1.4 Underutilized-neglected legumes

A variety of underutilized legumes are rich in protein, dietary fiber, and bioactive compounds (Hunter et al., 2019). They in powdered form can be added in flour, and dairy products etc. Legume crops are adapted to harsh climatic conditions and thus can support agricultural sustainability and food security. Too simple content!!!!.

4.1.5 Wild terrestrial plants and weeds

Many wild but edible plants are rich sources of phytochemicals, dietary fiber, and micronutrients. These plants have been increasingly studied for their remarkable role in functional foods. They can help in diversifying diets and can decrease dependence on conventional staples. The previous four sub-sessions 4.1.2–4.1.5 are too simple and need to add more contents.

4.2 Development of novel food products

The process of novel food product development is complex, dependent on series of various steps and aim at creating safe, nutritious, appealing, palatable and marketable food products (Ververis et al., 2020). Development of novel food can attain nutritional adequacy in population suffering from nutritional deficiencies. Although the process of development of novel foods is multi-faceted however the aftermaths are meaningful and robust in availing nutritional adequacy. Fig. 1 exhibits a schematic series of steps for developing novel food products. The novel food products can be classified based on food sources from which they produce, the type of technology used for their development and functional benefits they exert upon consumption.

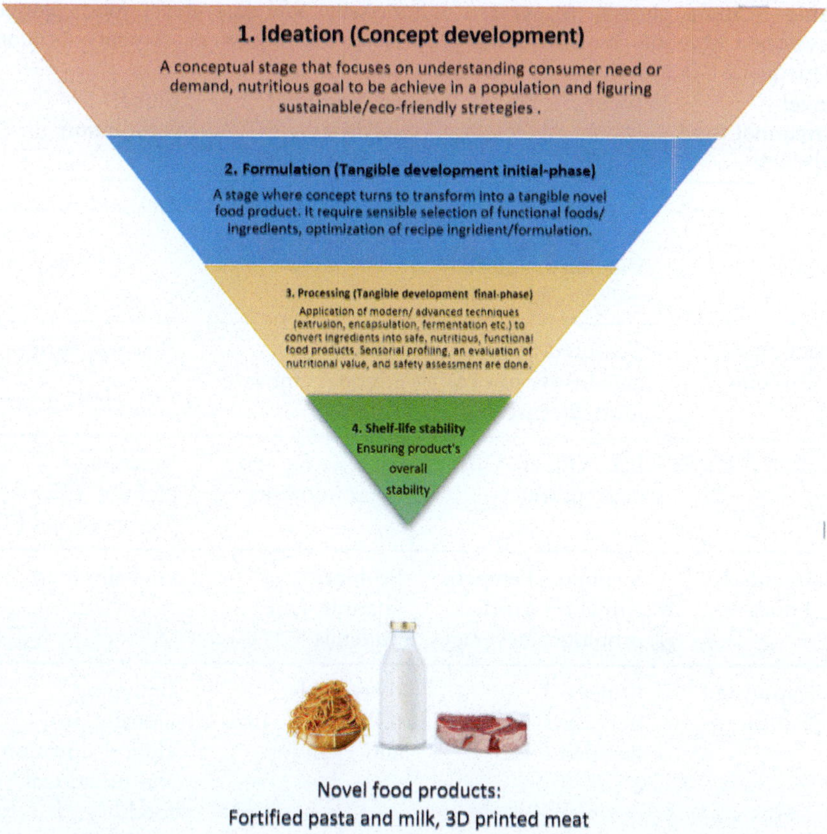

Fig. 1 Schematic series of steps involved in novel food product development.

4.3 Advanced techniques for novel food development

Food science is continuously evolving with the passage of time. It introduces meaningful advanced techniques through which novelty in the food products can be achieved. There are tremendous modern food applications that can promote nutritional adequacy through developing novel food products (Guiné et al., 2016). These innovative food processing techniques can create novel foods with improved nutritional adequacy, functionality and sustainability. Most of these modern food processing techniques are employed to enhance nutrient density and protein quality in novel foods, to improve the bioavailability of nutrients and to counter micronutrient deficiencies. Table 5 shows information regarding some globally recognized nutritionally adequate novel food

Table 5 Some globally recognized nutritionally adequate novel food products developed through modern food processing techniques to combat nutritional inadequacy [40 and 41].

Novel functional food products	Types	Key essential nutrients	Global recognition
Fortified Foods	Fortified cereals (brown rice), Vitamin D-fortified milk, Iodized salt	Vitamin A, D Iron and Iodine	Asia and Africa
Plant-Based Proteins	Textured plant-based (soy, pea etc.) meat alternatives	Proteins or essential amino acids	Asia and Africa
Cultured Foods	Lab cultivated meat, dairy products	Protein, fat and micronutrients	Emerging globally, US, Singapore and EU
Fermented Foods	Yoghurt, Tempeh, Kimchi, Fortified probiotics beverages	Probiotics, vitamins and minerals	Globally used
Encapsulated Foods	Omega-3 beverages, mineral-fortified foods	Fatty acids (Omega-3), Iron and vitamins	Emerging globally, in clinical nutrition and functional foods
Personalized 3D-printed Foods	Customized meals and Texture-modified meals	Protein, vitamins and minerals	Global experimental and clinical applications

products that have been developed through modern food processing techniques. However, some important advanced food techniques for developing novel food products ae discussed here.

4.3.1 Food fortification and enrichment

Food fortification is referred to as "addition of essential nutrients (mainly iron, zinc, vitamin A, D. B12 and $\omega-3$ fatty acids etc.) during food processing to combat nutritional deficiencies". Recently, keeping in view of the current health situation among the people throughout the world, the health-related departments and food fortification experts are developing new fortified products with the addition of micronutrients in staples such as flour, salt, rice, oil etc. (Santosh et al., 2023). Food enrichment is a process of restoration of the inherent nutrients in the final food products that are lost during food processing. The process of food enrichment is crucial in attaining nutritional adequacy in public health.

4.3.2 Alternative proteins innovation

Globally, food industries are gradually relying on protein alternatives for improving protein quality and its digestibility. Not even this merely, but economic shift and global security in relation to meat and meat-based foods are also main reasons for developing alternative proteins. Protein is a key macronutrient for performing a variety of vital functions in the body. It is important for people living in poor areas where food purchasing power is very low and people cannot buy meat or high-protein foods. Therefore, as a novel food alternative in the form of proteins is extracted from plant sources (soy proteins, pea and fava protein etc.), insects (crickets, grasshopper, mealworm etc.). Most of the proteins extracted from different non-animal sources are high-quality protein and have great potential to counter protein energy malnutrition in rural areas (Smetana et al., 2015).

4.3.3 Precision fermentation

Conventional food fermentation uses bacteria, yeast, fungi etc. to transform food into fermented foods (yoghurt, cheese, bread etc.) to improve food nutrient availability, digestibility and shelf life. This also helps in reducing antinutritional factors that interfere with the bioavailability of essential micronutrient. Precision fermentation is an advanced mode of conventional fermentation in which genetically engineered microorganisms are used as fermentation starters which is engineered to synthesize specific nutrients/functional compounds in fermented food (Lonchamp et al., 2022). Through precision fermentation a complete alternative protein (Quorn, a mycoprotein) has been developed without livestock from fungal starter (Lonchamp et al., 2020). This technique is in its emerging phase and has potential to combat situations of protein energy malnutrition in future.

4.3.4 Encapsulation

In encapsulation the nutrients are trapped inside a protective coating or capsule to protect them from hazards (heat, light, moisture, and oxygen) which improves nutrient stability until nutrients are released in the body for absorption (Maurya et al., 2023). Nano-encapsulation is a type of encapsulation in which nanoparticles are encapsulated. This technique has the potential to combat nutritional inadequacy. Essential micronutrients (iron, zinc, vitamin A, D. B12) may be encapsulated to falter the situation of micronutrient deficiencies.

4.3.5 3D and 4D food printing

Multidimensional food printing is an emerging technique in which food materials (food inks) are deposited into layers using a food printer like a conventional 3D printer for creating food of customized composition, shape, texture and flavors (Mantihal et al., 2020). It can be used as a tool for reducing nutritional inadequacy through fortifying food materials with vitamins, minerals, or probiotics during printing. Moreover, using nutritional profiling software or AI-driven systems nutrient concentration may be optimized as per individualized needs.

4.3.6 Cellular agriculture

The foods developed through cellular agriculture are referred to as cultivated foods. In this modern technique animal food products are grown from cells rather than whole animals. The muscle or fat cells from an animal are cultured in a nutrient-rich medium using bioreactors which allow them to grow into food produce like meat, dairy etc. (Smith et al., 2022). Bioreactors can precisely control or customize the nutrient content (protein, fat and micronutrients) in these cultured foods thus nutritional adequacy can be met.

5. Conclusion

Nutritional adequacy is fundamental to maintaining health, supporting growth and development, and ensuring normal physiological functions throughout life. Despite advances in agriculture, food production and nutrition science, nutritional inadequacy remains a major global concern. Food insecurity, poverty, limited dietary diversity, micronutrient deficiencies, infectious diseases, poor water and sanitation, sociocultural inequalities, conflicts, natural disasters and weaknesses within food systems can all limit access to nutritious diets. Addressing these challenges therefore requires approaches that consider both the availability and quality of food, rather than focusing only on energy intake.

A nutritionally adequate diet should provide appropriate amounts of macronutrients, micronutrients, dietary fiber, water and beneficial phytochemicals, while ensuring adequate nutrient density, diversity and bioavailability. Dietary assessment and nutritional indicators are important for identifying deficiencies and evaluating whether dietary requirements are being met. Within this framework, novel food products provide promising opportunities to improve nutritional quality and diversify food sources. Ingredients such as algae, edible insects, yeast, fungi, underutilized legumes and wild edible plants can contribute valuable nutrients and functional components to the diet. Furthermore, technologies including food fortification, alternative protein innovation, precision fermentation, encapsulation, multidimensional food printing and cellular agriculture offer new possibilities for developing nutritionally enhanced and sustainable foods.

Overall, the successful development of novel foods requires careful consideration of nutritional composition, safety, bioavailability, technological feasibility and consumer acceptance. Integrating innovative food ingredients and processing technologies with broader nutrition and food-system strategies can help address nutrient deficiencies, improve dietary quality and support sustainable approaches to achieving nutritional adequacy and better public health.

References

Agrawal, N., Singh, A., & Rana, S. (2020). Study of dietary intake of micro and macronutrients and comparison with the recommended daily allowance (RDA). *Magnesium, 340*(312.53) 89-27.

Alamirew, S. K., Lemke, S., Stadlmayr, B., & Freyer, B. (2023). Dietary behaviour and sociocultural determinants of dietary diversity among rural women of reproductive age: A case of Amhara region, Ethiopia. *Nutrients, 15*(15), 3369.

Ali, J., Mohiuddin, M., Wang, X., Hussain, Z., Irshad, M., Zia, M., & Ahmad, S. (2024). Spatial variation in iodine content with relation to soil physicochemical properties in lower Himalayan region. *Environmental Research, 251*, 118569.

Almodaifer, S., Alsibaie, N., Alhoumendan, G., Alammari, G., & Kavita, M. S. (2017). Role of phytochemicals in health and nutrition. *BAOJ Nutrition, 3*, 28–34.

Anderson, J. W., Baird, P., Davis, R. H., Jr, Ferreri, S., Knudtson, M., Koraym, A., & Williams, C. L. (2009). Health benefits of dietary fiber. *Nutrition Reviews, 67*(4), 188–205.

Ashraf, I., Bestman, P. L., Assiri, A. A., Kamal, G. M., Uddin, J., Luo, J., & Ishaqui, A. A. (2024). Status of inappropriate complementary feeding and its associated factors among infants of 9–23 months. *Nutrients, 16*(24), 4379.

Azeem, A., Khan, S., Haq, M. Z. U., Shafiq, S., Aslam, M. T., Munir, S., & Zaman, M. S. (2025). Biofortification strategies for enhancing crop nutritional value: A review of methods, challenges, and future directions. *Discover Plants, 2*, 185.

Barasa, B., Watsusi, C.W., Gudoyi, P.M., Lutaaya, N., & Akello, G. (2022). Desert locust swarm invasions and control: Determinants, duration, food security and effectiveness of control measures. Duration, Food Security and Effectiveness of Control Measures.

Basurko, C., Savy, M., Galindo, M. S., Gatti, C., Osei, L., Nacher, M., & Dramé, M. (2025). Prevalence of food insecurity during pregnancy in Latin American and the Caribbean Countries: A systematic review. *The Journal of Nutrition, 155*(1), 250–259.

Baugreet, S., Hamill, R. M., Kerry, J. P., & McCarthy, S. N. (2017). Mitigating nutrition and health deficiencies in older adults: A role for food innovation? *Journal of Food Science, 82*(4), 848–855.

Beal, T., Massiot, E., Arsenault, J. E., Smith, M. R., & Hijmans, R. J. (2017). Global trends in dietary micronutrient supplies and estimated prevalence of inadequate intakes. *PLoS One, 12*(4), e0175554.

Biadgilign, S., Ayenew, H. Y., Shumetie, A., Chitekwe, S., Tolla, A., Haile, D., & Gebre, B. (2019). Good governance, public health expenditures, urbanization and child undernutrition nexus in ethiopia: An ecological analysis. *BMC Health Services Research, 19*(1), 40.

Bitwell, C., Indra, S. S., Luke, C., & Kakoma, M. K. (2023). A review of modern and conventional extraction techniques and their applications for extracting phytochemicals from plants. *Scientific African, 19*, e01585.

Borowitzka, M. A. (2018). *Biology of microalgae. Microalgae in health and disease prevention.* Academic Press. 23–72.

Bray, G. A., & Siri-Tarino, P. W. (2016). The role of macronutrient content in the diet for weight management. Endocrinology and metabolism. *Clinics (Sao Paulo, Brazil), 45*(3), 581–604.

Bryant, C., & Barnett, J. (2020). Consumer acceptance of cultured meat: An updated review (2018-2020). *Applied Sciences, 10*(15).

Carroll, G. J., Lama, S. D., Martinez-Brockman, J. L., & Pérez-Escamilla, R. (2017). Evaluation of nutrition interventions in children in conflict zones: A narrative review. *Advances in Nutrition, 8*(5), 770–779.

Cassells, D., Costantini, L., Ashery, A.F., Gadge, S., Pires, D.L., Sánchez-Cortés, M.Á., & Omodei, E. (2024). A 72h exploration of the co-evolution of food insecurity and international migration. arXiv preprint arXiv:2407.03117.

Centre for Coordination of Agricultural Research and Development for Southern Africa (CCARDESA). (2025). Southern Africa in 2025: A deepening food insecurity crisis amid global progress. CCARDESA. (Statistics on undernourishment and inability to afford a healthy diet in Africa.).

EFSA NDA Panel. (2014). Scientific Opinion on the safety of dried Palmaria palmata as a novel food pursuant to regulation (EC) No 258/97. *EFSA Journal, 12*(4), 3632.

EFSA NDA Panel. (2018). Scientific opinion on the safety of Porphyra sp. protein products as a novel food pursuant to Regulation (EU) 2015/2283. *EFSA Journal, 16*(6), 5340.

EFSA NDA Panel. (2019). Scientific opinion on the safety of Ulva spp. as a novel food pursuant to regulation (EU) 2015/2283. *EFSA Journal, 17*(5), 5701.

EFSA NDA Panel. (2020). Safety assessment of Saccharina latissima as a novel food pursuant to regulation (EU) 2015/2283. *EFSA Journal, 18*(2), 6016.

Fan, Y., Yao, Q., Liu, Y., Jia, T., Zhang, J., & Jiang, E. (2022). Underlying causes and co-existence of malnutrition and infections: An exceedingly common death risk in cancer. *Frontiers in Nutrition, 9*, 814095.

Fanta. (2015). Validation of food-based recommendations developed using optifood for groups at nutritional risk in the western highlands of guatemala. Washington, DC: FHI 360/FANTA; p. 1–4.

Fao, IFAD, Unicef, Wfp, & WHO. (2024). The state of food security and nutrition in the world 2024. Financing to end hunger, food insecurity and malnutrition in all its forms24 July 2024.

Farahmandnia, H., Zarei, M., & Sahebi, A. (2024). Nutritional status among earthquake survivors: A systematic review and meta-analysis. *Public Health, 227*, 24–31.

Food and Agriculture Organization of the United Nations. (2025). Indicator 2.1.2 Prevalence of moderate or severe food insecurity in the population, based on the Food Insecurity Experience Scale.

Food and Agriculture Organization, International Fund for Agricultural Development, UNICEF, World Food Programme, & World Health Organization. (2024). The state of food security and nutrition in the world 2024. FAO.

Food and Agriculture Organization. (2004). Vitamin and mineral requirements in human nutrition. FAO/WHO.

Food and Agriculture Organization. (2025). Committee on World Food Security: Section on Governance and structural causes of hunger and malnutrition.

Frison, S., Kerac, M., Checchi, F., & Prudhon, C. (2016). Anthropometric indices and measures to assess change in the nutritional status of a population: A systematic literature review. *BMC Nutrition, 2*(1), 76.

Gassara, G., & Chen, J. (2021). Household food insecurity, dietary diversity, and stunting in sub-Saharan Africa: A systematic review. *Nutrients, 13*(12), 4401.

Gizaw, Z., & Worku, A. (2019). Effects of single and combined water, sanitation and hygiene (WASH) interventions on nutritional status of children: A systematic review and meta-analysis. *Italian Journal of Pediatrics, 45*(1), 77.

González, A., Cruz, M., Losoya, C., Nobre, C., Loredo, A., Rodríguez, R., & Belmares, R. (2020). Edible mushrooms as a novel protein source for functional foods. *Food & function, 11*(9), 7400–7414.

Guiné, P. F. R., CD Ramalhosa, E., & Paula Valente, L. (2016). New foods, new consumers: Innovation in food product development. *Current Nutrition & Food Science, 12*(3), 175–189.

Gupta, O. P., Singh, A., Pandey, V., Sendhil, R., Khan, M. K., Pandey, A., ... Singh, G. (2024). Critical assessment of wheat biofortification for iron and zinc: A comprehensive review of conceptualization, trends, approaches, bioavailability, health impact, and policy framework. *Frontiers in Nutrition, 10*, 1310020.

Habte, T. Y., & Krawinkel, M. (2016). Dietary diversity score: A measure of nutritional adequacy or an indicator of healthy diet. *Journal of Nutrition and Health Sciences, 3*(3), 303.

Hasan, M. M., Ahmed, S., Soares Magalhaes, R. J., Fatima, Y., Biswas, T., & Mamun, A. A. (2022). Double burden of malnutrition among women of reproductive age in 55 low-and middle-income countries: Progress achieved and opportunities for meeting the global target. *European Journal of Clinical Nutrition, 76*(2), 277–287.

Huang, L., Xu, S., Chen, X., Li, Q., Lin, L., Zhang, Y., & Yang, N. (2020). Delayed lactogenesis is associated with suboptimal breastfeeding practices: A prospective cohort study. *The Journal of Nutrition, 150*(4), 894–900.

Hunter, D., Borelli, T., Beltrame, D. M., Oliveira, C. N., Coradin, L., Wasike, V. W., & Tartanac, F. (2019). The potential of neglected and underutilized species for improving diets and nutrition. *Planta, 250*(3), 709–729.

Jacobs, D. R., Jr, & Tapsell, L. C. (2013). Food synergy: The key to a healthy diet. *Proceedings of the Nutrition Society, 72*(2), 200–206.

Jiang, B., Lin, K., Buys, N., Zhang, B., Qi, Y., & Sun, J. (2025). Trend and burden of suboptimal breastfeeding in children under five years of age in 1990–2021: A systematic analysis for the global burden of disease study 2021. *Nutrients, 17*(7), 1134.

Johnson, C. M., Sharkey, J. R., Lackey, M. J., Adair, L. S., Aiello, A. E., Bowen, S. K., & Ammerman, A. S. (2018). Relationship of food insecurity to women's dietary outcomes: A systematic review. *Nutrition Reviews, 76*(12), 910–928.

Katoch, O. R. (2022). Determinants of malnutrition among children: A systematic review. *Nutrition (Burbank, Los Angeles County, Calif.), 96*, 111565.

Kiani, A. K., Dhuli, K., Donato, K., Aquilanti, B., Velluti, V., Matera, G., & Bertelli, M. (2022). Main nutritional deficiencies. *Journal of Preventive Medicine and Hygiene, 63*(2 3), E93.

Kiosia, A., Dagbasi, A., Berkley, J. A., Wilding, J. P., Prendergast, A. J., Li, J. V., & Frost, G. (2024). The double burden of malnutrition in individuals: Identifying key challenges and re-thinking research focus. *Nutrition Bulletin, 49*(2), 132–145.

Kitchener, E., Thurecht, R., Connors, A., Dixson, B., & Kafer, G. (2026). Nutrient intake evidence in the Pacific: A scoping review of research coverage, challenges, and opportunities. *Food Science & Nutrition, 14*(1), e71360.

Llorca-Jaña, M., Martínez-Carrión, J. M., & Salvatore, R. (Eds.). (2025). *Health, nutrition and inequality in Latin America: An anthropometric history*. Taylor & Francis.

Lonchamp, J., Akintoye, M., Clegg, P. S., & Euston, S. R. (2020). Functional fungal extracts from the Quorn fermentation co-product as novel partial egg white replacers. *European Food Research and Technology, 246*(1), 69–80.

Lonchamp, J., Stewart, K., Munialo, C. D., Evans, L., Akintoye, M., Gordon, S., & Euston, S. R. (2022). Mycoprotein as novel functional ingredient: Mapping of functionality, composition and structure throughout the Quorn fermentation process. *Food Chemistry, 396*, 133736.

Lopes, S. O., Abrantes, L. C. S., Azevedo, F. M., Morais, N. D. S. D., Morais, D. D. C., Goncalves, V. S. S., & Priore, S. E. (2023). Food insecurity and micronutrient deficiency in adults: A systematic review and meta-analysis. *Nutrients, 15*(5), 1074.

Ma, C., Xia, S., Song, J., Hou, Y., Hao, T., Shen, S., & Jiang, X. (2024). Yeast protein as a novel protein source: Processing, functional properties, and potential applications in foods. *Innovative Food Science & Emerging Technologies, 93*, 103606.

Makwela, M. S., & Mashaba, R. G. (2025). Determinants of sub-optimal complementary feeding practices among caregivers of children aged 6–23 months in low-and middle-income countries: Scoping review. *Frontiers in Public Health, 13*, 1655685.

Mantihal, S., Kobun, R., & Lee, B. B. (2020). 3D food printing of as the new way of preparing food: A reviw. *International Journal of Gastronomy and Food Science, 22*, 100260.

Marphatia, A. A., Cole, T. J., Grijalva-Eternod, C., & Wells, J. C. K. (2016). Associations of gender inequality with child malnutrition and mortality across 96 countries. *Global Health, Epidemiology and Genomics, 1*, e6.

Maurya, V. K., Shakya, A., Bashir, K., Jan, K., & McClements, D. J. (2023). Fortification by design: A rational approach to designing vitamin D delivery systems for foods and beverages. *Comprehensive Reviews in Food Science and Food Safety, 22*(1), 135–186.

Miano, T. F., & Dogan, M. (2021). Novel foods: Sustainable food process. *International Journal of Ecosystems and Ecology Science, 11*(2), 311–314. https://doi.org/10.31407/ijees11.215.

Miazzi, M. M., Dellino, M., Fanelli, V., Mascio, I., Nigro, D., De Giovanni, C., & Montemurro, C. (2025). Novel foods in the European framework: Benefits and risks. *Critical Reviews in Food Science and Nutrition, 65*(29), 6460–6469.

Moore, L. L., Singer, M. R., Qureshi, M. M., Bradlee, M. L., & Daniels, S. R. (2012). Food group intake and micronutrient adequacy in adolescent girls. *Nutrients, 4*(11), 1692–1708.

Natekar, P., Deshmukh, C., Limaye, D., Ramanathan, V., & Pawar, A. (2022). A micro review of a nutritional public health challenge: Iron deficiency anemia in India. *Clinical Epidemiology and Global Health, 14*, 100992.

Newsom, A., Koli, M., & Sebesvari, Z. (2021). Technical Report: Locust Outbreak 2019-2021. Interconnected Disaster Risks 2020/2021. United Nations University - Institute for Environment and Human Security (UNU-EHS).

Nour, T. Y., & Altintaş, K. H. (2025). Obesity after natural disasters and associated risk factors: A systematic review. *Disaster Medicine and Public Health Preparedness, 19*, e8.

Ojo, T., Ruan, C., Hameed, T., Malburg, C., Thunga, S., Smith, J., & Peprah, E. (2022). HIV, tuberculosis, and food insecurity in Africa—a syndemics-based scoping review. *International Journal of Environmental Research and Public Health, 19*(3), 1101.

Onwezen, M. C., Bouwman, E. P., Reinders, M. J., & Dagevos, H. (2021). A systematic review on consumer acceptance of alternative proteins: Pulses, algae, insects, plant-based meat al ternatives, and cultured meat. *Appetite, 159*.

Perez-Puyana, V. M., Jiménez-Rosado, M., & Romero, A. (2024). Novel food and beverages: Production and characterization. *Foods, 13*(13), 2085.

Rani, J., Sharma, U. K., & Sharma, D. N. (2018). Role of adequate water intake in purification of body. *Environment Conservation Journal, 19*(1&2), 183–186.

Ribeiro, A. C. P., Esmerino, E. A., Tavares Filho, E. R., Cruz, A. G., & Pimentel, T. C. (2025). Unraveling the potential of co-creation on the new food product development: A comprehensive review on why and how listening to consumer voices. *Trends in Food Science & Technology, 159*, 104978.

Rizaldi, M. A., Ali, K., Rara, S. M. H., & Panjaitan, B. S. R. (2025). Water, sanitation and hygiene (WASH) and its association with stunting in developing countries in Asia: A systematic review. *Svāsthya: Trends in General Medicine and Public Health, 2*(2) e81.

Roberts, T. J., Carnahan, E., & Gakidou, E. (2013). Can breastfeeding promote child health equity? A comprehensive analysis of breastfeeding patterns across the developing world and what we can learn from them. *BMC Medicine, 11*(1), 254.

Saber, A. A. (2026). War and wasting: The role of precision nutrition in managing malnutrition and chronic disease in conflict-affected populations. *Precision Nutrition, 5*(1), e00133.

Santosh, O., Bajwa, H. K., Indira, A., Bano, S., Dolma, N., Bisht, M. S., & Chongtham, N. (2023). Novel food product development through food-to-food fortification with nutrient and bioactive compound-rich bamboo shoot. *Bamboo science and technology*. Singapore: Springer Nature Singapore. 67–87.

Scaffardi, L., & Formici, G. (2022). *Novel foods and edible insects in the European Union*. Springer International Publishing.

Schönfeldt, H. C., Pretorius, B., & Hall, N. (2016). Bioavailability of nutrients. *Encyclopedia of Food and Health, 1*, 401–406.

Shafiq, A., Hussain, A., Asif, M., Jameel, A., Sadiq, S., & Kanwel, S. (2021). Determinants of gender disparity in nutritional intake among children in Pakistan: Evidence from PDHS. *Children, 9*(1), 7.

Shahid, N., Salman, F., & Makhdum, M. (2021). Major factors responsible for child malnutrition: A review. *Journal of Pakistan Medical Association, 71*(2), 729–733.

Shergill-Bonner, R. (2017). Micronutrients. *Paediatrics and Child Health, 27*(8), 357–362.

Sidari, R., & Tofalo, R. (2019). A comprehensive overview on microalgal-fortified/based food and beverages. *Food Reviews International, 35*(8), 778–805.

Singh, B., Pavithran, N., & Rajput, R. (2023). Effects of food processing on nutrients: A review. *Current Journal of Applied Science and Technology, 42*(46), 34–49.

Smetana, S., Mathys, A., Knoch, A., & Heinz, V. (2015). Meat alternatives: Life cycle assessment of most known meat substitutes. *International Journal of Life Cycle Assessment, 20*(9), 1254–1267.

Smith, D. J., Helmy, M., Lindley, N. D., & Selvarajoo, K. (2022). The transformation of our food system using cellular agriculture: What lies ahead and who will lead it? *Trends in Food Science & Technology, 127*, 368–376.

Sodjinou, V. D., Mengouo, M. N., Douba, A., Tanifum, P., Yapi, M. D., Kuzanwa, K. R., & Masresha, B. (2022). Epidemiological characteristics of a protracted and widespread measles outbreak in the Democratic Republic of Congo, 2018-2020. *The Pan African Medical Journal, 42*, 282.

Soni, K., Samtiya, M., Krishnan, V., & Dhewa, T. (2022). Antinutritional factors: Nutrient bioavailability and health beneficial effects. *Conceptualizing plant-based nutrition: Bioresources, nutrients repertoire and bioavailability*. Singapore: Springer Nature Singapore. 157–179.

Tidiane Ndour, C., Diop, W., & Asongu, S. (2025). The effect of natural disasters on food security in Sub-Saharan Africa. *Social Responsibility Journal, 21*(1), 180–197.

United Nations Children's Fund & World Health Organization. (2025). Progress on household drinking water, sanitation and hygiene 2000–2024: special focus on inequalities.

United Nations Children's Fund (UNICEF). (2022). Malnutrition and food insecurity during the COVID-19 pandemic.

United Nations Children's Fund. (2025). Triple threat of malnutrition puts children in South Asia at risk, UNICEF warns.

Valença, S. E., Almeida, L. F., Castro, J. D. S., Almeida, P. H., Fernandes, D. P., Dos Santos, C. A., & Ribeiro, A. Q. (2025). The usual dietary recall as a dietary assessment method: A scoping review. *British Journal of Nutrition,* 1–14.

Ververis, E., Ackerl, R., Azzollini, D., Colombo, P. A., de Sesmaisons, A., Dumas, C., & Gelbmann, W. (2020). Novel foods in the European Union: Scientific requirements and challenges of the risk assessment process by the European Food Safety Authority. *Food Research International, 137*, 109515.

Wang, H. M. D., Li, X. C., Lee, D. J., & Chang, J. S. (2017). Potential biomedical applications of marine algae. *Bioresource Technology, 244*, 1407–1415.

Wei, C., Liu, Y., Li, Y., Zhang, Y., Zhong, M., & Meng, X. (2020). Evaluation of the nutritional status in patients with COVID-19. *Journal of Clinical Biochemistry and Nutrition, 67*(2), 116–121.

Wolfe, R. R., Cifelli, A. M., Kostas, G., & Kim, I. Y. (2017). Optimizing protein intake in adults: Interpretation and application of the recommended dietary allowance compared with the acceptable macronutrient distribution range. *Advances in Nutrition, 8*(2), 266–275.

World Food Programme. (2024). Global Report on Food Crises: acute hunger remains persistently high in 59 countries with 1-in-5 people assessed in need of critical urgent action.

World Health Organization & United Nations Children's Fund. (2025). Progress on household drinking-water, sanitation and hygiene 2000-2024: Special focus on inequalities 26 August 2025.

Wretlind, A. (1982). Standards for nutritional adequacy of the diet: European and WHO/FAO viewpoints. *The American Journal of Clinical Nutrition, 36*(2), 366–375.

Yaya, S., Hudani, A., Udenigwe, O., Shah, V., Ekholuenetale, M., & Bishwajit, G. (2018). Improving water, sanitation and hygiene practices, and housing quality to prevent diarrhea among under-five children in Nigeria. *Tropical Medicine and Infectious Disease, 3*(2), 41.

Zambrano-Villacres, R., Arteaga-Pazmiño, C., Guevara Castillo, W. D., Herrera-Fontana, M. E., Domínguez Brito, L. D., Becerra Granados, L. M., & Duran-Aguero, S. (2026). The seven methods for the evaluation of nutritional status—ABCDEFG: Narrative review. *Applied Sciences, 16*(2), 845.

Zavala, E., King, S. E., Sawadogo-Lewis, T., & Roberton, T. (2021). Leveraging water, sanitation and hygiene for nutrition in low-and middle-income countries: A conceptual framework. *Maternal & Child Nutrition, 17*(3), e13202.

Zhao, T., Liu, S., Zhang, R., Zhao, Z., Yu, H., Pu, L., & Han, L. (2022). Global burden of vitamin A deficiency in 204 countries and territories from 1990 to 2019. *Nutrients, 14*(5), 950.

Zheng, C., & Tako, E. (2024). Biochemical biomarkers of nutritional status. *Frontiers in Nutrition, 11*, 1397071.

CHAPTER SEVEN

Lipid and protein oxidation of foods and their implications on food quality, health, and diseases

E.D.N.S. Abeyrathne[a], X. Huang[b], and D.U. Ahn[c,*]
[a]Department of Animal Science, Uva Wellassa University, Badulla, Sri Lanka
[b]College of Food Science and Technology, Huazhong Agricultural University, Wuhan, P.R. China
[c]Department of Animal Science, Iowa State University, Ames, IA, United States
*Corresponding author. e-mail address: duahn@iastate.edu

Contents

Advances in Food and Nutrition Research, Volume 122
ISSN 1043-4526, https://doi.org/10.1016/bs.afnr.2026.07.003

Abstract

During food processing and storage, lipids and proteins are prone to oxidative degradation. This deterioration is caused by reactions with reactive oxygen species or with secondary compounds formed during oxidation. The mechanisms of lipid and protein oxidation share several similarities, and both processes can negatively influence the functional, nutritional, and sensory quality of food products. Lipid oxidation in foods has been recognized as a major quality issue. Still, protein oxidation in food systems has been largely ignored because it has been considered a secondary reaction to lipid oxidation. Both lipid and protein oxidation can influence food quality, including the production of off-odor and off-flavor, color and texture changes, and decreased protein functionality and consumer acceptance, and are involved in the pathogenesis of various diseases such as atherosclerosis, inflammation, cellular dysfunction, aging, neurodegenerative diseases, cancers, and other health conditions in humans and animals. The consumption of oxidized lipids and protein products has raised increasing concern due to their potential roles in the development of various chronic diseases in humans. Lipid and protein oxidation produce reactive oxygen species (ROS) and a variety of secondary oxidation products, such as aldehydes, ketones, and epoxides, which can interact with biological molecules and disrupt normal physiological functions. However, lipid oxidation products tend to promote protein oxidation, and their impact is more detrimental than that of protein oxidation. Consuming foods with oxidized lipids and proteins can increase oxidative stress in the body. Because the functions of proteins in biological systems are highly specific, oxidative modifications can compromise protein function and lead to pathological consequences. The oxidation processes, products, and mechanisms of lipids are well elucidated. However, details on protein oxidation, including peptide and amino acid oxidation, side chain degradation/cleavage and their end products, and the structural (primary, secondary, and tertiary) modifications of proteins on functionality, nutritional, digestibility, product quality changes, and the biological and health effects of consuming protein oxidation products need further study.

1. Introduction

Lipids are among the major nutrient sources for humans and animals and support many functional and regulatory processes in the body, including signal transduction, myelination, and synaptic plasticity. They are also involved in the structural developments of the human body (Zheng et al., 2019; Abeyrathne et al., 2021; Wang et al., 2023a). In food, lipid content and fatty acid composition are two critical congenital parameters that determine the susceptibility of food to oxidative changes. Lipid content and fatty acid composition in farm animals and fish vary significantly with diet (Baggio et al., 2002; Gao et al., 2021; Borys et al., 2012). On the other hand, fish is considered a source of ω-3 fatty acids, especially eicosapentaenoic acid (EPA) and docosahexaenoic acid (DHA) (Prato & Biandolino, 2012). However, they are highly susceptible to oxidative changes. Eggs are also rich in ω-3 fatty acids and are important in controlling cardiovascular diseases, inflammation, immune function, and mental health disorders (Xiao et al., 2020; Zhao et al., 2021). Not only animals but also plants contain fats and oils, which are important sources of nutrition for humans and are used in cooking. Plant oils are rich in linoleic, linolenic, oleic, and palmitic acids, and are also used in other industries for coatings, inks, plasticizers, lubricants, and paints (Gonçalves & Romano, 2021; Adekunle, 2015). However, these lipids can be easily oxidized over time under various conditions, and it is one of the major reasons for quality deterioration in food systems, particularly in products rich in unsaturated fatty acids such as fish, meat, dairy products, and vegetable oils. Lipid oxidation is a complex chain reaction involving initiation, propagation, and termination phases, resulting in the formation of hydroperoxides, the primary oxidation products, and the secondary oxidation products such as aldehydes, ketones, and alcohols (Choe & Min, 2007; Min & Ahn, 2005). These compounds are responsible for rancid flavors, off-odors, discoloration, and nutritional losses. In addition, some oxidation products pose health risks due to their potential toxicity (Shahidi & Zhong, 2010).

Protein oxidation is the chemical modification of proteins induced by reactive oxygen species or by indirect reactions involving secondary products of oxidative stress (Stadtman & Levine, 2000; Zhang et al., 2013; Kehm et al., 2021). The effects of protein oxidation in biological systems and in foods differ and should be discussed separately. In biological systems, protein oxidation compromises biological functions and causes cellular dysfunction, aging, neurodegenerative diseases, and various health conditions (Stadtman & Levine, 2000; Davies, 2016). However, studies on the health effects of oxidative stress in humans and animals are mainly focused on lipid oxidation products, and the biological effects of consuming oxidized proteins are very rare or nonexistent because lipids and proteins cannot be separated. Both lipid and protein oxidation can influence food quality, but lipid oxidation products tend to promote protein oxidation, and their impact is more detrimental than that of protein oxidation (Hematyar et al., 2019). Lipid oxidation causes off-odor and off-flavor generation, color and texture changes, and decreased protein functionality and consumer acceptance. Protein oxidation also negatively influences the functionality, nutritional quality, and eating quality of foods (Xiao et al., 2011a; Utrera & Estévez, 2013b; Nawaz et al., 2022). However, the degree of changes in food quality due to protein oxidation is less pronounced than that of lipid oxidation, and most of the functional and quality changes caused by protein oxidation in foods can be prevented using methods that prevent lipid oxidation. So, no separate methods to prevent protein oxidation are used, other than lipid oxidation in food systems.

2. Process of lipid and protein oxidation

2.1 Process of lipid oxidation: Initiation and propagation

Depending upon the reaction mechanisms and factors involved, lipid oxidation can be divided into autoxidation, photooxidation, and enzyme-catalyzed oxidation (Ahmed et al., 2016). Among them, autoxidation, the most common form of lipid oxidation in foods, proceeds through initiation, propagation, and termination stages. During the initiation step, free radicals abstract hydrogen atoms from the methylene group of a polyunsaturated fatty acid. Then, the fatty acid rearrangement (diene conjugation) stabilizes the fatty acid radical (Guéraud et al., 2010; Ayala et al., 2014). In the presence of oxygen, the conjugated dienes become peroxyl (lipid) radicals ($LOO^{\cdot}$) with high reactivity. The total number of bis-allylic carbons determines the rate of peroxyl radical formation in a fatty acid molecule. Peroxyl radicals are highly reactive and can abstract a hydrogen atom from an adjacent polyunsaturated fatty acid (Huang & Ahn, 2019), a process called propagation, and then form a hydroperoxide. When the propagation process is initiated, it continues until the termination step, when all unstable peroxyl radicals become stable non-radical species (Ayala et al., 2014). The initiation, propagation, and formation of hydroperoxides process is shown in Fig. 1.

2.2 Degradation of lipid hydroperoxides

Hydroperoxides, the primary products of lipid oxidation, are unstable and can decompose into the secondary lipid oxidation products. The degradation of hydroperoxides can be triggered by free radicals, metal catalysts, $Fe\text{-}O_2$ complexes (ferryl and perferryl), heat, and light. The presence of Fe^{2+} and oxygen is important because they can form $Fe\text{-}O_2$ complexes and generate free radicals, which are the major catalysts and the driving force for the further degradation of lipid hydroperoxides

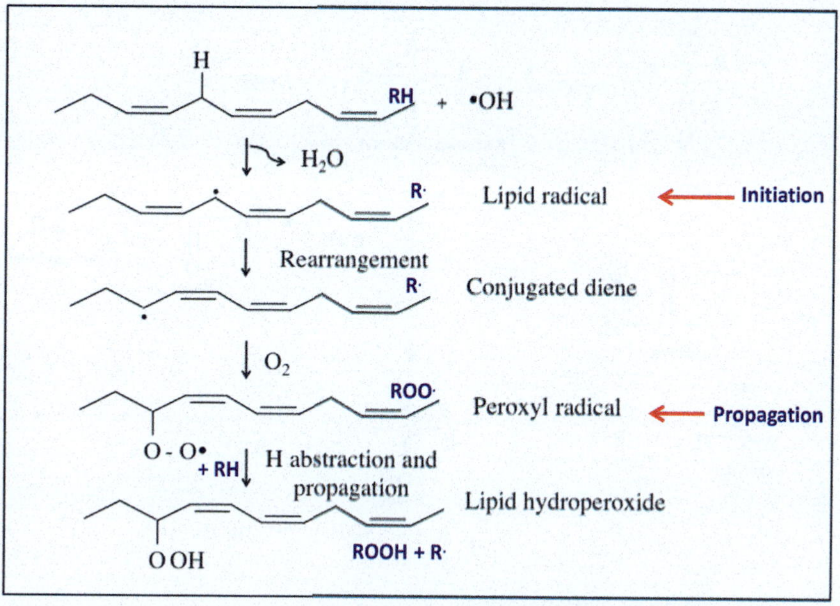

Fig. 1 Initiation and propagation processes of lipid oxidation.

(Fridovich, 1987; Halliwell & Gutteridge, 1989). Iron-oxygen complexes (ferryl and perferryl) and free radicals can be produced through the reactions of ferrous iron (Fe^{2+}) and oxygen (O_2). The binding of oxygen with ferrous iron produces ferryl, and the reaction intermediate of ferrous ion and oxygen is perferryl, as shown below:

$$Fe^{2+} + O_2 \rightarrow Fe^{2+} - O_2\,(\text{ferryl})$$

$$Fe^{2+} + O_2 \leftrightarrow Fe^{3+} - {}^{\cdot}O_2^{-}\,(\text{perferryl}) \leftrightarrow Fe^{3+} + {}^{\cdot}O_2^{-}$$

Lipid hydrogen peroxides are degraded into peroxyl and alkoxyl radicals through the catalytic activity of Fe-O_2 complexes, and the radicals are further degraded into secondary oxidation products by β-scission.

$$2LOOH \rightarrow LO^{\cdot}\,(\text{alkoxyl radical}) + LOO^{\cdot}\,(\text{peroxyl radical}) + H_2O \text{ iron complexes}$$

The degradation of lipid hydroperoxides leads to the formation of new free radicals, such as peroxyl and alkoxyl radicals, that can cause further reactions (Halliwell & Gutteridge, 1989). The peroxyl and alkoxyl radicals undergo further degradation via β-scission, producing various secondary oxidation products, including aldehydes, alcohols, ketones, hydrocarbons, and epoxy compounds (Fig. 2).

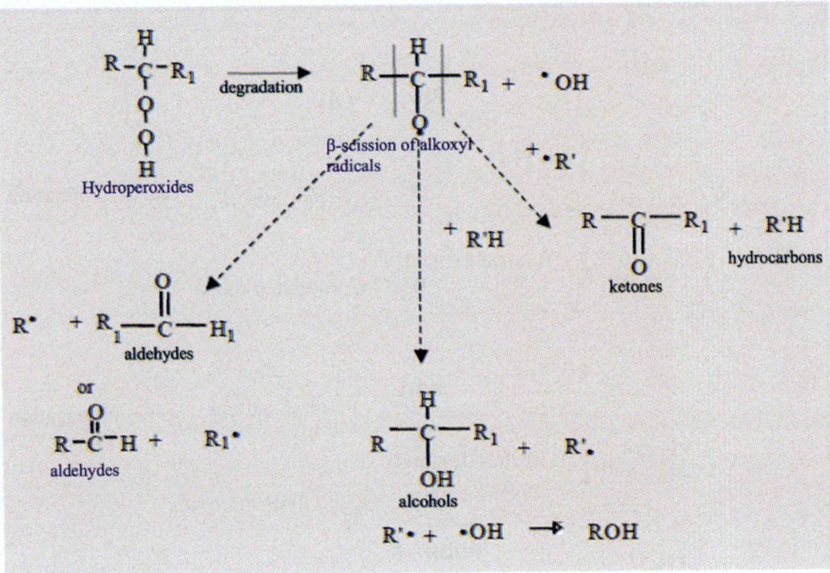

Fig. 2 Degradation of hydroperoxides.

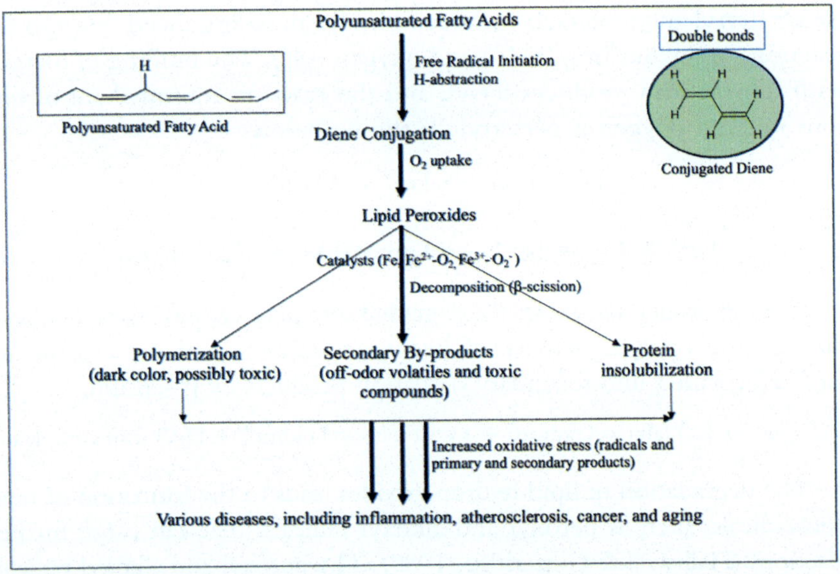

Fig. 3 A schematic diagram of the lipid oxidation (autoxidation) pathway and the effects of lipid oxidation products. *(Adopted from Abeyrathne et al., Antioxidants 2021, 10, 1587).*

2.3 Process of protein oxidation

Protein oxidation can be initiated by reactive oxygen species from internal and external sources, which can react with amino acids, altering protein structure. Protein oxidation in food systems can diminish functionality, such as protein solubility, texture, flavor, color, and water-holding capacity, and negatively influence product quality (Utrera & Estévez, 2013b; Al-Hijazeen et al., 2016; Domínguez et al., 2022; Wang et al., 2023a). Most of the reactive oxygen species that cause protein oxidation in foods are generated by the lipid oxidation processes, even though direct contact of proteins with external reactive oxygen species and oxidation catalysts occurs during production, processing, cooking, and storage (Xiao et al., 2011b; Hematyar et al., 2019; Nawaz et al., 2022; Geng et al., 2023). The main external sources of reactive oxygen species that contribute to protein oxidation in biological systems are irradiation (X-ray, γamma-ray, UV), air pollutants, and oxidized foods. The internal sources include inflammation (neutrophils, macrophages), endothelial cells (nitroxides from arginine), and mitochondria ($\cdot O_2^-$) (Huang & Ahn, 2019; Stadtman & Levine, 2000; Dragoev, 2024). Structural modifications of

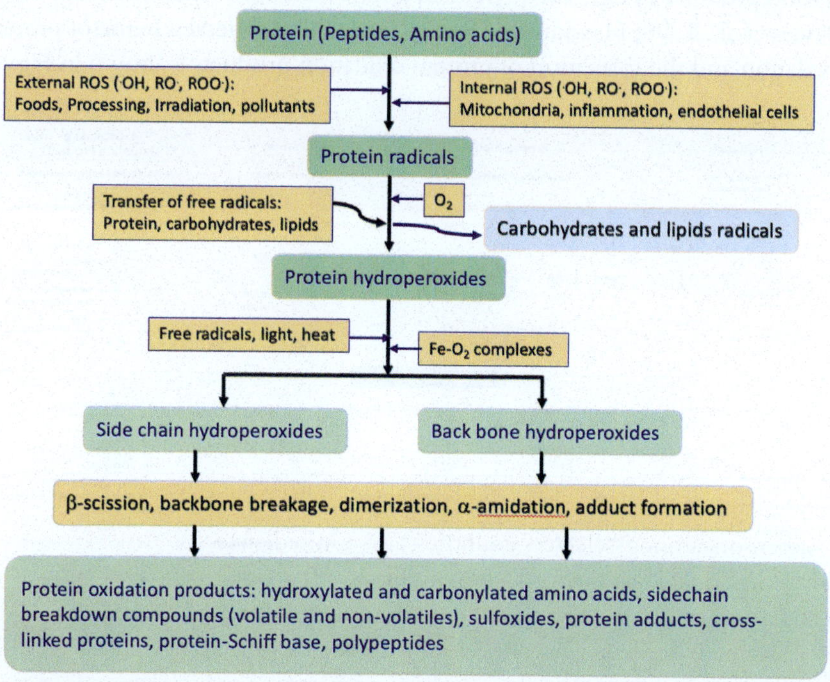

Fig. 4 The general scheme of protein oxidation and the formation of protein oxidation products.

proteins in biological systems compromise their functions and cause cellular dysfunction, aging, neurodegenerative diseases, and various health conditions (Reddy, 2023; Yang et al., 2024; Pooja et al., 2025) Figs. 3 and 5.

The process, principles, and mechanisms of protein oxidation are similar to lipid oxidation but do not follow the same pathway (Hawkins & Davies, 2001; Stadtman, 2006; Estevez, 2011; Hellwig, 2020). The pathway of protein oxidation is more complex, and the range of products is greater. However, its consequences in food systems are less noticeable to consumers than those of lipid oxidation (Domínguez et al., 2022). As in lipid oxidation, protein oxidation is initiated by the highly reactive oxygen and nitrogen species (e.g., ·OH, ROO·, RO·, $Fe^{2+}O_2$, NO·, and ONOO·) and the free radicals (amino acids, peptides, proteins) formed can transfer free radicals to other biomolecules and become hydroperoxides, which go through various degradation processes, including β-scission, dimerization, and α-amidation, producing various carbonylated, hydroxylated, sulfoxides, cyclic amino acids, and amino acid side-chain degradation products. These oxidation products can influence the functions of proteins, causing diseases, and damage DNA in biological systems and the functional, nutritional, and sensory characteristics in food products (Vogt, 1995; Gebicki & Gebicki, 1999; Luxford et al., 1999; Davies et al., 1999; Hawkins & Davies, 2001). The general scheme of protein oxidation and the formation of protein oxidation products is shown in Fig. 4.

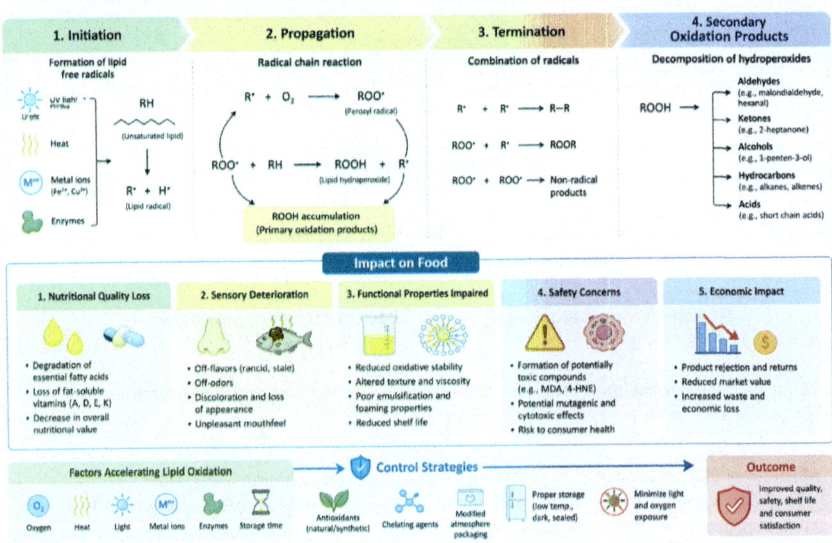

Fig. 5 Effects of lipid oxidation on foods.

3. Factors affecting lipid and protein oxidation in foods

Lipid oxidation is considered one of the main factors that directly affects the nutritional and functional quality, as well as the sensory and consumer acceptance, of foods. Various internal and external factors affect lipid oxidation in food products (Huang & Ahn, 2019; Domhnguez et al., 2022). The internal factors include fatty acid composition, antioxidants, and prooxidants. In contrast, the external factors include temperature, light, moisture level, atmospheric oxygen, additives such as anti- and prooxidants, processing methods, cooking, and storage conditions (Wu et al., 2017; Grabowski et al., 2019; Huang & Ahn, 2019).

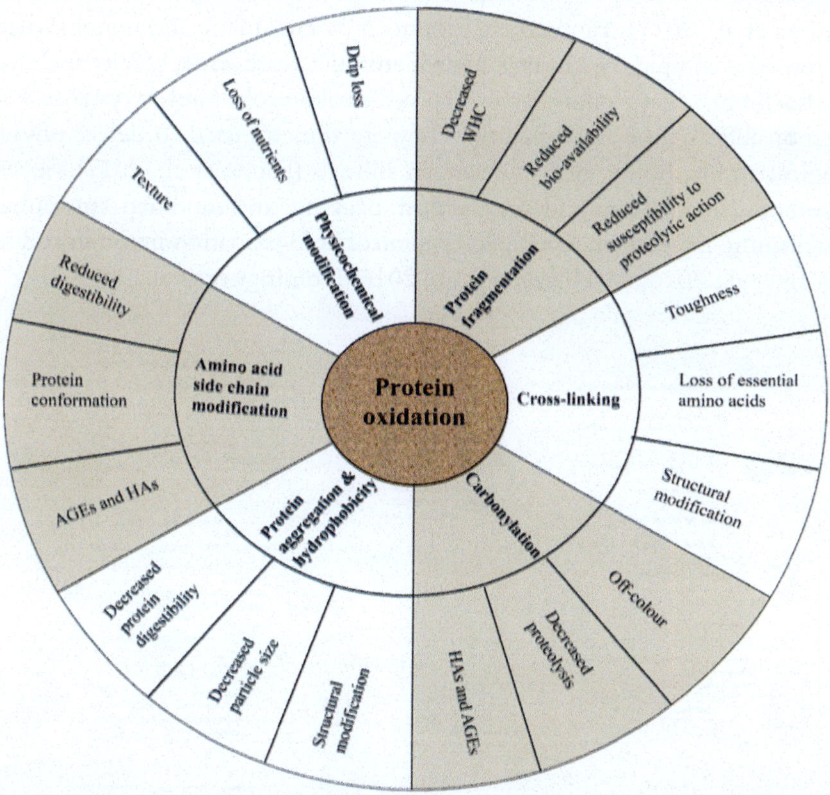

Fig. 6 Protein oxidation and its associated consequences during thermal processing of meat and meat-based products. *(Adopted from Nawaz et al., 2022. Food Res. Int. 157 (2022) 111322).*

Since lipids play a crucial role in food quality, and changes in lipids are the primary parameters in shelf-life analyses. The shelf-life of most meat and meat-based products is determined mainly by the degree of lipid oxidation (Domínguez et al., 2019; Amaral et al., 2018; Wazir et al., 2019). Rancidity is an undesirable flavor produced by the oxidation of lipids in raw meat. Warmed-over flavors are mostly developed by the secondary lipid oxidation products, such as hexanal, pentanal, pentanol, and nonanal (Min & Ahn, 2005; Huang & Ahn, 2019; Abeyrathne et al., 2021). Some lipid oxidation products, such as malondialdehyde, acrolein, 4-hydroxy-trans-nonenal, 4-hydroxy-trans-hexanal, and crotonaldehyde-like compounds, not only generate off-flavor but also cause diseases such as inflammation, aging, cancer, and atherosclerosis in humans (Wu et al., 2017). Also, they can interfere with the signaling pathways, leading to significant biomolecular damage in biological systems (Vieira et al., 2017; Papuc et al., 2017). Biomarkers produced by enzymatic and nonenzymatic hydrolysis of lipids, including hydroperoxides, carbonyls, aldehydes, alcohols, furans, keto-cholesterols, epoxy-cholesterols, and oxysterols, are responsible for the final quality. They are directly used to detect physiological and pathological conditions in animals (Papuc et al., 2017). Several methods, such as antioxidants, vacuum packing, and modified-atmosphere packaging, have been developed to control lipid oxidation in food products (Kim et al., 2016; Al-Hijazeen et al., 2016; Domínguez et al., 2019). Fig. 6.

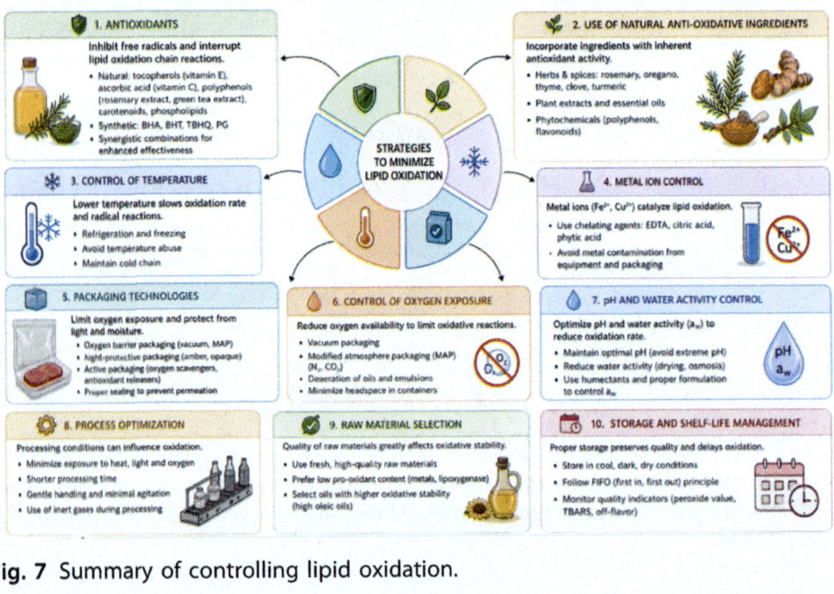

Fig. 7 Summary of controlling lipid oxidation.

As in lipid oxidation, oxygen availability is the most important factor affecting protein oxidation. However, other factors such as the presence of transitional metals (Fe and Cu), exposure to UV light, antioxidants, high temperature, extreme pH conditions, processing conditions, storage time and temperature, and environmental pollutants also play important roles in protein oxidation (Utrera et al., 2012; Huang & Ahn, 2019; Stadtman & Levine, 2000). Therefore, any preventive methods for lipid oxidation, including adding antioxidants, preventing oxygen contact, chelating metal catalysts, and storing at low temperatures, would be relevant to preventing protein oxidation in foods Fig. 7.

4. Effects of lipids and proteins on foods

Lipids and proteins are fundamental components of food systems and contribute significantly to nutritional value, sensory attributes, and overall food quality. Lipids are chemically unstable, particularly when they contain unsaturated fatty acids, and they are highly prone to oxidative and hydrolytic degradation. These degradation processes lead to extensive chemical transformations, including protein oxidation, that adversely affect the nutritional, sensory, and functional properties of foods. Among the chemical modifications, lipid and protein oxidation are the most significant, as it triggers complex reactions that produce both primary and secondary oxidation products. These compounds degrade food quality and affect the safety and consumer acceptability of food products (Du et al., 2001b; Wang et al., 2012) Table 1.

4.1 Changes in nutritional value and functional characteristics

The most critical consequences of lipid and protein oxidation in foods are the loss of nutritional quality and functional characteristics. Lipids are a major source of energy and provide essential fatty acids, linoleic and linolenic acids, which are susceptible to oxidative degradation due to their double bonds (Wang et al., 2023a; Frydrych et al., 2025). During lipid oxidation, highly reactive oxygen species usually attack the methylene group between two double bonds, abstract a hydrogen atom, and eventually form lipid hydroperoxides. As oxidation progresses, these hydroperoxides decompose into secondary products, such as aldehydes and ketones, leading to structural degradation of fatty acids and loss of

Table 1 Odor characteristics of major volatile compounds from lipid oxidation.

Volatile compound	**Odor characteristics**
Pentane, heptane, octane	Gasoline-like
Cyclohexane	Sweet, aromatic
Cyclohexanone	Sweet, peppermint
Acetone	Minty chemical, sweet
1-Penten-3-one	Rotten, sewer-like, fruity
2-Heptanone	Cured ham-like, toasted, nutty
1-Octen-3-one/1-octen-3-ol	Mushroom-like, dirty, dust, blood
Hexanol	Sweet, alcohol, woody
Acetaldehyde	Green, sweet, fruity
Hexanal	Green, grassy, fruity
2-Methylpropanal	Toasted, fruity, pungent
3-Methylbutanal	Fruity, almond-like, toasted
Octanal	Stew-like, boiled meat-like, rancid
Hydrogen sulfide	Boiled or rotten eggs, sewage
Methanethiol	Rotten eggs, meat or fish, cheesy
Dimethyl sulfide	Decayed cabbage

nutritional value (Abeyrathne et al., 2021; Wang et al., 2023a). Lipid oxidation products can interact with proteins, forming complexes that reduce protein digestibility and bioavailability, as well as tenderness, water-holding capacity, and juiciness of meat (Hematyar et al., 2019). Reactive compounds, such as malondialdehyde, can bind to amino acids, contributing to overall nutritional loss (Abeyrathne et al., 2021).

When proteins in food products are oxidized, various chemical changes, including carbonylation, carboxylation, and Schiff base formation, cause structural modifications and potential loss of functionality, nutritional value, digestibility, and bioavailability of proteins (Utrea & Estevez, 2013; Geng et al., 2023; Zhu et al., 2025; Sun et al., 2026). Most carbon-centered radicals ($^{\cdot}$CH-) rapidly react with oxygen (O_2) and produce protein hydroperoxides (amino acid and peptides, as well). Although reactive oxygen and nitrogen species can react with all cellular components, proteins scavenge the majority (50–75%) of these species (Kehm et al., 2021). Among amino acids, sulfur-containing (cysteine and methionine), aromatic (tyrosine, tryptophan, phenylalanine, histidine, and proline), and basic (lysine and arginine) amino acids are the most susceptible to oxidative modifications (Davies, 2016). The metal-catalyzed oxidation system oxidizes the side chains of susceptible amino acids, leading to protein aggregation, carbonylation, and changes in hydrophobicity and 3D structure (Amici et al., 1989; Nawaz et al., 2022). The major quality parameters of meat, including hydrophobicity, solubility, WHC, tenderness, and texture, are directly related to protein oxidation (Lund et al., 2011). Oxidation of amino acid side chains results in protein insolubility and aggregation (Rowe et al., 2004).

4.2 Changes in the color of foods due to oxidation

Color is a key quality attribute that inspires consumer perception. Lipid oxidation significantly affects the color of food products, especially in meat and fish. The primarily cause is the oxidation of myoglobin, the pigment responsible for the red color of fresh meat (Du et al., 2001a). In fresh meat, oxymyoglobin gives a bright red appearance. However, under oxidative conditions, oxymyoglobin is converted to metmyoglobin, resulting in a brown discoloration (Du et al., 2001a; Abeyrathne et al., 2022). Lipid oxidation and pigment oxidation are closely interconnected processes. Free radicals generated during lipid oxidation accelerate pigment degradation, leading to rapid discoloration (Wang et al., 2023b). In plant-based foods and edible oils, oxidation can degrade pigments, such as carotenoids and chlorophyll, leading to color fading or darkening and reducing visual appeal (Geng et al., 2023).

4.3 Changes in flavor/taste and other organoleptic properties of foods

Flavor deterioration is one of the most noticeable effects of lipid oxidation in foods. The breakdown of lipid hydroperoxides produces various volatile compounds, including aldehydes, ketones, alcohols, and hydrocarbons, which are responsible for off-odors and flavors (Wang et al., 2012). Aldehydes, such as hexanal and heptanal, are particularly important contributors to the oxidation off-odor (Table 1). These compounds produce undesirable sensory notes, often described as rancid, cardboard-like, metallic, or grassy (Abeyrathne et al., 2022; Wang et al., 2023b). Even at low concentrations, these compounds significantly impact food acceptability. In cooked meat products, lipid oxidation leads to the development of "warmed-over flavor" (WOF), which occurs during refrigerated storage and is associated with rapid formation of oxidation products (Fu et al., 2022). Lipid oxidation can also mask desirable flavors: in dairy products, oxidation leads to the loss of fresh flavors and the development of oxidized notes. At the same time, in edible oils, oxidation results in bitterness and sharp off tastes (Wang et al., 2023a). Lipid oxidation products can interact with proteins, leading to cross-link, increasing toughness, and reducing tenderness in meat products (Pagey et al., 2025; Aruchunan et al., 2025). In fried foods, lipid oxidation can lead to lipid polymerization and increased oil viscosity, affecting crispness and overall texture (Estévez & Xiong, 2021; Bazina et al., 2025). These changes reduce the sensory quality and consumer acceptance of food products.

Protein oxidation affects other organoleptic properties such as color, odor, texture, and overall mouthfeel. The backbone fragmentation, cleavage of amino acid side chains, disulfide bond formation, and α-amidation of proteins by protein oxidation produce various volatiles, as in lipid oxidation. Especially, sulfur compounds, such as hydrogen sulfide, sulfur dioxide, mercaptomethane, methyl sulfide, methyl thioacetate, dimethyl disulfide, and trimethyl sulfide, can be produced through the degradation of methionine and cysteine side chains by free radicals from irradiation, and play significant roles on the odor and flavor of oxidized food products (Jo & Ahn, 2000; Fan et al., 2002; Ahn, 2002, Ahn & Lee, 2002; Lee & Ahn, 2003; Ahn et al., 2016a; Wang et al., 2023b). The accumulation of volatile oxidation products from lipid and protein oxidation leads to unpleasant odors ranging from stale and rancid to sulfury (Lee & Ahn, 2003; Ahn et al., 2016a; Wang et al., 2023a; Pagey et al., 2025). However, irradiation-induced degradation of non-sulfur amino acid side chains produced various volatile compounds but had only marginal effects on the odor of oxidized samples (Ahn et al., 2016b). Texture changes are also significant, particularly in protein-rich foods. Overall, lipid and protein oxidation are interdependent and lead to simultaneous deterioration of multiple sensory attributes, ultimately reducing the shelf life, marketability, and consumer acceptability of food products (Estévez & Xiong, 2021; Wang et al., 2023a).

5. Effects of lipid and protein oxidation on biological systems

The consumption of oxidized lipids and protein products has raised increasing concern due to their potential role in the development of various chronic diseases in humans. Lipid and protein oxidation produce reactive oxygen species (ROS) and a variety of secondary oxidation products, such as aldehydes, ketones, and epoxides, which can interact with biological molecules and disrupt normal physiological functions (Vieira et al., 2017; Hu et al., 2021). One of the most studied compounds is malondialdehyde (MDA), a reactive aldehyde formed during lipid peroxidation. MDA can form adducts with proteins, DNA, and phospholipids, leading to structural and functional damage and contributing to mutagenesis and carcinogenesis (Jackson, & Penumetcha, 2019). Another important group of lipid oxidation products is 4-hydroxy-2-nonenal (4-HNE) and other α,β-unsaturated aldehydes. These compounds are highly reactive and can modify cellular proteins and enzymes, impair their normal function, and contribute to oxidative stress (Li et al., 2022). Oxidized lipids, particularly oxidized low-density lipoproteins (LDL), play a key role in the formation of atherosclerotic plaques. These plaques accumulate in arterial walls, restricting blood flow and increasing the risk of heart attacks and strokes (Li et al., 2022; Liu et al., 2025). Many of the oxidized lipid oxidation products, such as aldehydes and oxysterols, can be absorbed in the intestines, transported into blood stream, and be involved in the pathogenesis of various diseases, including atherosclerosis, Alzheimer's disease, neurodegenerative diseases, diabetes, hypertension, mutagenicity, cancer, and inflammatory bowel diseases in humans and animals (Hu & Xie, 2021).

Furthermore, oxidized lipids may contribute to inflammatory responses in the body. Many lipid oxidation products act as signaling molecules that trigger inflammation, which is a common underlying factor in several chronic conditions, including diabetes, neurodegenerative diseases, and cancer (Pooja et al., 2025). In addition to long-term health effects, consuming oxidized fats can also have immediate physiological consequences. These include reduced digestibility, gastrointestinal irritation, and decreased absorption of essential nutrients. The degradation of essential fatty acids and fat-soluble vitamins further exacerbates nutritional deficiencies (Li et al., 2022; Liu et al., 2025). Importantly, the extent of the health impact depends on factors such as oxidation level, frequency of consumption, and individual susceptibility. While the human body possesses natural antioxidant defense systems, excessive intake of oxidized lipids can overwhelm these defenses and lead to cumulative damage over time.

Oxidative stress occurs when the production of free radicals exceeds the antioxidant defenses, leading to cellular damage in biological systems. ROS and reactive nitrogen species can directly oxidize proteins, and oxidation products can then react with reducing sugars or carbohydrates (Lund et al., 2011). Consuming foods with oxidized lipids and proteins can increase oxidative stress in the body. Because the functions of proteins in biological systems are highly specific, oxidative modifications can compromise protein function and lead to pathological consequences, including DNA damage, mitochondrial dysfunction, inflammation, coronary vascular diseases, cancer, and aging in humans (Yang et al., 2024). Therefore, lipid and protein oxidation products accumulate more readily in biological systems than in food systems.

6. Detection methods of lipid and protein oxidation

Many methods are used to determine the level of oxidation in foods by detecting lipid oxidation products. The oxidation products are categorized into primary and secondary products. Peroxides and hydroperoxides are the primary oxidation compounds, and aldehydes, ketones, epoxides, alcohols, hydroxy compounds, oligomers, and polymers are the secondary products of lipid oxidation (Kunyaboon et al., 2021). Critical criteria for detecting lipid oxidation include accuracy, precision, sensitivity, and detection limits (Mas-Bargues et al., 2021). Either direct or indirect methods are used to determine lipid oxidation in foods. The most common methods for determining primary oxidation products include peroxide value, iodine value, and diene conjugation measurements (Jo & Ahn, 1999; Barriuso, & Astiasarán, 2013; Kunyaboon et al., 2021).

6.1 Methods used to detect the primary lipid oxidation products

The three most widely used methods for primary oxidation products in foods are the iodometric, ferric thiocyanate, and diene conjugation methods, which determine the amount of hydroperoxides formed by oxidation (Abeyrathne et al., 2021). The peroxide value (PV) is determined by titration or spectrophotometry, and the diene conjugation method quantifies all conjugated fatty acids, including lipid peroxides, in food products after lipid extraction. All three methods are used to determine the early stage of lipid oxidation, and their sensitivities are lower than those for measuring secondary products. However, the values of primary products indicate the degree of lipid oxidation at an early stage and do not reflect the food's sensory quality. However, they indicate the potential for further oxidation in the later stage, when oxygen becomes available.

The ferric thiocyanate method requires prior lipid extraction from the food sample before analysis. This assay is based on the conversion of ferrous ions (Fe^{2+}) to ferric ions (Fe^{3+}) through their reaction with lipid hydroperoxides (LOOH). During this process, hydroperoxides undergo homolytic decomposition, generating highly reactive lipid alkoxyl radicals (LO•). These radicals subsequently participate in further reactions with ferrous ions, hydroperoxides, and surrounding solvent molecules. Owing to its effectiveness in detecting primary lipid oxidation products, the ferric thiocyanate assay has been successfully applied to assess oxidative deterioration in insect-derived food products (Ojha et al., 2021; Tenyang et al., 2021). Conjugated dienes are formed to stabilize an oxidized fatty acid radical by moving a double bond to a neighboring carbon-carbon bond (Huang & Ahn, 2019). When oxygen is available, the PUFA peroxyl radical reacts with oxygen to form a hydroperoxide (lipid peroxide) and a conjugated PUFA. The diene conjugation values agree well with those of the peroxide and iodine values, but the absorption peak is distinct only when lipids are fully peroxidized (Buege & Aust, 1978; Corongiu & Milia, 1983; Recknagel et al., 1984).

6.2 Methods to determine the secondary oxidation products

While primary oxidation products are commonly used as indicators of oxidative deterioration due to their ease of determination, they are transient compounds that decompose over time into a diverse range of secondary (degraded) oxidation products. The fluorometric TBARS assay is based on the formation of three-carbon compounds, mainly malondialdehyde (MA), during the oxidation of unsaturated fatty acids. These compounds react with thiobarbituric acid (TBA) to form a colored complex that can be quantified. Nevertheless, the specificity of this method may be compromised because various food constituents, including proteins, amino acids, and carbohydrates, can also react with TBA, potentially interfering with the results (Jo & Ahn, 1999; Pérez-Palacios, & Estévez, 2020). Due to the limited specificity and sensitivity of traditional TBARS assays, more advanced analytical techniques have been developed to improve the accuracy of lipid oxidation measurements. Chromatographic methods have been developed to determine the primary and secondary oxidation products present in food (Kakuda et al., 1981). The method is highly sensitive compared with other methods, but it is time-consuming. Product identification and quantification can be easily achieved when a high-performance liquid chromatography (HPLC) unit is coupled to a mass spectrometer (Kato et al., 2018; Abeyrathne et al., 2021). A recently developed analytical technique, comprehensive two-dimensional liquid chromatography (LC × LC) coupled with an evaporative light-scattering detector (ELSD), has been employed to analyze lipid oxidation products. This approach enables the simultaneous detection of both polar and non-polar oxidation compounds (Steinhotst-Slikkerveer et al., 2000).

6.3 Sensory analysis method

Sensory evaluation is a method used to assess human responses to food properties. Through the participation of trained or untrained panelists, attributes such as aroma, flavor, and taste can be examined to characterize the sensory quality of food products (Fruhwirth et al., 2003). Although sensory evaluation is an inexpensive and versatile tool for assessing a variety of food products, its outcomes may vary considerably due to differences in panelists' preferences, experiences, cultural influences, and demographic characteristics, as well as evaluation conditions, which can affect the consistency of results.

6.4 Methods to detect both the primary and secondary oxidation products

The p-anisidine value (p-AV) and total oxidation value (TOTOX) are widely employed analytical indices for assessing lipid oxidation in food systems, particularly in edible oils. These methods provide information on both primary and secondary oxidation products, thereby offering a more comprehensive evaluation of oxidative deterioration (Cong et al., 2020). The p-anisidine assay specifically targets aldehydic compounds, including 2-alkenals and 2,4-dienals, which are generated during the decomposition of lipid hydroperoxides. The concentration of these oxidation products is determined spectrophotometrically by measuring absorbance at 350 nm after reaction with the p-anisidine reagent (Cong et al., 2020; Shahidi & Zhong, 2015). The p-anisidine value (p-AV) and total oxidation value

(TOTOX) are widely used to evaluate the oxidative stability and quality of edible oils, particularly those enriched in EPA and DHA. These methods are relatively straightforward to perform, require minimal technical expertise, and provide useful information regarding the extent of lipid oxidation (Shahidi & Zhong, 2015; Ismail et al., 2016). Despite these advantages, the p-anisidine assay has certain limitations, as it may not accurately assess oxidation in highly colored omega-3-rich oils or in products containing added flavoring agents, which can interfere with the analytical measurements (Ismail et al., 2016).

7. Detection methods of protein oxidation

Detecting protein oxidation is more difficult than detecting lipid oxidation. Proteins are much larger than lipids, and their oxidation pathways are more complex; thus, their end products are more diverse and complex than those of lipid oxidation. The oxidative modifications of proteins include oxidation of sulfur-containing amino acids, such as cysteine and methionine; oxidation of aromatic amino acids, such as tyrosine, tryptophan, phenylalanine, and histidine; and glycoxidation of basic amino acids, such as lysine and arginine. Among the protein oxidation markers, measuring protein carbonylation using the dinitrophenylhydrazine (DNPH) method is most commonly used in food systems (Lund et al., 2011). Carbonyls of proteins are produced by ROS via four pathways (i) direct oxidation of the side chains (lysine, threonine, arginine, tyrosine, cysteine, methionine, tryptophan, histidine, and proline) (Requena et al., 2001; Kehm et al., 2021), (ii) nonenzymatic glycation in the presence of reducing sugars (Akagawa et al., 2005); (iii) oxidative cleavage of the peptide backbone via the α-amidation pathway or via oxidation of glutamyl side chains (Berlett & Stadtman, 1997), and (iv) covalent binding to non-protein carbonyl compounds such as 4-hydroxy-2-nonenal (HNE) or malondialdehyde (MDA) (Feeney et al., 1975) and are used as protein oxidation markers. The carbonylated proteins are derivatized with 2,4-dinitrophenylhydrazine and measured by fluorescence spectroscopy (excitation at 450 nm and emission at 530 nm), immunochemistry (anti-DNPH), and mass spectrometry (DNPH fragments in LC-MS) (Kehm et al., 2021).

7.1 Detection of protein carbonylation

Carbonyls are formed in proteins through the deamination of alkaline amino acids such as lysine, arginine, and proline. The 2,4-dinitrophenylhydrazine

(DNPH) assay is the simplest, most convenient, and most inexpensive method commonly used in muscle foods (Domínguez et al., 2022). In this method, carbonyl groups in proteins react with DNPH under acidic conditions to form hydrazones, which are quantified at 370 nm using a spectrophotometer (Zhang et al., 2013; Soladoye et al., 2015; Hellwig, 2020). The results are expressed as nanomoles of carbonyl/mg of protein. However, this method lacks specificity and overestimates carbonyls (Lorido et al., 2016). To address the drawbacks of the DNPH assay, Enzyme-Linked Immunosorbent Assays (ELISA) and immunoblot methods have been developed. These methods are highly sensitive but complex, time-consuming, and expensive compared with other techniques (Schaich, 2016). Alternatives to the DNPH assay, a fluorescence microscopy assay that uses fluorescent labeling (coumarin-hydrazide), have been successfully tested in muscle foods (Cropotova & Rustad, 2019).

7.2 Detection of loss of sulfhydryl groups and loss of tryptophan

The thiol groups of amino acids are very sensitive to oxidation by ROS, leading to the formation of oxidized products (fulfenic and sulfinic acids) and disulfide cross-links. To determine oxidized sulfur amino acids, the Ellman reagent, 5,5'-dithiobis (2-nitrobenzoate) (DTNB), is used as a derivatization agent to form a disulfide bond with free thiol groups. This reaction releases a thiolate ion with a maximum absorbance at 412 nm, which determines the total thiol content. So, the free thiol content should be determined in unreduced proteins. The loss of thiol groups is then calculated by difference (Soladoye et al., 2015). The loss of thiol groups can also be determined with 4,4-dithiodipyridine (4-DPS) at 324 nm, which has better solubility and smaller molecular volume than DTNB. Both DTNB and 4-DPS assays have been widely applied in muscle foods.

Tryptophan has an aromatic ring that is responsible for the natural fluorescence observed at 350 nm upon excitation at 280 nm. Changes in the fluorescence of this amino acid are used to assess the oxidative degradation of tryptophan in proteins. The loss of tryptophan is evaluated by fluorescence spectroscopy, which offers several advantages, such as simplicity and rapidity, and is a solvent-free procedure that has been successfully used to evaluate protein oxidation in muscle foods (Utrera et al., 2012).

7.3 Assessment of cross-linking and fragmentation of proteins

Free radicals produced during protein oxidation can react with peptide chains, leading to protein fragmentation (Heinonen et al., 2020). Specifically, cross-linking, fragmentation, and disruption of protein quaternary structure cause changes in molecular weight patterns, which can be easily evaluated by separating proteins on polyacrylamide gel electrophoresis with sodium dodecyl sulfate (SDS-PAGE) (Gebicki & Gebicki, 1999; Soladoye et al., 2015). This technique is the most commonly used to evaluate this manifestation of protein oxidation and has been employed in various meats.

Schiff bases can be formed from the reaction of alkaline amino acids (e.g., lysine) and protein carbonyls (mainly amino acids). Schiff bases emit fluorescence. So, they can be easily measured by fluorescence spectroscopy with excitation at 350 nm and emission at 450 nm. Results are expressed as fluorescence intensity units emitted by protein oxidation products at 450 nm (Lorido et al., 2016; Carvalho et al., 2017).

8. Control of lipid and protein oxidation in foods

Given its significant impact on food quality and safety, controlling lipid and protein oxidation is a critical step in food processing, preservation, and storage. Various strategies, ranging from antioxidants to advanced packaging technologies, have been developed to minimize lipid and protein oxidation. One of the most effective approaches to minimizing lipid oxidation is the addition of antioxidants. Antioxidants interrupt oxidation chain reactions by scavenging free radicals, chelating metals, generating reducing conditions, or stabilizing heme pigments (Shahidi & Zhong, 2010; Durand et al., 2025). A range of synthetic antioxidants, including butylated hydroxyanisole (BHA), butylated hydroxytoluene (BHT), tertiary-butylhydroquinone (TBHQ), and propyl gallate, have been widely adopted in the food industry owing to their effectiveness in inhibiting oxidation and their cost-efficient nature (Choe & Min, 2006). Therefore, the antioxidant radicals lose their reactivity and cannot propagate free radicals to new polyunsaturated fatty acids, thereby stopping chain reactions (Huang & Ahn, 2019). TBHQ is particularly effective in stabilizing frying oils. However, safety concerns and consumer preference for natural products have led to reduced acceptance of synthetic antioxidants. Natural antioxidants are increasingly preferred due to their safety and added health benefits. Common examples include tocopherols (vitamin E), ascorbic acid (vitamin C), and polyphenols found in plant sources (Pokorný, 2007; Chibuye et al., 2024; Shahidi & Zhong, 2010; Abeyrathne et al., 2022).

8.1 Control of oxygen exposure

Oxygen plays a crucial role in lipid oxidation by reacting with lipid radicals and promoting the formation of peroxides. Consequently, reducing oxygen availability is an effective strategy for slowing oxidative reactions and minimizing lipid deterioration in food products. There are many methods for controlling the amount of oxygen that the food encounters. Among them, vacuum packing, modified-atmosphere packing, and the use of oxygen scavengers and barrier materials are common industrial techniques.

Vacuum packaging is a widely used technique to control lipid oxidation by physically removing air from the package before sealing, thereby significantly reducing oxygen concentration available for oxidative reactions (Lee & Ahn, 2003). Since oxygen is a component in lipid peroxidation, its removal stops the initiation and propagation phases of oxidation. This method is particularly effective for high-fat foods, such as meat, fish, dairy products, and nuts, where oxidative rancidity can rapidly compromise quality (Oh et al., 2023; Porter, 2025; Robertson, 2016).

Modified atmosphere packaging (MAP) involves changing the air inside a package with a controlled gas mixture, naturally reducing oxygen levels while increasing inert gas concentrations, such as nitrogen (N_2) or carbon dioxide (CO_2) (Khan & Mittal, 2020). In modified atmosphere packaging (MAP), nitrogen is commonly used as an inert filler gas to replace oxygen and maintain package integrity by preventing collapse. Carbon dioxide contributes to product preservation through its antimicrobial activity, which suppresses the growth of spoilage microorganisms. The reduction of oxygen within the package environment helps retard oxidative reactions involving lipids and proteins, thereby enhancing the sensory attributes, nutritional value, and overall shelf life of food products (Floros & Matsos, 2005; Robertson, 2016; Khan & Mittal, 2020).

Oxygen scavengers and barrier materials are advanced packaging solutions designed to minimize oxygen exposure further and enhance the oxidative stability of lipid-containing foods (López-de-Dicastillo et al., 2023). Oxygen scavengers are typically small sachets or incorporated compounds that chemically react with and remove residual oxygen present within the package after sealing (Dey, & Sudarsan, 2019; López-de-Dicastillo et al., 2023). In parallel, barrier materials such as multilayer films made from polymers like polyethylene terephthalate (PET) and ethylene vinyl alcohol (EVOH), or aluminum laminates, are used to prevent the ingress of oxygen from the external environment during storage. These materials have low oxygen transmission rates, making them highly effective in protecting sensitive foods over extended periods (Han, 2014; Türksever et al., 2025).

8.2 Temperature control

Temperature control plays a crucial role in minimizing lipid oxidation, as oxidation reactions are highly temperature-dependent and are generally accelerated at higher temperatures. When food is exposed to elevated temperatures, the kinetic energy of molecules increases, leading to a higher rate of free radical formation and faster progression of oxidative chain reactions. Refrigeration and freezing are essential preservation methods used to control lipid oxidation because lowering the temperature of food systems slows down chemical reaction rates. At reduced temperatures, molecular mobility decreases, limiting interactions between oxygen and lipid molecules and the formation and propagation of free radicals responsible for oxidation (Suárez-Medina et al., 2024). Refrigeration (typically 0–5 °C) is effective for short-term storage, particularly for perishable foods such as meat, fish, and dairy products, while freezing (−18 °C or lower) is more suitable for long-term preservation (Aung & Chang, 2014; Fellows, 2017). Despite its effectiveness in slowing lipid oxidation, freezing has certain limitations that can negatively impact food quality. One major issue is the formation of ice crystals within the food matrix, which can rupture cell membranes and disrupt cellular structures (Wang et al., 2023a; Yang et al., 2025).

Additionally, repeated freeze–thaw cycles can exacerbate this damage, further accelerate oxidative deterioration, and lead to off-flavors and textural degradation (Zhu et al., 2025). In fatty fish and meat products, this can result in noticeable rancidity even after frozen storage (Hu & Xie, 2021). Maintaining consistent temperature conditions throughout storage, transportation, and distribution is critical in controlling lipid oxidation. Fluctuations in temperature can accelerate oxidative reactions by intermittently increasing molecular mobility and promoting the breakdown of lipid structures (Aung & Chang, 2014; Choe & Min, 2006).

8.3 Control of light exposure

Light exposure, particularly in the ultraviolet (UV) and visible regions, plays a significant role in initiating and accelerating lipid oxidation in food systems through photooxidation mechanisms. When food is exposed to light, photosensitizers absorb energy, leading to the formation of highly reactive species that can rapidly attack unsaturated lipids (Csapó et al., 2019). Unlike thermal oxidation, photooxidation can occur at relatively low temperatures and proceeds rapidly once initiated (Wang et al., 2023a). Therefore, controlling light exposure during processing, storage, and retail display is essential to maintaining product stability and extending shelf life (Min & Boff, 2002; Tao, 2015).

Light-induced oxidation occurs primarily through the action of photosensitizers - naturally occurring compounds in foods, such as chlorophyll, porphyrins, myoglobin, riboflavin, bilirubin, erythrosine, etc., that absorb light energy and transfer it to molecular oxygen (Baptista et al., 2017; Tsubone et al., 2019; Fujii et al., 2023). The rate of this process depends on factors such as light intensity, wavelength, presence of sensitizers, and oxygen availability. Foods like milk and green vegetables are particularly susceptible due to their natural riboflavin and chlorophyll content. Because singlet oxygen-mediated oxidation is much faster than conventional autoxidation, even minimal light exposure can significantly impact food quality (Min & Boff, 2002; Tsubone et al., 2019).

Protective packaging is a key strategy for minimizing light-induced lipid oxidation by limiting the transmission of harmful wavelengths of light to food products. Materials that are opaque, pigmented, or specifically designed to block ultraviolet and visible light can significantly reduce photooxidation (Sani et al., 2024). In addition, advanced packaging technologies may incorporate light-filtering additives that selectively block wavelengths that activate photosensitizers (Tsubone et al., 2019; Akhila et al., 2023). By effectively reducing light penetration, protective packaging helps preserve sensory attributes, nutritional value, and overall product quality throughout storage and distribution (Robertson, 2016).

8.4 Effects of prooxidants

Prooxidants, particularly transition metal ions, such as iron (Fe^{2+}/Fe^{3+}) and copper (Cu^{+}/Cu^{2+}), play a crucial role in accelerating lipid oxidation by catalyzing the decomposition of lipid hydroperoxides into highly reactive free radicals (Rodrigo et al., 2010; Ghazizadeh et al., 2020). These radicals further propagate the oxidation chain reaction, leading to rapid deterioration of food quality. Metal ions can come from raw materials, water, processing equipment, or environmental contamination (Mani et al., 2021; Shahidi & Zhong, 2010; Mani et al., 2021).

Chelating agents are compounds that bind metal ions and form stable, inactive metal complexes, thereby preventing them from participating in oxidation reactions. Common food-grade chelators include citric acid, phosphates, and ethylenediaminetetraacetic acid (EDTA), which are widely used in food systems to inhibit metal-catalyzed oxidation (Kim et al., 2019; Franz, 2013). Their effectiveness depends on factors such as pH, concentration, and specific metal types. When used appropriately, chelating agents can significantly enhance the oxidative stability and overall quality of food products (Pokorný, 2007).

Processing controls are essential in minimizing the introduction and activity of prooxidant metals during food production and handling. The use of non-reactive materials, such as stainless steel for equipment, storage tanks, and pipelines, helps prevent metal leaching that could otherwise catalyze oxidation reactions. In contrast, the use of reactive metals, like iron or copper, should be avoided or properly coated (Gordon et al., 2020). Minimizing mechanical damage during processing is also important, as it can release endogenous metal ions from food tissues. By implementing these controls, the overall exposure of lipids to catalytic metal ions is reduced, thereby slowing oxidation and preserving the sensory and nutritional quality of foods (Shahidi & Zhong, 2010).

8.5 Modification of food composition

The susceptibility of foods to lipid oxidation is largely governed by their intrinsic composition. Factors including the degree of fatty acid unsaturation, the balance between prooxidative and antioxidative constituents, and the physical characteristics of the food matrix significantly affect oxidative stability. Foods enriched with unsaturated lipids are generally more prone to oxidative degradation, particularly in the presence of naturally occurring pigments and metal ions that can catalyze oxidation reactions (Gupta et al., 2003; Junchniewicz & Harasym, 2026). Fatty acid composition is one of the most critical determinants of lipid oxidation, as the degree of unsaturation directly affects lipid susceptibility to oxidation. Polyunsaturated fatty acids (PUFAs), which contain multiple double bonds, are highly prone to oxidation because these double bonds provide reactive sites for oxygen attack (Lottenberg et al., 2012; Abeyrathne et al., 2021). Therefore, modifying the fatty acid profile through blending oils, hydrogenation (where appropriate), or selecting more stable lipid sources can enhance oxidative stability while maintaining nutritional value (Choe & Min, 2006).

In emulsified food systems, such as milk, mayonnaise, and salad dressings, lipid oxidation primarily occurs at the oil–water interface, where lipids, oxygen, and prooxidants interact. Emulsifiers and stabilizers play a crucial role in controlling this interface and can significantly influence oxidative stability (Xiao et al., 2024). Properly selected emulsifiers can form protective interfacial layers around oil droplets, reducing the accessibility of oxygen and metal ions to the lipids. Some emulsifiers may also possess antioxidant properties or work synergistically with added antioxidants (Ghelichi et al., 2023; McClements & Decker, 2000; Ghelichi et al., 2023).

Water activity (a_w) is another important factor influencing lipid oxidation, as it affects the mobility of reactants and the availability of oxygen and prooxidants within the food system. Interestingly, lipid oxidation tends to be minimal at very low and very high-water activity levels but is often maximized at intermediate levels. At low water activity, limited molecular mobility restricts oxidation reactions, while at high water activity, dilution effects and reduced oxygen solubility can inhibit oxidation (Vu et al., 2020; Gumus & Decker, 2021; Labuza, 1971; Mora et al., 2025).

8.6 Processing techniques

Processing techniques have a significant impact on lipid oxidation by altering the chemical composition, physical structure, and lipid exposure to prooxidative conditions. Mechanical, thermal, and biochemical processes can accelerate or inhibit oxidation depending on their application (Ganjeh et al., 2024; Mora et al., 2025). Fermentation, blanching, mixing, homogenization, smoking, curing, marination, and irradiation are important food processing methods that help reduce or delay lipid oxidation in foods. Fermentation can inhibit oxidation by producing organic acids, antioxidant peptides, and beneficial microorganisms that reduce oxidative reactions in food systems (Kanimozhi & Sukumar, 2025; Sawant et al., 2025; Park & Mannaa, 2025). Blanching helps inactivate oxidative enzymes such as lipoxygenase and peroxidase, thereby slowing lipid deterioration in fruits and vegetables (Zafar et al., 2024; Fellows, 2017; Xiao et al., 2017). Proper mixing uniformly distributes antioxidants, salts, and preservatives throughout the food matrix, thereby improving oxidative stability. At the same time, homogenization reduces fat globule size and creates stable emulsions that can minimize localized oxidation when properly controlled. Smoking contributes antioxidant phenolic compounds from wood smoke, which retard rancidity and improve shelf life in meat and fish products. Curing with salt, nitrites, and nitrates inhibits microbial growth and oxidative reactions by stabilizing pigments and limiting free radical formation (Shahidi & Zhong, 2010; Choe & Min, 2007; Pokorný, 2007; Mora et al., 2025).

8.7 Emerging technologies

Encapsulation is an advanced technique widely used to protect sensitive lipids and antioxidants from adverse environmental conditions such as oxygen, light, moisture, and heat. In this process, the active compounds are enclosed within a protective coating or matrix, which can be made from proteins, polysaccharides, or lipids (Shishir et al., 2018; Klojdová et al., 2023). Encapsulation not only enhances the stability of bioactive compounds but also enables their controlled release during digestion or processing, thereby improving their functional and nutritional benefits. Encapsulation is particularly effective in preserving sensitive nutrients and bioactive lipids in complex food systems (Desai & Park, 2005).

Nanotechnology has emerged as a powerful tool in improving the delivery, stability, and bioavailability of antioxidants in food systems. Nanoemulsions, emulsions with droplet sizes in the nanometer range, offered enhanced physical stability and increased surface area, thereby improving the solubility and dispersion of hydrophobic antioxidants in aqueous systems (Nile et al., 2020; Agnihotri, 2025). This increased surface area also enhanced the interactions between antioxidants and free radicals, thereby improving their antioxidant efficiency. Additionally, nanocarriers can improve the absorption and bioavailability of nutrients, making them more effective at lower concentrations (Arshad et al., 2021; McClements, 2012).

Unlike traditional packaging systems that primarily serve as protective barriers, active packaging is designed to perform additional functions through deliberate interactions with the food or package atmosphere, thereby improving product stability and extending storage life. In the context of lipid oxidation, active packaging systems can incorporate antioxidants, oxygen scavengers, or other functional additives that are gradually released into the food or headspace during storage (Han, 2014; Amit et al., 2017; Fadiji et al., 2023). These systems help inhibit oxidative reactions by reducing oxygen availability or directly neutralizing free radicals. Common materials used include polymer films embedded with natural antioxidants such as tocopherols, polyphenols, or essential oils. Active packaging not only enhances oxidative stability but also improves sensory attributes such as flavor, aroma, and color over time (Panou & Karabagias, 2023).

8.8 Good storage and handling practices

Good storage and handling practices are critical for minimizing lipid oxidation, a major cause of quality deterioration in fat-containing foods. One of the most important factors is controlling oxygen exposure, since oxygen is a primary initiator of oxidative reactions (Wang et al., 2023a). Food products should be stored in airtight, vacuum-sealed, or nitrogen-flushed packaging to limit contact with oxygen. Reducing headspace oxygen in containers and using oxygen-impermeable packaging materials can significantly slow down oxidation (Longanathan et al., 2022).

Refrigeration or freezing can greatly slow down oxidative processes by reducing molecular activity and the rate of free radical formation (Oh et al., 2023). Temperature fluctuations should also be avoided, as repeated warming and cooling can accelerate oxidation and promote condensation, which may further accelerate degradation. Maintaining a consistent low temperature is particularly important for highly unsaturated lipids, which are more prone to oxidation (Wang et al., 2023a; Longanthan et al., 2023).

Light control is equally important, as exposure to light, especially ultraviolet (UV) light, can initiate photooxidation. This is why many oils and fat-containing products are packaged in opaque or amber-colored containers that block light penetration (Roy et al., 2024). The combination of light and oxygen exposure can significantly accelerate the formation of hydroperoxides and other primary oxidation products, which eventually break down to secondary oxidation products, generating oxidation flavor (Csapó et al., 2019).

Effective hygiene during handling and storage plays a significant role in controlling lipid oxidation. Certain metal ions, particularly iron and copper, can promote oxidative reactions by catalyzing the formation of free radicals. Consequently, the use of equipment contaminated with these metals should be avoided to reduce the risk of oxidation (Wang et al., 2023a). Using clean, non-reactive materials, such as stainless steel, and avoiding prolonged contact with prooxidant surfaces can help reduce oxidative reactions. In addition, minimizing mechanical stresses, such as excessive agitation, grinding, or exposure to air during processing, can prevent the formation of free radicals (Pagey et al., 2025). The addition of natural or synthetic antioxidants, when permitted, can further enhance stability by scavenging free radicals and interrupting oxidation chains (Fellows, 2017; Longanthan et al., 2023).

9. Future research on lipid and protein oxidation

During food processing and storage, lipids and proteins are particularly prone to oxidative degradation. This deterioration is caused by reactions with reactive oxygen species or with secondary compounds formed during oxidation. The mechanisms of lipid and protein oxidation share several similarities, and both processes can negatively influence the functional, nutritional, and sensory quality of food products. Lipid oxidation in foods has been recognized as a major quality issue and has been studied extensively. However, protein oxidation in food systems has been largely ignored because it has been considered a secondary reaction to lipid oxidation. There are a few reasons why protein oxidation in food systems has not been largely ignored: (1) most of the reactive oxygen species that cause protein oxidation in foods are generated by lipid oxidation processes even though a direct contact of proteins with external reactive oxygen species and the results of lipid and protein oxidation in food systems are difficult to separate; (2) the effects of lipid oxidation to the functional, nutritional, and sensory qualities of foods are much greater than protein oxidation; (3) the molecular size of proteins is much bigger than lipids, and the reaction pathways and the products of protein oxidation are more complicated and diverse than lipid oxidation. Thus, the methods to determine the degree of protein oxidation and the oxidation processes are not well established; and (4) most of the functional, nutritional, and quality changes by protein oxidation in foods can be prevented using the methods that prevent lipid oxidation. It is well established that oxidative stress is closely linked to various biological functions, including inflammation, cellular dysfunction, aging, neurodegenerative diseases, cancers, and other health conditions in humans. Also, various biomarkers and detection methods for protein oxidation in biological systems have been established, but those in food systems require further work. However, most studies on the health effects of consuming oxidized food/feed in humans and animals are also primarily focused on lipid oxidation products. Protein oxidation and the production of oxidation products from individual proteins in model systems are available, but not in foods. Also, some model system studies that used amino acid homopolymers focused only on volatile oxidation products (Ahn, 2002; Ahn et al., 2016a, 2016b), and non-volatile compounds were not even discussed.

Another important area that needs attention is the structural changes in proteins at different stages of oxidation. It is known that free radicals can break the side chains and peptide bonds. However, peptide bond breakage may occur after the secondary and tertiary structures are disrupted. The initial stage of protein oxidation begins at the side chains of hydrophilic amino acids, destabilizing the tertiary structure, and then the secondary structure. Unless the primary structure of proteins is exposed by a denaturing environment (heat, pH, etc.), peptide bond breakage may not occur. Therefore, depending on the degree of protein oxidation or the conditions under which proteins are stored, the oxidation processes and products can vary significantly. In this sense, cooked meat or foods can be more susceptible to protein oxidation than raw products, as in lipid oxidation, even if the effects of antioxidant enzymes or lipid oxidation products are not considered.

Therefore, more studies on protein oxidation, independent of lipid oxidation, in foods are needed. The best model to suggest for protein oxidation study is using egg white with irradiation (electron beam, X-ray, and gamma-ray) treatment. Unlike other foods or food ingredients, egg white contains virtually no lipids, so that any lipid-related reactions can be prevented. Using irradiation, hydroxyl ($^{\cdot}OH$) radicals that can initiate protein oxidation can be produced, and complications from lipid oxidation-related products can be prevented. Using this system, the protein oxidation process, the effects of oxygen, the action of metal catalysts (Fe and Cu), the control of protein oxidation, the intermediate and end products produced, and the functional quality, nutritional, and sensory quality, and the biological functions of protein oxidation products, can be determined. Earlier studies elucidated many of the processes and mechanisms of protein oxidation. They discussed protein oxidation products arising from β-scission, dimerization, and α-amidation reactions of protein hydroperoxides, the reactions and factors involved, the primary and end products, and the consequences of protein oxidation in foods and biological systems. However, details on the protein (peptides, amino acids) or side chain degradation/cleavage and their end products, including carbonylated, hydroxylated, sulfoxides, cyclic amino acids, volatile and non-volatile compounds, and amino acid side-chain degradation products, the structural (primary, secondary and tertiary) modifications of proteins and the functionality, nutritional, digestibility, and product quality changes, and the biological and health effects of consuming protein oxidation products, are not well known, and further studies are needed in these areas.

References

Abeyrathne, E. D. N. S., Nam, K., & Ahn, D. U. (2021). Analytical methods for lipid oxidation and antioxidant capacity in food systems. *Antioxidants, 10*, 1587.

Abeyrathne, E. D. N. S., Nam, K., Huang, X., & Ahn, D. U. (2022). Plant- and animal-based antioxidants' structure, efficacy, mechanisms, and applications: A review. *Antioxidants, 11*, 1025.

Adekunle, K. F. (2015). A review of vegetable oil-based polymers: Synthesis and applications. *Open Journal of Polymer Chemistry, 5*, 34–40.

Agnihotri, S. (2025). Nanoemulsions as advanced delivery systems of bioactive compounds for sustainable food preservation applications. *Biocatalysis and Agricultural Biotechnology, 68*, 103702.

Ahmed, M., Pickova, J., Ahmad, T., Liaquat, M., Farid, A., & Jahangir, M. (2016). Oxidation of lipids in foods. *Sarhad Journal of Agriculture, 32*, 230–238.

Ahn, D. U. (2002). Production of volatiles from amino acid homopolymers by irradiation. *Journal of Food Science, 67*, 2565–2570.

Ahn, D. U., & Lee, E. J. (2002). Production of off-odor volatiles from liposome-containing amino acid homopolymers by irradiation. *Journal of Food Science, 67*, 2659–2665.

Ahn, D. U., Lee, E. J., Feng, X., Zhang, W., Lee, J. H., Jo, C., & Nam, K. C. (2016a). Mechanisms of volatile production from sulfur-containing amino acids by irradiation. *Radiation Physics and Chemistry, 119*, 80–84.

Ahn, D. U., Lee, E. J., Feng, X., Zhang, W., Lee, J. H., Jo, C., & Nam, K. C. (2016b). Mechanisms of volatile production from non-sulfur amino acids by irradiation. *Radiation Physics and Chemistry, 119*, 64–73.

Dey, A., & Sudarsan, N. (2019). Oxygen scavengers for food packaging applications: A review. *Trends in Food Science & Technology, 90*, 26–34.

Akagawa, K., Sasaki, D., Kurota, Y., & Suyama, K. (2005). Formation of α-aminoadipic and γ-glutamic semialdehydes in proteins by the Maillard reaction. *Annals of the New York Academy of Sciences, 1043*, 129–134.

Akhila, K., Ramakanth, D., Rao, L. L., & Gaikwad, K. K. (2023). UV-blocking biodegradable film based on flaxseed mucilage/pectin impregnated with titanium dioxide and calcium chloride for food packaging applications. *International Journal of Biological Macromolecules, 239*, 124335.

Al-Hijazeen, M., Lee, E. J., Mendonca, A., & Ahn, D. U. (2016). Effects of tannic acid on lipid and protein oxidation, color, and volatiles of raw and cooked chicken breast meat during storage. *Antioxidants, 5*, 19.

Amaral, A. B., da Silva, M. V., & da Silva Lannes, S. C. (2018). Lipid oxidation in meat: Mechanisms and protective factors: A review. *Food Science and Technology, 38*, 1–15.

Amici, A., Levine, R. L., Tsai, L., & Stadtman, E. R. (1989). Conversion of amino acid residues in proteins and amino acid homopolymers to carbonyl derivatives by metal--catalyzed oxidation reactions. *Journal of Biological Chemistry, 264*, 3341–3346.

Amit, S. K., Uddin, M. M., Rahman, R., Islam, S. R., & Khan, M. S. (2017). A review on mechanisms and commercial aspects of food preservation and processing. *Agriculture & Food Security, 6*, 51.

Arshad, R., Gulshad, L., Haq, I. U., Farooq, M. A., Al-Farga, A., Siddique, ... Karra, E. (2021). Nanotechnology: A novel tool to enhance the bioavailability of micronutrients. *Food Science & Nutrition, 9*, 3354–3361.

Aruchunan, U., Henry, C. J., & Sim, S. Y. J. (2025). Role of protein–lipid interactions for food and food-based applications. *Food Hydrocolloids, 160*, 110715.

Aung, M. M., & Chang, Y. S. (2014). Temperature management for quality assurance of perishable food supply chains. *Food Control, 40*, 198–207.

Ayala, A., Munoz, M. F., & Argüelles, S. (2014). Lipid peroxidation: Production, metabolism, and signaling mechanisms of malondialdehyde and 4-hydroxy-2-nonenal. *Oxidative Medicine and Cellular Longevity, 2014*, 360438.

Baggio, S. R., Vicente, E., & Bragagnolo, N. (2002). Cholesterol oxides, cholesterol, total lipid and fatty acid composition in turkey meat. *Journal of Agricultural and Food Chemistry, 50*, 5981–5986.

Baptista, M. S., Cadet, J., Mascio, P. D., Ghogare, A. A., Greer, A., Hamblin, M. R., ... Yoshimura, T. M. (2017). Type I and type II photosensitized oxidation reactions: Guidelines and mechanistic pathways. *Photochemistry and Photobiology, 93*, 912–919.

Barriuso, B., & Astiasarán, I. (2013). A review of analytical methods measuring lipid oxidation status in foods: A challenging task. *European Food Research Technology, 236*, 1–15.

Bazina, N., Ahmed, T., Almdaaf, M., Hallalah, H. M. O. A., & Jibia, S. (2025). Chemical changes in deep-fat frying: Reaction mechanisms, oil degradation, and health implications. *Food Science & Nutrition, 13*, e70969 *2025*.

Berlett, B. S., & Stadtman, E. R. (1997). Protein oxidation in aging, disease, and oxidative stress. *Journal of Biological Chemistry, 272*, 20313–20316.

Borys, B., Oprządek, J., Borys, A., & Przegalińska-Gorączkowska, M. (2012). Lipid profile of intramuscular fat in lamb meat. *Animal Science Papers and Reports, 30*, 45–56.

Buege, J. A., & Aust, S. D. (1978). Microsomal lipid peroxidation. *Methods in Enzymology, 52*, 302–310.

Carvalho, R. H., Ida, E. I., Madruga, M. S., Martínez, S. L., Shimokomaki, M., & Estévez, M. (2017). Underlying connections between the redox system imbalance, protein oxidation and impaired quality traits in pale, soft and exudative (PSE) poultry meat. *Food Chemistry, 215*, 129–137.

Chibuye, B., Singh, I. S., Ramasamy, S., & Maseka, K. K. (2024). Natural antioxidants: Sources, mechanisms, and applications. *Next Research, 1*, 100086.

Choe, E., & Min, D. B. (2007). Chemistry of deep-fat frying oils. *Journal of Food Science, 72*, R77–R86.

Choe, E., & Min, D. B. (2006). Mechanisms and factors for edible oil oxidation. *Comprehensive Reviews in Food Science and Food Safety, 5*, 169–186.

Cong, S., Dong, W., Zhao, J., Hu, R., Long, Y., & Chi, X. (2020). Characterization of the lipid oxidation pocess of robusta green coffee beans and shelf-life prediction during accelerated storage. *Molecules (Basel, Switzerland), 25*, 1157.

Corongiu, F. P., & Milia, A. (1983). An improved and simple method for determining diene conjugation in autoxidized polyunsaturated fatty acids. *Chemico-Biological Interactions, 44*, 289–297.

Cropotova, J., & Rustad, T. (2019). A novel fluorimetric assay for visualization and quantification of protein carbonyls in muscle foods. *Food Chemistry, 297*, 125006.

Csapó, J., Prokisch, J., Albert, C., & Sipos, P. (2019). Effect of UV-light on food quality and safety. *Acta University Sapientae, Alimentaria, 12*, 21–41.

Davies, M. J. (2016). Protein oxidation and peroxidation. *The Biochemical Journal, 473*, 805–825.

Davies, M. J., Fu, S., Wang, H., & Dean, R. T. (1999). Stable markers of oxidant damage to proteins and their application in the study of human disease. *Free Radical Biology and Medicine, 27*, 1151–1163.

Desai, K. G. H., & Park, H. J. (2005). Recent developments in microencapsulation of food ingredients. *Drying Technology, 23*, 1361–1394.

Domínguez, R., Pateiro, M., Gagaoua, M., Barba, F. J., Zhang, W., & Lorenzo, J. M. (2019). A comprehensive review on lipid oxidation in meat and meat products. *Antioxidants, 8*, 2–31.

Domínguez, R., Pateiro, M., Munekata, P. E. S., Zhang, W., Garcia-Oliveira, P., Carpena, M., ... Lorenzo, J. M. (2022). Protein oxidation in muscle foods: A Comprehensive Review. *Antioxidants, 11*, 60.

Dragoev, S. G. (2024). Lipid peroxidation in muscle foods: Impact on quality, safety and human health. *Foods, 13*, 797.

Du, M., Hur, S. J., Nam, K. C., Ismail, H., & Ahn, D. U. (2001a). Volatiles, color, and lipid oxidation of broiler breast fillets. *Poultry Science, 80*, 1748–1753.

Du, M., Nam, K. C., & Ahn, D. U. (2001b). Cholesterol and lipid oxidation products in cooked meat. *Journal of Food Science, 66*, 1396–1401.

Durand, E., Laguerre, M., Bourlieu-Lacanal, C., Lecomte, J., & Villeneuve, P. (2025). Lipid oxidation and antioxidation: Evaluation methods and emerging approaches. *Progress in Lipid Research, 97*, 101317.

Estevez, M. (2011). Protein carbonyls in meat systems: A review. *Meat Science, 89*, 259–279.

Estévez, M., & Xiong, Y. L. (2021). Protein oxidation in foods: Mechanisms, consequences, and antioxidant solutions. *Foods, 10*, 2346.

Fadiji, T., Rashvand, M., Daramola, M. O., & Iwarere, S. A. (2023). Antimicrobial packaging for extending shelf life of food. *Processes, 11*, 590.

Fan, X., Sommers, C. H., Thayer, D. W., & Lehotay, S. J. (2002). Volatile sulfur compounds in irradiated precooked turkey breast analyzed with pulsed flame photometric detection. *Journal of Agricultural and Food Chemistry, 50*, 4257–4261.

Feeney, R. E., Blankenhorn, G., & Dixon, B. F. (1975). Carbonyl-amine reactions in protein chemistry. *Advances in Protein Chemistry, 29*, 135–203.

Fellows, P. J. (2017). *Food processing technology: Principles and practice* (4th ed.). Cambridge: Woodhead Publishing.

Floros, J. D., & Matsos, K. I. (2005). Introduction to modified atmosphere packaging. In J. H. Han (Ed.). *Innovations in food packaging* (pp. 159–172). Academic Press.

Franz, K. J. (2013). Broadening the notion of metal chelators in medicine. *Current Opinion in Chemical Biology, 17*, 143–149.

Fridovich, I. (1987). The biology of oxygen radicals: General concept. In B. Halliwell (Ed.). *Proceedings of a Brook Lodge Symposium, Augusta.*

Fruhwirth, G. O., Wenzl, T., El-Toukhy, R., Wagner, F. S., & Hermetter, A. (2003). Fluorescence screening of antioxidant capacity in pumpkin seed oils and other natural oils. *European Journal of Lipid Science and Technology, 105*, 266–274.

Frydrych, A., Kulita, K., Jurowski, K., & Piekoszewski, W. (2025). Lipids in clinical nutrition and health. *Foods, 14*, 473.

Fu, L., Du, L., Sun, Y., Fan, X., Zhou, C., He, J., & Pan, D. (2022). Effect of lentinan on lipid oxidation and quality change in goose meatballs during cold storage. *Foods, 11*, 1055 2022.

Fujii, J., Soma, Y., & Matsuda, Y. (2023). Biological action of singlet molecular oxygen. *Molecules (Basel, Switzerland), 28*, 4085.

Ganjeh, A. M., Gomes, A., Barreira, M. J., Pinto, C. A., Casal, S., & Saraiva, J. A. (2024). Pressure-based technologies and lipid oxidation. *Food Chemistry, 461*, 140768.

Gao, F., Ben-Amotz, D., Yang, Z., Han, L., & Liu, X. (2021). Complementarity of FT-IR and Raman spectroscopies for species discrimination of meat and bone meals. *Food Chemistry, 345*, 128754.

Gebicki, S., & Gebicki, J. M. (1999). Cross-linking of DNA and proteins induced by protein hydroperoxides. *Biochemistry Journal, 338*, 629–636.

Geng, L., Liu, K., & Zhang, H. (2023). Lipid oxidation in foods and implications on proteins. *Frontiers in Nutrition, 10*, 1192199.

Ghazizadeh, H., Saberi-Karimian, M., Aghasizadeh, M., Sahebi, R., Ghazavi, H., Khedmatgozar, H., ... Ghayour-Mobarha, M. (2020). Prooxidant–antioxidant balance (PAB) as a prognostic index. *Obesity Medicine, 19*, 100272.

Ghelichi, S., Hajfathalian, M., Yesiltas, B., Sørensen, A. M., García-Moreno, P. J., & Jacobsen, C. (2023). Oxidation and oxidative stability in emulsions. *Comprehensive Reviews in Food Science and Food Safety, 22*, 1864–1901.

Gonçalves, S., & Romano, A. (2021). *Aromatic oils from forest and their application. Non-timber forest products.* Chambridge: Springer19–34.

Gordon, A., DeVlieger, D., Vasan, A., & Bedard, B. (2020). Technical considerations for the implementation of food safety and quality systems in developing countries. In A. Gordon (Vol. Ed.), *Food safety and quality systems in developing countries: III*, (pp. 1–40).

Grabowski, R., Ilowska, J., Chrobak, J., Szmatola, M., Drabik, J., & Trzos, M. (2019). Methods of modification of vegetable oils to obtain base oils. *Tribologia, 6*, 23–29.

Guéraud, F., Atalay, M., Bresgen, N., Cipak, A., Eckl, P. M., Huc, L., ... Uchida, K. (2010). Chemistry and biochemistry of lipid peroxidation products. *Free Radical Research, 44*, 1098–1124.

Gumus, C. E., & Decker, E. A. (2021). Oxidation in low moisture foods. *Foods, 10*(4), 860.

Gupta, K., Jain, V., Jain, S., Dhawan, K., & Talwar, G. (2003). Analysis of food. In B. Caballero, L. Trugo, & P. Finglas (Eds.). *Encyclopedia of food sciences and nutrition* (pp. 213–222). London: Academic.

Halliwell, B., & Gutteridge, J. M. C. (1989). *Lipid peroxidation: A radical chain reactionFree radicals in biology and medicine* (2nd ed.). London: Clarendon press, 188–276.

Han, J. H. (2014). *Innovations in food packaging* (2nd ed.). London: Academic Press.

Hawkins, C. L., & Davies, M. J. (2001). Generation and propagation of radical reactions on proteins. *Biochimica et Biophysics Acta, 1504*, 196–219.

Heinonen, M., Gürbüz, G., & Ertbjerg, E. (2020). Oxidation of proteins. In D. B. Rodriguez-Amaya, & J. Amaya-Farfan (Eds.). *Chemical Changes during Processing and Storage of Foods* (pp. 85–123). Elsevier Inc.

Hellwig, M. (2020). Analysis of protein oxidation in food and feed products. *Journal of Agricultural and Food Chemistry, 68*, 12870–12885.

Hematyar, N., Rustad, T., Sampels, S., & Dalsgaard, T. K. (2019). Relationship between lipid and protein oxidation in fish. *Aquaculture Research, 50*, 1393–1403.

Hu, C., & Xie, J. (2021). Freeze–thaw cycles and fish quality. *Foods, 10*, 1350.

Hu, Y., Zhao, G., Zhang, M., Zhou, D., & Zhu, B. (2021). Potential adverse health effects of dietary lipid oxidation products. *Journal of Food Bioactives, 15*, 51–62.

Huang, X., & Ahn, D. U. (2019). Lipid oxidation and its implications to meat quality and human health. *Food Science and Biotechnology, 28*, 1275–1285.

Ismail, A., Bannenberg, G., Rice, H. B., Schutt, E., & MacKay, D. (2016). Oxidation in EPA- and DHA-rich oils: An overview. *Lipid Technology, 28*, 3–4 *2016*.

Jackson, V., & Penumetcha, M. (2019). Dietary oxidized lipids and health consequences. *International Journal of Food Science and Technology, 54*, 1981–1988.

Jo, C., & Ahn, D. U. (2000). Production of volatiles from irradiated oil emulsion systems prepared with amino acids and lipids. *Journal of Food Science, 65*, 612–616.

Jo, C., & Ahn, D. U. (1999). Fluorometric analysis of 2-thiobarbituric acid reactive substances in turkey. *Poultry Science, 77*, 475–480.

Junchniewicz, S., & Harasym, J. (2026). Oxidation mechanism in Bigels and Emulgels-challenges and solutions. *Molecules, 31*, 970.

Kakuda, Y., Stanlry, D. W., & van deVoort, F. R. (1981). Determination of TBA number by high performance liquid chromatography. *Journal of American Oil Chemists Society, 58* 773-755.

Kanimozhi, N. V., & Sukumar, M. (2025). Probiotic fermentation and phytochemicals. *Food Wellness, 1*, 100018.

Kato, S., Shimizu, N., Hanzawa, Y., Otoki, Y., Ito, J., Kimura, F., ... Eitsuka, T. (2018). Determination of triacylglycerol oxidation mechanisms in canola oil using liquid chromatography-tandem mass spectrometry. *NPJ Science of Food, 2018*, 1.

Kehm, R. T., Baldensperger, J., Raupbach, & Höhn, A. (2021). Protein oxidation - Formation mechanisms, detection and relevance as biomarkers in human diseases. *Redox Biology, 42*, 101901.

Khan, M. A. M., & Mittal, A. (2020). Modified atmosphere packaging technique: An overview. International conference on Recent Advances in Engineering and Science (ICRAES-2020) held on 11–12 January 2020. 1–4. University Polytechnic, AMU.

Kim, J. H., Ahn, D. U., Eun, J. B., & Moon, S. H. (2016). Antioxidant effect of extracts from the coffee residue in raw and cooked meat. *Antioxidants, 5*, 21.

Kim, J.-J., Kim, Y.-S., & Kumar, V. (2019). Heavy metal toxicity: An update of chelating therapeutic strategies. *Journal of Trace Elements in Medicine and Biology, 54*, 226–231.
Klojdová, I., Milota, T., Smetanová, J., & Stathopoulos, C. (2023). Encapsulation strategies. *Pharmaceuticals, 16*, 362.
Kunyaboon, S., Thumanu, K., Park, J. W., Khongla, C., & Yongsawatdigul, J. (2021). Evaluation of lipid oxidation and quality of silver carp during storage. *Foods, 10*, 495.
Labuza, T. P. (1971). Kinetics of lipid oxidation in foods. *Critical Reviews in Food Technology, 2*, 355–405.
Lee, E. J., & Ahn, D. U. (2003). Antioxidants and lipid oxidation in turkey meat. *Journal of Food Science, 68*, 1631–1638.
Li, Y., Zhao, T., Li, J., Xia, M., Li, Y., Wang, X., ... Sun, P. (2022). Oxidative stress and 4-HNE. *Journal of Immunology Research, 2022*, 2233906.
Liu, S., Liu, J., Wang, Y., Deng, F., & Deng, Z. (2025). Oxidative stress: Signaling pathways and functions. *MedComm, 6*, e70268 *2025*.
Longanathan, L., Saravanakumar, S. S., & Murugan, R. (2022). Investigation of physico-mechanical, thermal, morphological, optical and biodegradation properties of polyvinyl alcohol films reinforced with alkali treated Limonia acidissima shell powder. *Polymer Composite, 43*, 3544–3559.
López-de-Dicastillo, C., López-Carballo, G., Vázquez, P., Schwager, F., Aragón-Gutiérrez, A., Alonso, J. M., ... Gavara, R. (2023). Oxygen scavenger multilayer systems. *Polymers, 15*, 3899 *2023*.
Lorido, L., Ventanas, S., Akcan, T., & Estévez, M. (2016). Effect of protein oxidation on the impaired quality of dry-cured loins produced from frozen pork meat. *Food Chemistry, 196*, 1310–1314.
Lottenberg, A. M., Afonso, M. S., Lavrador, M. S. F., Machado, R. M., & Nakandakare, E. R. (2012). Dietary fatty acids and metabolic syndrome. *Journal of Nutritional Biochemistry, 23*, 1027–1040.
Lund, M. N., Heinonen, M., Baron, C. P., & Estevez, M. (2011). Protein oxidation in muscle foods: A review. *Molecular Nutrition & Food Research, 55*, 83–95.
Luxford, C., Morin, B., Dean, R. T., & Davies, M. J. (1999). Histone H1- and other protein- and amino acid-hydroperoxides can give rise to free radicals which oxidize DNA. *Biology Journal, 344*, 125–134.
Mani, S., Swargiary, G., Tyagi, S., Singh, M., Jha, N. K., & Singh, K. K. (2021). Nanotherapeutic approaches. *Life Sciences, 281*, 119773.
Mas-Bargues, C., Escrivá, C., Dormant, M., Borrás, C., & Viña, J. (2021). Lipid peroxidation measured by malondialdehyde. *Archives of Biochemistry and Biophysics, 709*, 108941.
McClements, D. J. (2012). Nanoemulsions for improving bioavailability of hydrophobic foods. *Food & Function, 3*, 362–370.
McClements, D. J., & Decker, E. A. (2000). Lipid oxidation in oil-in-water emulsions. *Journal of Food Science, 65*, 1270–1282.
Min, B. R., & Ahn, D. U. (2005). Mechanism of lipid peroxidation in meat and meat products – A Review. *Food Science and Biotechnology, 14*, 152–163.
Min, D. B., & Boff, J. M. (2002). Chemistry and reaction of singlet oxygen in foods. *Comprehensive Reviews in Food Science and Food Safety, 1*, 58–72.
Mora, R. L., Vanare, S. P., & Pegg, R. B. (2025). Lipid oxidation in low-moisture snacks. *European Journal of Lipid Science and Technology, 127*, e70044 *2025*.
Nawaz, R., Aadil, M., Kumar, M., Wang, M., Chen, F., & Cheng, K.-W. (2022). Protein oxidation in muscle-based products: Effects on physicochemical properties, quality concerns, and challenges to food industry. *Food Research International, 157*, 111322.
Nile, S. H., Baskar, V., Selvaraj, D., Nile, A., Xiao, J., & Kai1, G. (2020). Nanotechnologies in food science. *Nano-Micro Letters, 12*, 45.

Oh, W. Y., Kim, M. J., & Lee, J. (2023). Lipid oxidation in oil matrices. *Food Science and Biotechnology, 32*, 1805.

Ojha, S., Bußler, S., Psarianos, M., Rossi, G., & Schlüter, O. K. (2021). Edible insect processing pathways and implementation of emerging technologies. *Journal of Insects Food and Feed, 7*, 877–890.

Pagey, S. V., Landge, B. P., & Nakhale, S. (2025). Lipid oxidation in food systems: Mechanisms, detection methods, and prevention strategies. *International Journal of Applied Research, 11*, 82–88.

Panou, A., & Karabagias, I. K. (2023). Biodegradable packaging materials. *Coatings, 13*, 1176.

Papuc, C., Goran, G. V., Predescu, C. N., & Nicorescu, V. (2017). Mechanisms of oxidative processes in meat. *Comprehensive Reviews in Food Science and Food Safety, 16*, 96–123.

Park, I., & Mannaa, M. (2025). Fermented foods and gut health. *Foods, 14*, 2292.

Pérez-Palacios, T., & Estévez, M. (2020). Chapter 13—Analysis of lipids and lipid oxidation products. In A. K. Biswas, & P. K. Mandal (Eds.). *Meat quality analysis – Advanced evaluation methods, techniques, and technologies* (pp. 217–239). Cambridge, MA, USA: Academic Press.

Pokorný, J. (2007). Are natural antioxidants better? *European Journal of Lipid Science and Technology, 109*, 629–642.

Pooja, G., Shweta, S., & Patel, P. (2025). Oxidative stress and free radicals in disease pathogenesis: a review. *Discover Medicine, 2*, 104.

Porter, N. A. (2025). Chemical mechanisms of lipid peroxidation. *Redox Biochemistry and Chemistry, 12*, 100054.

Prato, E., & Biandolino, F. (2012). Total lipid content and fatty acid composition of fish species. *Food Chemistry, 131*, 1233–1239.

Recknagel, R. O., Glende, E. A., & Jr (1984). Spectrophotometric detection of lipid conjugated dienes. *Methods in Enzymology, 105*, 331–337.

Reddy, V. P. (2023). Oxidative stress in health and disease. *Biomedicines 2023, 11*(11), 2925.

Requena, J. R., Chao, C.-C., Levine, R. L., & Stadtman, E. R. (2001). Glutamic and aminoadipic semialdehydes are the main carbonyl products of metal-catalyzed oxidation of proteins. *Proceedings of the National Academy of Sciences. USA, 98*, 69–74.

Robertson, G. L. (2016). *Food packaging: Principles and practice* (3rd ed.). Boca Raton: CRC Press.

Rodrigo, R., Vinay, J., Castillo, R., Cereceda, M., Asenjo, R., Zamorano, J., ... Larraín, E. (2010). Vitamins C and E therapy. *International Journal of Cardiology, 138*, 221–228.

Rowe, L. J., Maddock, K. R., Lonergan, S. M., & Huff-Lonergan, E. (2004). Influence of early postmortem protein oxidation on beef quality. *Journal of Animal Science, 82*, 785–793.

Roy, S., Ramakrishnan, R., Goksen, G., Singh, S., & Łopusiewicz, Ł. (2024). UV-light barrier packaging films. *Innovative Food Science & Emerging Technologies, 91*, 103550.

Sani, M. A., Khezerlou, A., Tavassoli, M., Abedini, A. H., & McClements, D. J. (2024). UV-screening packaging materials. *Trends in Food Science & Technology, 145*, 104366.

Sawant, S. S., Park, H.-Y., Sim, E.-Y., Kim, H.-S., & Choi, H.-S. (2025). Microbial fermentation in food. *Fermentation, 11*, 15.

Schaich, K. M. (2016). Analysis of lipid and protein oxidation in fats, oils, and foods. In M. Hu, & C. Jacobsen (Eds.). *Oxidative stability and shelf life of foods containing oils and fats* (pp. 133–156). The Netherlands: Elsevier Inc. Amsterdam.

Shahidi, F., & Zhong, Y. (2010). Lipid oxidation and improving oxidative stability. *Chemical Society Reviews, 39*, 4067–4079.

Shahidi, F., & Zhong, Y. (2015). Measurement of antioxidant activity. *Journal of Functional Foods, 18*, 757–781.

Shishir, M. R. I., Xie, L., Sun, C., Zheng, X., & Chen, W. (2018). Nano-encapsulation of bioactive compounds. *Trends in Food Science & Technology, 78*, 34–60.

Soladoye, O. P., Juárez, M. L., Aalhus, J. L., Shand, P., & Estévez, M. (2015). Protein oxidation in processed meat: Mechanisms and potential implications on human health. *Comprehensive Review in Food Science and Food Safety, 14*, 106–122.

Stadtman, E. R. (2006). Protein oxidation and aging. *Free Radical Research, 40*, 1250–1258.

Stadtman, E. R., & Levine, R. L. (2000). Protein oxidation. *Annals of the New York Academy Sciences*, 191–208.

Steinhotst-Slikkerveer, L., Hans-Gerd Janssen, L., & Bauer-Plank, C. (2000). Analysis of nonvolatile lipid oxidation products in vegetable oils by normal-phase high-performance liquid chromatography with mass spectrometric detection. *Journal of American Oil Chemists' Society*, 77, 837–845.

Suárez-Medina, M. D., Sáez-Casado, M. I., Martínez-Moya, T., & Rincón-Cervera, M.Á. (2024). Low temperature storage and lipid quality. *Foods, 13*, 1097.

Sun, G., Wang, M., Chen, T., Ge, Z., Zhang, Y., Wang, H., ... Li, Y. (2026). Synthesis, structural modulation, and functional applications of Schiff bases in food science. *Advances in Colloid and Interface Science, 354*, 103903.

Tao, L. (2015). Oxidation of polyunsaturated fatty acids. *Advances in Food Technology and Nutrition Sciences, 1*, 135–142.

Tenyang, N., Tiencheu, B., Mamat, A., Mawamba, L. A., & Ponka, R. (2021). Effect of cooking methods on the nutritive value and lipid oxidation of two Cricket species consumed in Cameroon. *European Journal of Nutrition and Food Safety, 13*, 11–23.

Tsubone, T. M., Baptista, M. S., & Itri, R. (2019). Membrane remodeling by lipid oxidation. *Biophysical Chemistry, 254*, 106263.

Türksever, C., Ko, B., & Esmer, O. K. (2025). Sustainable alternatives in multilayer packaging: Storage stability of pudding powder under accelerated storage conditions. *Foods, 14*, 3806..

Utrea, M., & Estevez, M. (2013a). Impact of Trolox, quercetin, genistein and gallic acid on the oxidative damage to myofibrillar proteins: The carbonylation pathway. *Food Chemistry, 141*, 4000–4009.

Utrera, M., & Estévez, M. (2013b). Oxidative damage to poultry, pork, and beef during frozen storage through the analysis of novel protein oxidation markers. *Journal of Agricultural Food Chemistr, 61*, 7987–7993.

Utrera, M., Rodríguez-Carpena, J. G., Morcuende, D., & Estévez, M. (2012). Formation of lysine-derived oxidation products and loss of tryptophan during processing of porcine patties with added avocado byproducts. *Journal of Agricultural and Food Chemistry, 60*, 3917–3926.

Vieira, S. A., Zhang, G., & Decker, E. A. (2017). Biological implications of lipid oxidation products. *Journal of the American Oil Chemists' Society, 94*, 339–351.

Vogt, W. (1995). Oxidation of methionyl residues in proteins: Tools, targets, and reversal. *Free Radical Biology and Medicine, 18*, 93–105.

Vu, T. P., He, L., McClements, D. J., & Decker, E. A. (2020). Lipid stability in low moisture foods. *Food Research International, 130*, 108844.

Wang, D., Xiao, H., Lyu, X., Chen, H., & Wei, F. (2023a). Lipid oxidation in food science and nutritional health: A comprehensive review. *Oil Crop Science, 8*, 35–44.

Wang, H., Suo, R., Wang, Y., Sun, J., Liu, Y., Wang, W., & Wang, J. (2023b). Effects of electron beam irradiation on protein oxidation and textural properties of shrimp (*Litopenaeus vannamei*) during refrigerated storage. *Food Chemistry, X20*, 101009.

Wang, J., Jin, G., Zhang, W., Ahn, D. U., & Zhang, J. (2012). Effect of curing salt content on lipid oxidation and volatile flavor compounds of dry-cured turkey ham. *LWT - Food Science and Technology, 48*, 102–106.

Wazir, H., Chay, S. Y., Zarei, M., Hussin, F. S., Mustapha, N. A., Abdullah, W. Z. W., & Saari, N. (2019). Effects of storage on lipid oxidation. *Antioxidants, 8*, 486.

Wu, H., Yin, J., Zhang, J., & Richards, M. (2017). Factors affecting lipid oxidation in meat. *Journal of Agricultural and Food Chemistry, 65*, 8011–8017.

Xiao, H. W., Pan, Z., Deng, L. Z., El-Mashad, H. M., Yang, X. H., Mujumdar, A. S., ... Zhang, Q. (2017). Thermal blanching developments. *Information Processing in Agriculture, 4*, 101–127.

Xiao, N., Zhao, Y., Yao, Y., Wu, N., Xu, M., Du, H., & Tu, Y. (2020). Biological activities of egg yolk lipids. *Journal of Agricultural and Food Chemistry, 68*, 1948–1957.

Xiao, S., Zhang, W. G., Lee, E. J., Ma, C. W., & Ahn, D. U. (2011a). Lipid and protein oxidation of chicken breast rolls as affected by dietary oxidation levels and packaging. *Journal of Food Science, 76*, C612–C617.

Xiao, S., Zhang, W. G., Lee, E. J., Ma, C. W., & Ahn, D. U. (2011b). Effects of diet, packaging, and irradiation on protein oxidation, lipid oxidation, and color of raw broiler thigh meat during refrigerated storage. *Poultry Science, 90*, 1348–1357.

Xiao, T., Ma, X., Hu, H., Xiang, F., Zhang, X., Zheng, Y., ... Wu, X. (2024). Progress in understanding oxidative stress, aging, and aging-related diseases. *Antioxidants, 13*, 394.

Yang, J., Luo, J., Tian, X., Zhao, Y., Li, Y., & Wu, X. (2024). Progress in understanding oxidative stress, aging, and aging-related diseases. *Antioxidants, 13*, 394.

Yang, Z., Xiong, Q., Yin, T., You, J., Xiong, S., & Liu, R. (2025). Freeze-induced deterioration in surimi. *Food Chemistry, 493*, 146083.

Zafar, A., Shaheen, M., Tahir, A. B., da Silva, A. P. G., Manzoor, H. Y., & Zia, S. (2024). Edible crickets in food applications. *Trends in Food Science & Technology, 143*, 104254.

Zhang, W., Xiao, S., & Ahn, D. U. (2013). Protein oxidation: Basic principles and implications for meat quality. *Critical Review in Food Science and Nutrition, 53*, 1191–1201.

Zhao, Y. C., Shi, H. H., Wang, C. C., Yang, J. Y., Xue, C. H., Jiang, X. M., ... Wang, Y. U. (2021). Enrichment of eggs with DHA and EPA. *Food Chemistry, 346*, 128958.

Zheng, L., Faith, M., Giuffrida, F., O'Neill, B. V., & Schneider, N. (2019). Dietary polar lipids and cognitive development. *Advances in Nutrition, 10*, 1163–1176.

Zhu, L., Fan, Z., Li, W., & Shan, Y. (2025). Goat milk exhibits a higher degree of protein oxidation and aggregation than cow milk during cold storage. *Foods, 14*, 852.

Zhu, Y., Zhu, J., Shi, X., & Fan, M. (2025). Effects of freezing, frozen storage and thawing on the water status, quality, nutrition and digestibility of Meat: A review. *Food Science & Nutrition, 13*, e70774.

CHAPTER EIGHT

Nexus of innovative packaging, food quality and sustainability

Evans Frimpong Boateng[a,*], Mustapha Muhammad Nasiru[b], Rita Dankwa[a], Anthony Pius Bassey[c], James Owusu-Kwarteng[a], Paa Kwesi Bordoh[d], and Fawze Alnadari[e]

[a]Department of Food Science and Technology, University of Energy and Natural Resources, Sunyani, Bono, Ghana
[b]College of Food Science and Engineering, Bohai University, Jinzhou, Liaoning, P.R. China
[c]State Key laboratory of Meat Quality Control and Culture Meat Development, MOST: College of Food Science and Technology, Nanjing Agriculture University, Nanjing, Jiangsu, P.R. China
[d]Department of Horticulture and Crop Production, Postharvest Unit, University of Energy and Natural Resources, Sunyani, Bono, Ghana
[e]Functional Protein Laboratory, Research and Development Center, Jiangsu Tianmeijian Nature Bioengineering Co., Ltd., Nanjing, Jiangsu, P.R. China
*Corresponding author. e-mail address: evans.boateng@uenr.edu.gh; nanafrimpong94@yahoo.com

Contents

Advances in Food and Nutrition Research, Volume 122
ISSN 1043-4526, https://doi.org/10.1016/bs.afnr.2026.03.005

Abstract

Packaging research is one of the primary fields of study in the food ecosystem that go beyond conventional containment and protection. Innovative packaging recognizes the importance of food shelf-life stability, minimizes environmental effect, promotes health, and supports the circular economy. Highlighting the innovations in food packaging is crucial to ensuring that consumers enjoy healthful food. Consequently, insights into the incorporation of agents like antioxidants, antimicrobial, oxygen/moisture absorbers directly into the packaging material to achieve desired functionalities are underpinned. Food waste is also decreased by using sensors like biosensors and time-temperature indicators to track food parameters like freshness or temperature exposure and to give retailers and customers visible data in real time. Modern food packaging systems consider the use of barcodes, radio frequency identification tags, fast response codes, artificial intelligence, and internet of things devices to enhance food product monitoring and ensure food safety and quality throughout the supply chain. While packaging innovations are promising, challenges include higher production costs for sophisticated food packages development and comprehensive risk assessment of food contact chemicals. Hence, the need for better recycling infrastructure, and ensuring that sustainable materials can provide the same level of synergistic impact of barrier protection and functionalities of consumer quality food demands. Future developments in edible packaging, coatings based on nanotechnology, and food contact material packaging systems mechanisms call for the development of packaging that is readily recyclable, compostable, or reusable to harness sustainability issues.

1. Introduction

Food packaging provides food products with protective barriers functionalities ranging from safety, easy handling, confinement, transporting, exporting, distribution, and convenient. Packaging significantly affects food product quality and marketability to mitigate challenges associated with sustainable food management by reducing food waste (Alamri et al., 2021; Pathiraja & Munaweera, 2024). To an extent consumer aesthetic satisfaction (color, forms, sizes, labels, and shape) for a food packaging material influences food product purchase and consumption (Gupta, 2023). Food packages make available the food product or food industry's important information's (manufacture date and expiry, ingredients and composition, trademark, manufacturer name and address, preparation steps, details etc.) to consumers per food supply chain regulations to curb food related risks. Food packages influence food products shelf-life extension and confer a positive impact on postharvest losses including insect pest damage control, enhanced physical and nutritional qualities. In a broader perspective, intrinsic factors such as food biological structure, pH, nutrient composition, water content, respiratory rate and extrinsic indexes including food storage temperature, microbes, light, relative humidity, heat, dust, pressure, composition of surrounding gas, and pests are dependent factors of food preservation and spoilage control which are relative to food packages. Thus, food packaging encompasses science, art, and technology targeted to safeguard food product and sustain its integrity (Dobrucka, 2013; Shekarchizadeh & Nazeri, 2020). A paradigm shift has occurred recently from the traditional packaging objectives of protection, convenient, containment, and communication to the advent of effective food packaging approaches aimed at guaranteeing food product safety and checking the increased food product complexities (Singh et al., 2023). Food packaging materials include foils, textiles, rubber, laminate, polymers, wood, metals, blister packs, fibrous materials, paper, glass, and plastic. These packaging materials have been employed in developing food packages of diverse assortment of shapes and dimensions (sizes, forms etc.) such as paper packaging, film packaging, carton packaging, containers, textile packaging, foam packaging, plastic packaging etc. Active, interactive, indications, smart, and intelligent innovative packaging strategies are not only important to the food ecosystem but also integral for logistic management, quality control, waste reduction, and premium pricing. Therefore, to mitigate the legislative concerns and circular economy's techno feasibility challenges related to carbon footprints, functional packaging is a progressive innovative concept of packaging aimed at addressing food safety and consumer's concern. This chapter sought to delve into the advancement of food packaging systems innovations and their synergic effect on food quality. This would complement available literatures on food packaging from a more ecofriendly, cost effective, biodegradable, and smart packages view-point worthy of contributing to food quality, safety, sustainability, and global food security.

1.1 Functional application of packaging on food

The food sector has recently become interested in innovative food packaging because of its unique technological applications and functional qualities. According to Ndwandwe et al. (2025), functional packaging provides advantages beyond protection with improved attributes such as the monitoring the quality of food through real-time indexes like temperature or pH, and enhanced food safety and traceability in the food supply chain. Food is protected against physical damage, biological contamination, and chemical hazards yet, the recent packaging development technologies explored in minimizing the utilization of non-biodegradable fossil resources and petroleum-based plastics as food packaging materials have been a great deal of addressing sustainable packaging alternatives. The potential health risk associated with the displacement of microplastic from plastic-based packaging films into foods and its accumulation in the human organs upon consumption is a contributing factor spurring interest in functional food packaging solutions (Zhang et al., 2025). Consequently, the environmental related issues of synthetic, petroleum derived polymers, and non-degradable plastic packaging materials coupled with the influx of agro waste pollution effect on the environment has necessitated the exploitation of valorizing agro waste as a bio polymer or bio composite materials in functional packaging fabrication (Boateng et al., 2025; Sun et al., 2025). Amidst the potential of sustainable packaging effects by some promising biopolymer-based alternative materials such as, aliphatic polyesters, protein, and polysaccharides secured from renewable sources, there seems to be some related compromised performance in food preservation. These limitations are barrier properties, mechanical strength, and moisture sensitivity. However, diverse functionalization approaches have been examined to address these physicochemical characteristics, barriers, thermal, hydrophobicity, mechanical, and processing properties of these biopolymer packaging films (Syed et al., 2025). Siddiqui et al. (2024) revealed that anticorrosion, antimicrobial, antioxidation, oxygen (O_2) and moisture barrier, color protection, and sealing of food during preservation are key functional requirements of packaging. Also, the eco-conscious innovation nature of biopolymer-based packaging systems is characterised by the affordability, renewability, and ability to enhance polysaccharides functionality (Vardhan et al., 2025). Therefore, the ability to block various permeation pathways (water vapour, ultraviolet rays, carbon dioxide (CO_2), light, volatile compounds, O_2 etc.) and liquid (oil, water), improve microbial inhibition, emit antioxidant agent, among others is functional properties of cellulose derivatives merged with natural active compounds utilized in fabricating biopolymer-based packaging materials (Fan et al., 2025; Li et al., 2025). There have been several food packaging innovative technologies in recent years secured from the synergic impacts of utilizing biodegradable polymers (chitosan, alginate, pectin, zein, soy protein isolate, cellulose, gelatine, chitin, etc.) and other valuable bioactive ingredient modification approaches aimed at achieving food packaging key application purposes (Wei et al., 2025). The technologies of food packaging systems suitable of sensing and adjusting to stimuli likewise edible packaging films have unfold as auspicious intelligent and sustainable means of enhancing food safety, quality, and supply properties.

1.2 Impact of packaged food products on marketing, quality, safety, and value addition to the food sector

Over the course of time, the advent of research and technological innovations has garnered the market-driven goal of package food to become one of the paramount sectors in the food or consumer's market. The convenience and access for food processing, packaging, and delivery to consumer revolutions have impacted the marketing of package food products. The food industry has witness value addition of products ranging from ready-to-eat food products through to meeting the quality and safety indexes of food products during dissemination process. Globalization, industrialization, and the fast-paced economic demands of people are the driven-force affecting the package food market trajectory.

According to Mengozzi et al. (2025) the global packaged produce market is estimated at $1.9 trillion in 2020 and forecasted to get to $3.4 trillion by 2030. Consequently, a "packaged food market size and share analysis: growth trends and forecast (2026–2031)" detail by Mordor Intelligence, (2026) revealed that the package food market was segmented by distributional channels, category, geography, and product type. In that the product type segment comprise of condiments, baby food, ready meal, dairy, confectionery, sauce, food spread, snacks, meat, bakery, breakfast cereals, and many others. The distributional channel category segment includes the online retail stores, hypermarkets or supermarkets, convenience stores, etc. And the market segment of category centers on organic and conventional. Mordor Intelligence, (2026) revealed that the packaged product market was priced at $6.34 trillion in 2025 and expected to get to $8.15. trillion by 2031 marking a steady rise at 4.28 % compound annual growth rate (CAGR). Globally, the processed meat is the most consumed packaged food product and dominate a substantial market approximated to rise from $523 billion to $737 billion from 2020 to 2026 (Mengozzi et al., 2025). Conversely, there seem to be health and environmental concerns associated with packaged food products. Ayvar-Gama et al. (2025) and Sharma et al. (2025) posited that package foods contribute to unhealthy diet due to their energy dense and high saturated and trans-fat, free and added sugar, and sodium composition that led to chronic non-communicable aetiologies from epidemiological studies. In the event of extending the shelf stability of packaged foodstuff, additives and preservatives are employed which later causes adverse effect on consumer health. Further, a revelation regarding the relationship of packaged food with malnutrition concerns is inferred to be due to the displacement of healthier nutrient options like fruit and vegetables with low nutrient constituents and high calorie containing products (Sharma et al., 2025). Consequently, there is a complexity among the packaged foods impacts on the multifaceted areas of the economy, health, and environment. Herein, in as much as packaged food product accessibility and convenience attributes are related to the soaring of marketing and economic drive, pollution from food package materials remain a key environmental concern among the food industries and governments across the world. Waste generation in the environment is extensively related to packaged food products. According to Siciliano et al. (2025) waste is caused because of failure to adhere to product expiration date, over buying, and poor management of product portion. Thus, consumers play a vital part in product waste reduction. The production, manufacturing of packaging materials, processing, and distribution of food influence greenhouse gas emission to adversely impact the environment.

Growing population, economic and lifestyle dynamics, environmental and social influences, demographic shift, consumers need and wants, and aging population stem the driven factors in the major swift changes in agriculture and industry. Packaging is among the numerous characteristics of place, time, functionality, ease of possession, and quality that add value to food products coupled with their increased international market value and price enhancement in global economies and trend (Sharif et al., 2018). Therefore, package food transforms the food product into ready-to-eat and convenient forms thereby resulting in a rise in perceived value. On one hand, packaging food products influence product quality, enhance consumer satisfaction, and develop trust and confidence of products. On the other hand, the production of packages for use generate employment which impact economic development, growth, and improved living standards. Therefore, to safeguard consumer's trust in the patronage of packaged food products because of their variable options, affordability, availability, and convenience, sustainable, innovative and creative food packaging possessing safety, quality, and traceability features are recommended. Public and environmental health regulatory policies on sustainable packaging, the use of innovative functional ingredients in conventional composite packaging materials development of biodegradable, edible films, active, and smart or intelligent packages underscores innovative and sustainable packaging technologies (Boateng et al., 2025).

2. Food packaging materials

The design process is not only related to the important criteria's in selecting packaging material, instead, packaging management is commendable. This is because packaging materials directly contribute to mechanisms and functionalities such as the level of protection, containment, and compatibility with the food product to keep track of and preserve food quality and safety (Ardestani et al., 2026). Packaging material development and adoption in the food industries over the years have transcended beyond the simple concept of materials choice to present a critical analysis of each material class. These classes of materials include polymers, paper, metals, glass, bioplastics, and aluminum (Al) foil. The packaging materials research encompasses parameters such as evaluating their inherent properties, high cost, manufacturing processes, high mechanical strength, and good barrier properties. Also, most packaging materials multifaceted impact on food quality parameters such as shelf-life, safety, nutritional value, and sensory attributes that suit regulatory standards are underpinned to address sustainability concerns. Concisely, packaging materials progress from the most utilized conventional materials to emerging and innovative alternatives, affirming the narrative that reflects the industry's evolution towards high-performance packaging materials to provide sustainable solutions.

2.1 Polymers

Generally, synthetic and semi-synthetic polymers constitute the most prevalent category of contemporary food packaging materials of which petroleum-based and bio-based/non-degradable plastics are the commonly employed packaging materials (Geueke et al., 2018; Lei et al., 2025). These long-chain molecules, primarily derived from petrochemicals which consist of single polymers (large molecular structure made of repeating units), thus classified into thermoplastics (e.g., polyethylene-PE, polypropylene-PP, polyethylene terephthalate-PET, polypropylene (PP), polyvinyl chloride-PVC, high- and low-density polyethylene-HD/LDPE, polystyrene-PS, polyamide (nylon, aramids), ethylene-vinyl alcohol-EVOH) and thermosets, with their molecular structure dictating processing methods and final properties (Alojaly & Benyounis, 2022; Guzman-Puyol et al., 2022). Based polymers packaging materials are fabricated with the incorporation of chemical additives (lubricant, plasticizers, heat stabilizers, antioxidants, ultraviolet barriers etc.) to achieve desired chemical and physical functionalities (Singh et al., 2022). Characterized by versatile barrier properties against oxygen, water vapor, and light, along with customizable mechanical strength and thermal stability, polymers enable diverse packaging formats from flexible films to rigid containers. Flexible films (mono or multi-layer films) have unique attributes which include printable nature, portability, resilience, and adaptability (Tamizhdurai et al., 2024). As mentioned by Kaiser et al. (2017) linear-low-density polyethene-LLDPE, LPDE, ethylene-vinyl acetate-EVA, ionomers (ethylene acrylic acid (EAA), ethylene methacrylic acid (EMAA)), and PE plastomers are standardly utilized as flexible plastic packaging with ionomers exhibiting a remarkable grease and oil resistance. The impact of polymers on food quality is dual-natured. While effectively prolonging food product shelf-life and inhibiting contamination via optimized barrier characteristics, potential risks emerge from chemical migration and flavor scalping in achieving the various desired special functions (Zhang et al., 2025). The advent of multilayer plastics has garnered due to enhanced barrier performances and long-term reliability among foodstuff (Tamizhdurai et al., 2024). Advanced multi-layer co-extruded structures represent paramount innovation. Thus, integrating different polymers during manufacturing secure performance unattainable by single materials (Mousania & Atkinson, 2025). Conversely, the environmental profile remains challenging, with current recycling systems struggling to manage complex composites and persistent waste issues related to the carbon footprint of polymeric materials. Herein, the proliferation of driving research into improved technologies (polymer reduction, reuse, and recycling) and alternative packaging materials (Schyns & Shaver, 2021). Although, low cost, availability, and accessibility of the resources to produce plastics primarily meet the physical, mechanical, and chemical essential objectives of packaging yet polymers are non-biodegradable (Zhou et al., 2022). Therefore, the application of biopolymers such as starch, cellulose, pectin, and gums possessing high compatibility, low toxicity, natural antimicrobial, eco-friendly, and natural antioxidant exhibits desired sustainable features compared to polymers (Zhou et al., 2022). Consequently, the adoption of composite biopolymers results in improved barrier (water, temperature, light, humidity, gas (CO_2, N_2), hydrocarbons), strong structural integrity, thermal stability, mechanical strength among other characteristics of packages and extensively as a key means of managing the economic and environmentally sustainable use of polymers (Jat et al., 2025; Schyns & Shaver, 2021).

2.2 Paper and paperboard

As one of the oldest packaging materials, paper and paperboard consist of matted or felted sheet, or interwoven cellulose (plant) fibers derived from wood pulp or recycled content, distinguished primarily by thickness and rigidity (Riley, 2012). Other fibrous materials (cotton, sugar cane, lignin, and straw) are utilized in developing paper and paperboard. These renewable materials (natural source, biodegradability, and readily recyclability) offer excellent mechanical protection and cushioning and are considered the most eco-friendly packaging material with applications ranging from flexible bags, rigid or semi-rigid to corrugated shipping containers (Chen et al., 2013; Shirzad & Joodaky, 2024). According to Podd (2022) paper product used in food packaging range from tissue to paperboard (cardboard, board, boxboard, carton board). However, paper and paperboard inherent hygroscopic nature and poor barrier properties against gases, moisture, and greases limit direct food contact applications (Poddar et al., 2025). The physical properties, structure of the paper (substrate, absorbent hygiene products, heterogenous, porous), fatty-based, and dry nature of paper and paperboard are innovatively required to secure aqueous-based foods effectively. Innovative coatings and laminations with bio-based polymers or nanocomposites have transformed their functionality, creating effective barriers that enable packaging of liquid and fatty foods in food packaging application since the 17th century (Adibi et al., 2023). The sustainability profile of coatings employed in enhancing barrier properties should be particularly favorable, featuring high recyclability, biodegradability, and renewable sourcing, though composite structures incorporating synthetic coatings present end-of-life management hurdles (Khule et al., 2025). Current innovations focus on developing bio-based barrier benign barrier coatings that maintain the material's environmental advantages while expanding its functional applications (Ghaffar et al., 2025). The phenomenon hydrophilic nature of paper and paperboard makes it vulnerable to moisture and water. Regulations are advised for consideration in the paper industries to prevent the use of substances harmful to consumers' health when substances are used to enhance the physical, chemical, and hygienic characteristics of paper for food packaging because of the changing economic, marketing, supply and value chain, and sustainable goals in meeting commercial and consumer demand (Podd, 2022). Additional relevant material test issues of the packages made of paper and paperboard are often employed in the food industry for both home storage and business delivery. However, the applied compression load may compromise the paper and paperboard packaging' structural integrity. The thin wall of the paperboard package is distorted by the external force or the stacking of boxes one on top of the other. Thus, research has been conducted to seek the best material and structural qualities for paper use in food product packaging (Zaheer et al., 2018).

2.3 Metals and cans

Metal packaging materials (retort pouches, foil wraps, cans) and packages primarily including steel, stannous or tin plates, and Al, provide virtually absolute barrier protection against gases, light, moisture, and microorganisms, making it exemplary for prolong storage of food products (Deshwal & Panjagari, 2020). Manufactured through processes like drawing, wall-ironing, and soldering or welding into two-piece or three-piece cans, these materials offer mechanical robustness and exceptional suitability for thermal processing, safety, and suitable for transporting food in the value chain (Akram et al., 2023). Al and steel are the most abundant metal material employed in metal packaging due to its recyclability property. The use of metal packaging is adaptive for the manufacturing of metal packages of different aesthetic shape, size, and style to enhance the market value of solid, liquid, and powdered food products. Structural durability and long-term storage functionality are associated with metal packaging. Canned foods are cost effective and convenient (Wu et al., 2024). According to Debeaufort et al. (2021) and Geueke (2016) there are three primary categories of metallic materials on the market for use in food packaging including stainless steel plates, coated plates, and Al. The tinplate's ability to withstand corrosion from acidic food items depends on the chemical makeup of the steel. Light metal packaging is made from relatively pure Al. Three technological disciplines are used in the creation of Al packaging such as shape, surface treatment, and coatings. Dhuey et al. (2025) revealed that it was recognized that both volatile and non-volatile substances can cause corrosion in metal cans used to package specific food items. A chemical reaction (electromagnetic) between a metal substrate and its surroundings is known as corrosion (Wu et al., 2024). In the food industry, this kind of chemical reaction is problematic because it can lead to three things: (1) the migration of metals or other corrosion products from the packaging material into the food; (2) the loss of hermeticity due to the formation of leaks in the metal package; and (3) appearance changes and taste of the contained food (Dhuey et al., 2025). The migration of bisphenol A (industrial endocrine disrupting compounds), lead (Pb), cadmium (Cd), mercury (Hg), Al, iron (Fe), nickel (Ni), can bulging, tin dissolving, blackening, and corrosion are the main health and product safety issues associated with metal packing (Huang et al., 2017). Because metals are not inert to food products, they are coated with protective lacquers to stop metal components from migrating and interacting with food. Metal packing materials offer lower global warming potential and increased recyclability due to their magnetic characteristics which help in simpler segregation (Akram et al., 2023; Deshwal & Panjagari, 2020). Also, soldering is unapplicable in can foods due to concerns attributed to the migration of Pb or tin (Sn) into foods thereby making welding suitable in the food industries (Deshwal & Panjagari, 2020). Critical to food quality preservation is internal protective enamels, coatings, and lacquers that prevent corrosive interactions between metal surfaces and food constituents, particularly with acidic or sulfur-containing products (Barilli et al., 2003). This ensures sensory quality maintenance while preventing metal uptake (Ninčević et al., 2007). Continuous innovation focuses on lightweighting through advanced material grades and easy-open end mechanisms. The sustainability advantage lies in permanent material characteristics enabling infinite recyclability without quality degradation, with recycled content significantly reducing the environmental footprint of production as a perfect fit in a circular economy (Bher & Auras, 2024).

2.4 Glass

Glass represents an amorphous, inert packaging material manufactured from silica (SiO_2, quartz), soda ash (Na_2CO_3), and limestone ($CaCO_3$), offering complete impermeability and chemical inertness (Grayhurst, 2012). By melting and rapid cooling, the amorphous inorganic material is developed into a rigid and non-crystalline state to be extensively employed in the packaging of food products as of its outstanding barrier characteristics, high thermal resistance, and chemical stability (Gomes et al., 2025). The malleability, durability, transparency, and excellent optical properties ensure absolute protection against gaseous exchange and chemical interactions, perfectly preserving food product taste, aroma, and composition (Saud AlMulaifi, 2024). The manufacturing process involves high-temperature melting followed by forming operations, resulting in packaging characterized by transparency, non-toxic, easy to sterilize, and consumer appeal but challenged by brittleness (fragile, susceptible to break), expensive, and high weight (heavy) (Boz, 2024). Soda is sometimes combined with silica to lower the melting point of the material to enhance the physical and chemical qualities of glass, making it more controllable during the manufacturing process. Another popular ingredient that gives glass stability, keeps it from dissolving in water, and strengthens its structural integrity is lime (CaO). Alumina (Al_2O_3) is added to strengthen the material's resistance to chemical corrosion for enhancing durability, while magnesium oxide (MgO) and potassium oxide (K_2O) are used to improve the mechanical qualities of the glass, particularly its ability to withstand thermal stress (Saud AlMulaifi, 2024). According to Grayhurst (2012) adding calcium (Ca) compounds makes this material insoluble. Glass surfaces may also be coated with titanium (Ti), Al or zirconium (Zr) chemicals to strengthen their stability and allow thinner and lighter containers. Glass has a good resilience to heat shock and minimal thermal expansion thanks to boron compounds (boric oxide (B_2O_3), borax). Other compounds help lower the duration and temperature necessary for melting as well as the presence of gas bubbles. In addition to these compounds, decolorizers such as Ni, cobalt (Co) or selenium (Se), are used to hide any color created by impurities such as Fe. Glass clarity is enhanced by Pb compounds. Conversely, glass is a superb barrier to everything except for light. It additionally ensures the absence of transmission, absorption, or migration activities between the food and the packaging (Molina-Besch & Olsson, 2022). Concisely, technological advancements address glass packages related limitations through lightweighting techniques that reduce mass while maintaining strength, and surface coatings that enhance lubricity and impact resistance. Also, from an environmental perspective, glass offers infinite reusability and recyclability in closed-loop systems, though the high energy input required for primary production and transportation due to weight presents sustainability challenges that are mitigated through increased recycling rates and refillable system implementation.

2.5 Bioplastics

Bioplastics or bio-based plastics encompass a diverse family of materials defined by either their biological origin, biodegradability, or both, representing the frontier of sustainable packaging innovation derived from natural substances (cellulose, sugars, and starch) (Huang et al., 2025; Mahmud et al., 2025). The biopolymer-derived formulations of bioplastics include poly (lactic acid) (PLA) incorporated with additives for processing and coloration, starch mixed with plasticizer, and (natural) fiber reinforced poly (3-hydroxybutyrate) (P3HB) (Lackner et al., 2023). Clear differentiation is essential between bio-based plastics (e.g., PLA from corn starch) and biodegradable plastics (e.g., polyhydroxyalkanoates (PHA) from microorganisms), as these characteristics do not always coincide. Yet, bioplastics are obtained from renewable biological origin and offer sustainable features that mitigate food packaging related health and environmental concerns (Sharma et al., 2025). Current material properties often show limitations in barrier performance and thermal stability compared to conventional polymers, restricting applications primarily to short-shelf-life products and fresh products where gas permeability can be advantageous (Ciawi et al., 2025). The environmental proposition centers on renewable resource utilization and potential end-of-life options including industrial composting and anaerobic digestion. For food packaging applications, several bioplastics have been proven to be very compatible with a variety of antimicrobial agents and have been assessed for their effectiveness against pathogenic microbes and spoiling under various test settings (Santana et al., 2025). Bio-based plastics are created using a variety of chemical and biological processes, such as hydrolysis, microbial fermentation, acidification, and others, employing renewable resources including corn, potatoes, and vegetable oil (Kowser et al., 2025). However, realization of these benefits depends on appropriate waste management infrastructure and clear consumer education, with evolving and ongoing research focusing on improving performance properties while maintaining sustainable credentials (renewability, affordability, biodegradability) (Kumar Chandra et al., 2024).

2.6 Aluminum foil

Aluminum is a slender, silver (Ag)-white metal utilized widely in the production of foil, laminated paper, and plastic packaging materials. It is extracted from bauxite ore, which normally consists of O_2 and alumina. To boost its strength functionality, Al is frequently alloyed with Mg and Mn (Nerlekar et al., 2026). Al foil, produced through progressive rolling of Al ingots into thin, flexible leaves less than 0.2 mm thick, of yield strength ranging from 200 to 600 MPa, and provides exceptional barrier properties against all gases, light, and moisture (Kerry, 2012). Majority of consumer food, drinks, pharmaceuticals, health care goods and cosmetics are bottled for delivery and sealed utilizing Al foil liner that is coated with a plastic coating (Liu & Naidu, 2025). Performance characteristics directly correlate with thickness, with thinner gauges exhibiting increased pinhole formation. Al foil provides aseptic properties (Hsu & Chang, 2007; Kerry, 2012). In favor of the idea that thermal insulation packaging is vital to conserving food safety and quality throughout cold food chain management, materials with low emissivity, such as Al foil, which are often employed in packaging applications, have high reflectivity and can productively reflect infrared radiation (Mwenya et al., 2025). Rarely used alone, Al foil serves as a critical component in laminated structures, combined with polymer films and papers to create high-performance flexible packaging for extremely sensitive food products. Its absolute barrier qualities make it indispensable in aseptic packaging systems and portions requiring complete protection from environmental factors (Noviello et al., 2020). The environmental assessment balances the high energy intensity of primary Al production against excellent recyclability and significant energy savings when recycled. In composite structures, foil's minimal thickness contributes negligible weight while providing maximum barrier protection, optimizing the environmental footprint of complex packaging systems (Fig. 1).

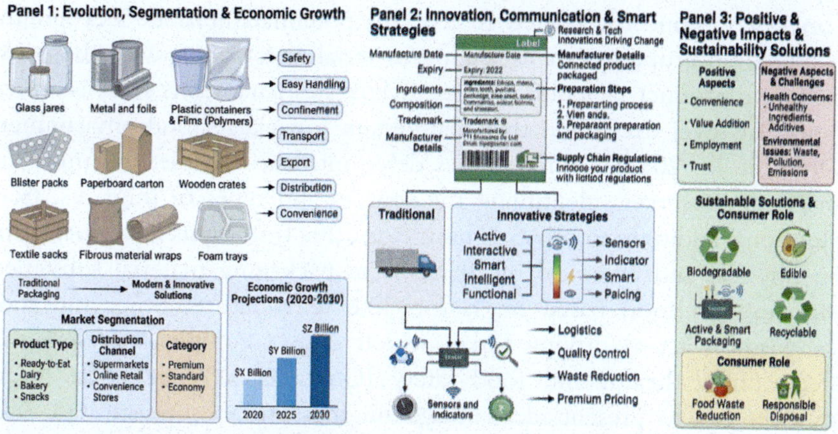

Fig. 1 Comprehensive roles and innovations in food packaging.

3. Synergistic interaction of food with packaging materials

Conventionally, the basic purpose of food packaging is related to the provision of inert containment and protection from external environmental factors for an enhanced food safety and quality (Radan et al., 2025). However, a more nuanced understanding reveals that packaging is rarely a passive bystander. Instead, it engages in a continuous and dynamic synergistic interaction with the food product it contains. This interplay is governed by the fundamental principles of mass transfer, where chemical substances from the packaging material can migrate into the food, and conversely, food components can be absorbed by the packaging. These interactions are not merely theoretical concerns, but they have direct, measurable consequences for food safety, sensory quality, shelf-life, and ultimately, consumer health. According to Dejene, (2025a) intercalation composite technology, which combines biopolymers and nanomaterials, has demonstrated promise in enhancing packaging materials qualities including bioactivity, thermal stability, and rigidity while preserving flexibility and biodegradability. Also, the incorporation of metal and metal oxide nanoparticles, like TiO_2, zinc oxide (ZnO), and Ag has evolved and improved packaging materials (biopolymer films) mechanical and barrier qualities. Hence, the synergic effect of nanoparticles in packaging materials on food can convert biopolymer sheets into intelligent packaging solutions that provide real-time food quality monitoring (Boateng et al., 2025).

Conversely, concerns related to toxicity potentials emanate beyond the macroscopic properties of packaging materials interaction with the packaged food product. Scientific report investigates parameters such as critical molecular-level exchanges that manifest in a package food environment (Omer et al., 2025). Other studies delve into the complex chemistry of packaging materials, identifying the sources and types of migrants, from intentional additives to NIAS (Tumu et al., 2024). Insights into elucidating the physical and chemical mechanisms of packaging material substances migration on food based on factors like temperature, time, and food composition act as driving forces which pose a complex problem that affects public health and food safety (Gupta et al., 2024). In recent advancement to present ideal mitigation measures, a sophisticated categorization of migration phenomena, reframing it not solely as a detrimental process to be minimized is appreciated. But as a spectrum of effects ranging from negative impacts on safety and quality to the deliberate, functional use of migration in active packaging technologies to enhance food preservation continue to be explored. Understanding these synergistic interactions is paramount for designing next-generation packaging solutions that not only provide protection but also actively accord to maintaining and enhancing food quality from production to consumption.

3.1 Packaging material chemical/substance composition

Modern food packaging materials contain complex chemical compositions that extend beyond their primary polymer, cellulosic, or metallic structures. Food contact articles are composed of food contact chemicals (FCCs) and one or more separate food contact materials. Food manufacturing and supply also depend on food contact items like filling lines, processing machinery (equipment), kitchen and tableware, and storage containers in the value chain (Muncke et al., 2020). Food contact materials are utilized to create food contact items that interact with food throughout processing, packing, transportation, storage, and consumption (serving) (Groh et al., 2021; Qian et al., 2018). These food chemical materials and food contact items include compounds termed as "food contact chemicals," which can contaminate food when they move from the materials into the food. Under some circumstances, low molecular weight (<10,000 Dalton) chemicals from printing inks, adhesives, and packaging raw materials (coating monomers, cleaning and washing chemicals, and fountain solvent) migrate to food, possibly introducing hazardous elements evidenced from biomonitoring (Muncke et al., 2020; Seref & Cufaoglu, 2025; Yenidoğan et al., 2023). Consumers exposure to intentional additives such as plasticizers, stabilizers, antioxidants, slip agents, and pigments, along with NIAS including oligomers, catalyst residues, and degradation products is synch with both food safety or public health (toxicology and carcinogens) and environmental concerns (Bignardi et al., 2017). Monomers, catalysts, solvents, antioxidants, antistatic, antifogging agents, slip additives, plasticizers, heat stabilizers, dyes, and pigments are among the compounds that can migrate from plastic polymers. These substances that interact with the product in the package and produce off-odors, off-flavors, or other hazardous substances could migrate and raise regulatory issues (Santos et al., 2024). The chemical profile varies significantly between material types: plastics may contain bisphenols and phthalates; paper packaging often incorporates per- and polyfluoroalkyl substances (PFAS) in grease-resistant coatings; metal cans employ epoxy phenolic lacquers; and glass typically has the simplest composition, primarily inorganic oxides with minimal additives (Muncke et al., 2020). FCCs (synthetic chemicals) are found in food, medications, household and personal care items, and environmental contaminants. Consequently, the rising incidence of non-communicable diseases has been linked to several of these substances (Geueke et al., 2025). Further, the effect of chemical migration from food contact materials to human exposure has been thoroughly researched for specific FCCs, such as bisphenol F (BPF), di (2-ethylhexyl) phthalate (DEHP), perchlorate, dibutyl phthalate (DBP), diethyl phthalate (DEP), and bisphenol A (BPA), while accounting for the existence of additional exposure sources. BPA is frequently tested in food contact items even though it is prohibited in various parts of the world in certain food contact applications, such as baby bottle. Due to their potential for endocrine disruption, bioaccumulation, reprotoxicity, mutagenicity, and carcinogenicity, and/or persistence, many chemicals that come into contact with food pose a risk to consumer health (Geueke et al., 2025; Gupta et al., 2024; Muncke et al., 2023; Zimmermann et al., 2022). Only certain substances used to produce food contact materials are examined, usually just for genotoxicity, while endocrine disruption and other risk qualities are ignored. Current regulatory regulations do not adequately protect the public's health from hazardous FCCs (Muncke et al., 2023). Hence, grasping the complex chemistry of packaging materials chemical is fundamental to predicting potential interactions with packaged foods, as these substances can migrate and impact both food safety and quality parameters. To ensure stringent safety assessment and regulatory compliance, laws play a crucial role in giving regulatory guidelines on quality assurance systems and confirming their application (Yenidoğan et al., 2023) (Fig. 2).

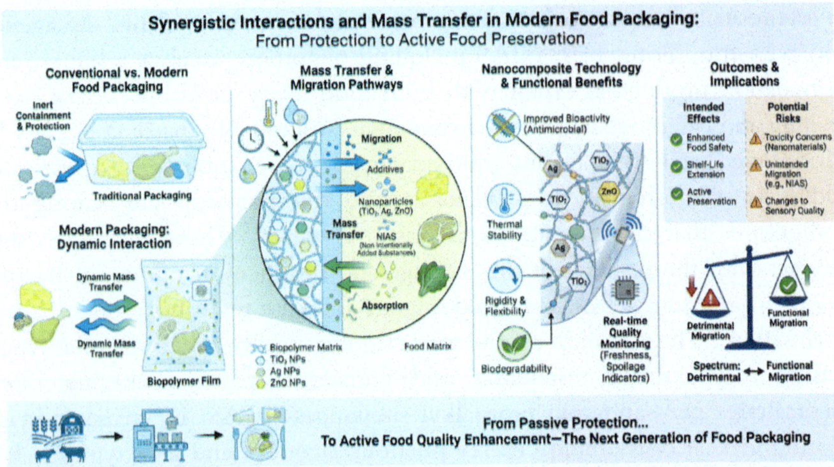

Fig. 2 Synergistic interaction and mass transfer of food with packaging materials.

3.2 Mechanisms of packaging material chemical/substance migration on food quality

Substances that have not been added for technical reasons during the development of food-contact materials and items are known as non-intentionally added substances (NIAS) (Geueke, 2018). According to Bradley and Coulier (2007), there are two classes of NIAS that leach into food matrices. The unintentional category is substance such as isomers, intermediates, and major technical impurities. The substances that are undetected or unidentified and so are unknown is related to the second category. Herein, reaction and breakdown products along with minor impurities often fall into this second category of unknown NIAS. The different migration mechanisms of gas phase migration, penetration migration, condensation/distillation migration, set-off migration, and contact migration are extensively discussed by Gupta et al. (2024); Spinei and Oroian (2025). Further, the major types of migrating compounds are nanoparticles, light stabilizers, monomers, thermal stabilizers, printing inks, oligomers, antioxidants, plasticizers (softness and fillers), and adhesives, pose potential health risks because of their connection with endocrine disruption and carcinogenic effects (Gupta et al., 2024; Spinei & Oroian, 2025). Chemical migration occurs through multiple mechanisms or chemistry of the food and food article, primarily diffusion, partitioning, and convection (permeation), influenced by temperature, contact time, and food properties (Bignardi et al., 2017). Although all diverse classifications of food contact material may express chemical migration phenomenon, yet the threshold of the chemical leaching and the exact type of migration differ greatly (Muncke et al., 2020). The diffusion mechanism follows Fick's and Henry's laws, where molecules move from regions of high concentration (packaging) to low concentration (food), with rates dependent on molecular size, polymer crystallinity, and temperature (Gupta et al., 2024; Kato & Conte-Junior, 2021). Muncke et al. (2020) revealed that the quantity of migrating compounds was the packaging's surface area to volume ratio. Because the increased surface-area-to-volume ratio as chemicals migrate from all kinds of food contact materials and articles, there is a proportionately larger migration effect from smaller container sizes. Chemical migration from packaging materials into foods under refrigerated conditions is both slower and smaller than that occurs at higher temperatures (Cozzini et al., 2022). Higher surface area to volume ratios in food packaging cause increased chemical migration into the food which is particularly true for meals meant for children (e.g. tiny size containers) (Vaferi et al., 2025). Partitioning governs the distribution equilibrium between packaging and food, significantly affected by food composition-particularly fat content, polarity, and pH (Gupta et al., 2024). These migration processes can directly compromise food quality through flavor scalping (absorption of food aromas by packaging), off-flavor transfer, nutrient degradation, and color changes (Kumari et al., 2024). Furthermore, migrated substances may catalyze oxidative reactions or interact with food components to form novel compounds with unknown safety profiles. Food and packaging materials can interact in several ways, including migration, permeation, and sorption. When food stuff and packaging come into direct contact as opposed to indirect contact, there is increased migration. The migration is comparatively slower in indirect contact because the food and packing material interact with the aid of a gaseous medium. Permeation involves transport of gases such as O_2 and carbon-dioxide (CO_2). By changing the gas composition of packaging materials, gas movement might jeopardize food quality. When components are moved between the two surfaces by either adsorption or absorption, sorption takes place at the food-packaging contact. Aroma and flavor modifications brought forth by sorption cause an organoleptic shift in the finished product (Nagalapur & Byadagi, 2024).

3.3 Categories and functionalities of packaging material chemical/substance migration phenomenon

Chemical migration from packaging materials can be measured in a controlled and consistent manner by using simulants that mimic the properties of different food types. A standardized substance that mimics the qualities of several food groups, such as fatty, moisture, and acidic foods, is called a food simulant (Gupta et al., 2024). To assess the dangers of chemicals that leach into food, migration must be closely monitored. The materials and chemicals used in food packaging provide health concerns that should be properly evaluated and closely watched. Such carcinogens must be removed to avoid contact and possible migration of carcinogenic chemical substances into food. The widespread of health concerns have been generated by risks associated with the presence of chemical residues in food. Subsequently, chemical poisoning may be indicated by both acute and chronic symptoms, such as light-headedness, nausea, vomiting, diarrhoea, loss of appetite, sleep problems, and decreased fertility (Alamri et al., 2021). To safeguard the food contents from outside threats like contaminant and to avoid unintentional food container chemical exposure to consumers, the food package materials chosen, the structural design, and the closure mechanisms are all crucial. Therefore, it is acknowledged that chemicals from packaging and other materials that come in touch with food might seep into the food and be consumed by the user. Tracking this migration has become essential to maintaining food safety (de Fátima Poças & Hogg, 2007). Risk assessment (scientific evaluation of hazards and their exposure), risk management (selection of control measures based on assessment outcomes), and risk communication (transparent exchange of information among stakeholders) are the three interrelated components that make up the process of risk analysis (Vaferi et al., 2025). It also describes how information and opinions on risks, hazards, risk-related factors, and perceptions are exchanged interactively throughout the whole risk analysis process (Coja & Steinwider, 2022; Maxim et al., 2021). Migration phenomena can be classified into three primary classes based on their impact on food quality. Negative migration encompasses the transfer of harmful substances, including endocrine disruptors, heavy metals, and NIAS, which pose safety concerns and may catalyze nutrient degradation or oxidative rancidity (Wrona & Nerin, 2019). Neutral migration involves substances with no significant impact on food safety or quality, typically including certain oligomers and mineral pigments that remain below sensory detection thresholds. Positive/functional migration represents an innovative approach where intentionally incorporated active substances migrate to extend shelf-life, including oxygen scavengers, antimicrobial agents, and antioxidants that actively improve food preservation and quality maintenance. Understanding these categories enables strategic material selection and the development of next-generation functional packaging systems that leverage beneficial interactions while minimizing negative impacts. To address concerns about exposure to one or more chemical, physical, or biological agents that could endanger human wellbeing and the environment, risk assessment is a thorough, systematic, interdisciplinary, and iterative procedure. Over time, risk assessment has developed into a specialized scientific field that involves analyzing and reviewing scientific data to determine the likelihood that exposure to hazardous substances can result in adverse consequences (Risk = Hazard × Exposure) (D'Amore et al., 2025). According to Vaferi et al. (2025), the partition coefficient (K_{pf}), a crucial metric for measuring chemical transfer from packing materials into food, is frequently used to assess food contact chemical migration. The chemical concentration in the food contact material is divided by the chemical concentration in the food composition to determine this factor. Hence, experimental approaches are frequently used to examine the effects of important variables on K_{pf}, including as temperature, material type, and dietary qualities. The expensive and time-consuming nature of conducting such an experiment is probably the reason why large FCCs have little migration data. The use of intelligent or mathematical models for the K_{pf} determination is a result of the complexity of the experimental investigation of chemical migration into foods. Using a few explanatory variables, these models can be used to predict the chemical migration from packaging materials into food (Vaferi et al., 2025). In summary, the use of analytical techniques for migration testing, and the identification of possible migrants are necessary to evaluate food simulants and food packaging/contact materials safety.

4. Advanced packaging systems

The evolution of food packaging has progressed beyond its traditional roles of containment and protection into a sophisticated discipline that actively enhances and monitors food quality throughout the supply chain. Over the years, there have been a paradigm shift from passive containment systems to advanced packaging technologies that interrelate dynamically with food products and their environment (Boateng et al., 2025). These cutting-edge solutions actively participate in quality preservation, real-time monitoring, and communication while traditional packaging acts as a barrier (Alnadari et al., 2023a; Alnadari et al., 2023b). According to Kumar and Gaikwad (2023), with the development of technology, food processing, preservation, and packaging have changed. In that, by altering its shape, most food found on earth may be ingested in space and appropriate packaging materials is employed to enhance the food variety, quality, acceptability, and shelf stability. To achieve specific preservation functions, emitters are specialized packaging components or incorporated systems that deliver active substances into the package environment in a regulated and sustained manner. Emitters actively deliver beneficial agents (such as antioxidants, CO_2, ethanol, humidity regulators, volatile organic actives, O_2 etc.) into the package headspace or, in certain cases, directly onto the product exterior, in contrast to scavengers, which passively remove undesirable compounds like O_2, moisture, or ethylene (Nuamduang & Leelaphiwat, 2025). In recent years food packaging research explorations cascade are four transformative categories that represent the forefront of food packaging innovation. Active packaging systems deliberately incorporate components that emit or absorb substances to extend expiration and sustain nutritional quality (Beikzadeh et al., 2025; Pateiro et al., 2019). Smart packaging technologies employ indicators, sensors, and data carriers to monitor and transmit critical information about food safety and freshness. The amalgamation of these intelligent systems into value chain operations enables unprecedented levels of food quality control and traceability (quick detection and recall of potentially dangerous food items) (Kumar et al., 2025). Finally, edible packaging represents a sustainable alternative that eliminates packaging waste entirely while facing unique challenges in consumer acceptance and regulatory compliance (Satheesh et al., 2026). Together, these advanced systems represent a fundamental reimagining of packaging's role in the food ecosystem, transforming it from a passive container to an active intelligent partner in quality storage, waste reduction, and consumer safety. The complexity and dynamics in packaging systems advancement provide a comprehensive analysis of their mechanisms, applications, and the technological convergence worthy in driving the future of food packaging.

4.1 Active packaging applications and food quality monitoring

Active packaging represents a transformative approach where the package, the product, and the environment interact dynamically to extend shelf stability, enhance safety, and preserve quality (Jiang et al., 2025). Active packaging effect on food quality and shelf-life extension monitoring exhibits good antioxidant, antibacterial, and antifungal properties. These systems integrate active components that deliberately eject or absorb substances through engineered migration and scavenging mechanisms. Oxygen plays a role in the oxidation of lipids, enzymatic processes, and the proliferation of food spoilage microorganisms, given that these gases are directly linked to the rate at which aerobic spoilage microbes multiply, the consumption of product substrates, and the start of metabolite oxidation processes, which are quite pertinent. All of which can lower the quality of food and eventually shorten its shelf stability (Escobar et al., 2023). Oxygen scavenging systems, which come in the form of trays, sheets, pads, labels, sachets, contain oxygen scavengers that can react chemically or biologically (enzymatically) with free oxygen molecules to scavenge oxygen and prevent free radical reactions from starting (Guo et al., 2024). Oxygen scavengers (primarily iron-based or organic compound systems) prevent oxidative rancidity and nutrient degradation, while carbon dioxide emitters suppress microbial growth in fresh products (Rozo et al., 2024). Even at trace amounts, ethylene, a naturally occurring plant hormone, speeds up the ripening and senescence of horticulture products. When left unchecked, ethylene buildup in packaging can rapidly impact harvested product quality and safety negatively. To prolong shelf stability and preserve postharvest quality, ethylene control in fresh produce packaging is crucial (Mahajan et al., 2025). Consequently, ethylene absorbers are a functional attribute of active packaging, particularly potassium permanganate ($KMnO_4$) based systems, metal organic frameworks, nanomaterials and photocatalysts (TiO_2, palladium (Pd), Ag), delay ripening and senescence in climacteric fruits and vegetables (Hong et al., 2025; Mahajan et al., 2025). One of the main causes of food product deterioration is an imbalance in moisture between food products and their surroundings. Food can become moist under several conditions, including the condensation of water vapor brought on by temperature changes, the presence of melting ice, the presence of tissue fluid exudates from cut meat products, and the respiration of fruits and vegetables (Bumbudsanpharoke & Boonsiriwit, 2025; Guo et al., 2024). Moisture regulators, including desiccants (silica gel, molecular sieves, zeolites, calcium oxide (CaO), porous clay, absorbent polymers, humectant salt etc.) and humidity controllers, maintain optimal water activity levels to prevent microbial proliferation and texture deterioration (Bumbudsanpharoke & Boonsiriwit, 2025). Recent innovations include antimicrobial packaging incorporating natural agents (nisin, essential oils, silver nanoparticles) and antioxidant-releasing systems that migrate preservatives directly to the food surface, providing targeted protection where most needed (Alnadari et al., 2023a; Alnadari et al., 2023b; Boateng et al., 2025). Conversely, the use of synthetic additives in the fabrication of active packaging systems is a threat to consumers or public health (Belasli et al., 2025). Therefore, it is admirable that natural and safe additives are used to establish consumer safety and ecological protection. Furthermore, it is praiseworthy to investigate sustainable alternatives where the threshold of the active ingredients, the regulation of the active agents' emissions, and the quick depletion and decline in the functional activity of the active package materials meet users demand (Li et al., 2026) (Fig. 3).

4.2 Smart food packaging technologies

According to Habiba et al. (2025a) the emergence of Industry 4.0, which is attributed by the unification of web-based platforms and smart technology, the packaging industry has undergone a considerable digital change known as packaging 4.0. Packaging 4.0 encompasses four major areas which include packaging production, packaging product relationship, packaging consumer interrelation, and packaging's role in sustainability. This transformation is a comprehensive framework that consolidate ecological, economic, and social factors. Smart packaging technologies rely on biosensors or chemical sensors to check, detect, and communicate data about food quality and safety

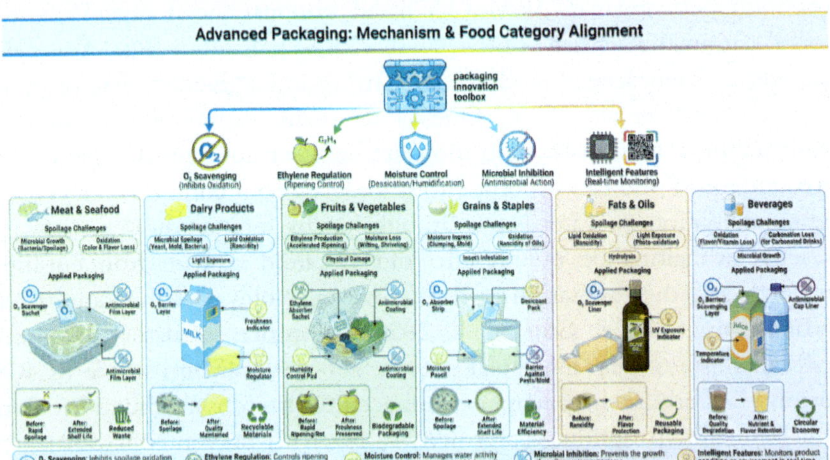

Fig. 3 Advanced packaging: mechanism and food category alignment.

throughout the supply chain (Schaefer & Cheung, 2018). The development of sophisticated preservation systems is given an unparalleled boost by the emerging science of artificial intelligence (AI). Therefore, by providing tailored postharvest freshness techniques for a variety of product, smart precision packaging directly addresses quality deterioration causes such atmospheric variation, temperature and humidity changes, and microbial infection (Du et al., 2026; Mobahi et al., 2025). Smart food packaging systems encompass primary and advance technological categories such as indicators, quick response (QR) codes, radio frequency identification (RFID) tags, barcode, sensors, and information carriers to deliver real-time statistics and information on the packaged food. Enhanced safety and quality control, increased traceability, increased consumer interaction, and decreased food waste are all benefits of smart packaging (Müller & Schmid, 2019). Smart food packaging has the capacity to completely change how consumers package, allocate, convey, and consume food and its products by leveraging these technologies in conjunction with the internet of things (IoT) (Putra & Kuswandi, 2025; Sobhan et al., 2025). In the food value chain, temperature regulation is required for preserving shelf-life, safety, and quality. Time-temperature indicators (TTIs) are intelligent packaging (IP) gadget that may accomplish this. These integrators take a pivotal part in tracking the food's exposure to temperature and storage duration. The TTIs are grouped as complete history indicators, critical temperature indicators, and critical temperature-time indicators. The ideas of polymerization, photochromic, redox, diffusion, or nanoparticles may serve as the foundation for chemical TTIs (Gurunathan, 2024). Further, TTIs provide visual evidence of cumulative temperature exposure, with diffusion-based, enzymatic, and polymer-based systems offering different activation energies and response ranges (Pavelková, 2013; Yar et al., 2025). Freshness indicators detect specific metabolites (volatile amines, hydrogen sulfide (H_2S), CO_2) produced by food spoilage, utilizing colorimetric dyes, pH-sensitive polymers, or enzymatic reactions that change color in response to quality deterioration progression (Müller & Schmid, 2019). Gas indicators, particularly oxygen sensors employ redox dyes like methylene blue to monitor package integrity and modified atmosphere conditions. Sensors are electrical devices that use a receptor to detect, identify, or quantify an alteration. A transducer then converts the signal into a signal that reflects a chemical or physical change. One kind of sensor pertaining to biological responses is a biosensor, in which a transducer transforms a biochemical signal into an electronic response that can be measured. Pathogens, allergies, and other hazardous substances can be found using biosensors (Sobhan et al., 2025). Advanced biosensors utilize antibodies, deoxyribonucleic acid (DNA) probes, or molecularly imprint polymers to detect specific pathogens or toxins with high sensitivity (Meliana et al., 2024; Rodrigues et al., 2021; Soundara Rajan & Wani, 2025). The RFID tags and near field communication (NFC) systems allow real-time tracing and condition monitoring, creating digital twins of physical products throughout the distribution chain (Soundara Rajan & Wani, 2025). In summary, smart food packaging technologies highlight interrelated important roles in the food industry, value chain (transparency and traceability), new sustainability trends (reduce waste and comply with regulations), and circular economy (economic growth based on marketers and customer experience).

4.3 Intelligent food packaging integration in food quality operations

The merger of intelligent packaging networks into food quality operations represents a paradigm shift toward Industry 4.0 in the food sector (Yu et al., 2025). According to Madhu (2025) and Özoğul (2025) by applying the acumen, learning, experience, and brainpower of human to computers, machines, or systems, AI can recognize novel issues that are predictable or unpredictable and take the appropriate response. A vast range of computational technologies that mimic human cognitive processes including learning, pattern identification, and decision-making are together referred to as AI (Jia et al., 2026). Traditional packaging is evolving from passive containment to a smart adaptable part of food quality control due to the use of AI into food packaging systems. Without directly affecting the food product itself, intelligent packaging uses technologies that sense, track, and share data about the product's freshness, preservation conditions, or spoiling status (Madhu, 2025). AI in food science could completely change how food packaging affects food items, customers, food quality and safety, sustainable demands, resource scarcity, and growing global food demand (Mittal et al., 2025). To enable continuous monitoring and enable dynamic, adaptive, and predictive capabilities that transform packaging systems from reactive tools into autonomous agents efficient of real-time monitoring, diagnosis, and responses, AI algorithms process the gathered data of food product storage patterns and the package in real time. AI encompasses several computational subclasses whereby machine learning helps packaging systems to forecast shelf stability or deterioration based on real information. The field of deep learning supports more intricate pattern identification such as visual anomaly sensing. Also, the computer vision self-operating visual assessment of package integrity and labeling. Whereas the natural language processing enables intelligent interactivity with consumers and regulatory content through multilingual intellect labels or QR-based interfaces (customer interaction and feedback) (Jia et al., 2026; Oyegbade et al., 2025). These technologies permit real-time quality monitoring all around the value chain, providing data-driven insights for dynamic shelf-life projection and waste reduction. Blockchain is utilized in the food sector to ameliorate traceability and transparency, streamline the supply chain, and do away with middlemen in the agriculture sector. Blockchain implementation in the agri-food sector can improve supply chain traceability and transparency, do away with the need for third parties, and foster consumer trust (Baku et al., 2024; Hasan et al., 2024). Digital traceability systems, combining QR codes, near field communication (NFC), and blockchain technology, create transparent and immutable records from production to consumption, enhancing both quality assurance and consumer trust. Smart labels integrated with cloud platforms enable dynamic expiration dating per the actual storage conditions rather than conservative estimates, potentially minimizing food waste (Ganeson et al., 2023). Packaging solutions that are extremely sensitive and environmentally friendly have been developed based on the advancements in nanotechnology, bio-based materials, and smart polymers. Nanotechnology, microfluidics, 3D printing, and the utilization of electromagnetic techniques are just a few of the many cutting-edge techniques used in intelligent packaging materials. Customized and intelligent packaging with integrated sensors and creative structural components can be made because of 3D printing's exceptional versatility in packaging design. This method reduces the amount of material used while improving packaging performance (Mkhari et al., 2025). Food packaging now has a temporal dimension due to 4D printing, which enables materials to actively alter their color or shape in reaction to environmental stimuli like pH, humidity, temperature, or light. More sophisticated and inventive printing methods, such as 5D and 6D printing, have lately started to appear, creating new opportunities for food production (Yu et al., 2025). Although, the advanced systems incorporating machine learning algorithms analyze data from multiple sensors to predict remaining packaged food shelf-life with increasing validity, allowing for optimized inventory management and markdown strategies. Yet, the intelligent packaging implementation challenges include cost considerations, standardization issues, and regulatory compliance, particularly regarding sensor materials and data privacy.

4.4 Edible films: Consumer safety and perception considerations

Edible films are thin edible coverings that are safe to eat and can be put to food surfaces to offer an extra layer of protection for marketing, distribution, and preservation (Walayat et al., 2024; Falguera et al., 2011). According to Usman et al. (2025) any material that is a necessary component of a food product that is safe to eat alongside the food itself is referred to as edible packaging. These materials may consist of films, thin layers, and surface coatings made of dietary biopolymers and food-grade additives that meet safety regulations (generally recognized as safe (GRAS)) and appropriate manufacturing methods (Gupta et al., 2024). Edible films and plant-based materials, which are formed from renewable resources including fruits, vegetables, and crops, have shown promise as sustainable and eco-friendly replaces for traditional packaging materials (Olawade et al., 2024; Zhou et al., 2022). Consequently, edible packaging represents a sustainable substitute to conventional materials, utilizing biopolymers derived from proteins (whey, collagen, zein, gelatin), polysaccharides (chitosan, pectin, chitin, carboxymethylcellulose, starch, alginate) (Wei et al., 2025; Boateng et al., 2025; Kumar & Gaikwad, 2023), and lipids (jojoba oil, paraffin wax, rice bran wax, beeswax, resins, candelilla wax) to create thin layers consumed with the product (Falguera et al., 2011; Gupta et al., 2024; Usman et al., 2025). Edible films advantages include significant reduction of packaging waste, carrier functionality for active compounds (antioxidants, antimicrobials, flavors), and improved product integrity through partial barrier properties (Arshad et al., 2025; Walayat et al., 2024). However, consumer perception presents complex challenges, with studies revealing dichotomous attitudes, while environmental benefits are appreciated, concerns regarding sensory alterations, safety of novel materials, and psychological barriers to "eating packaging" persist (Shao et al., 2025; Taylor et al., 2024). Also, edible film biopolymer associated disadvantages including brittleness, poor mechanical qualities, and low thermal stability are revealed by Pan et al. (2025). Therefore, using nanotechnology to incorporate nanoparticles (such as ZnO or Ag nanoparticles) improves the films' mechanical stability, antibacterial qualities, and other functionalities. Safety considerations encompass potential allergenicity of protein-based films, regulatory status of novel biopolymers, and microbial stability during storage. Future development focuses on multilayer edible films combining complementary properties of different biopolymers, nanotechnology enhancements for improved barrier properties, and clear labeling strategies to address consumer information needs and build trust (Gunwant et al., 2025).

5. The synergistic impact of nanotechnology on novel food packaging

Xuan et al. (2023) reveal that the milli-micron, one of the units of length, is shortened to "nano." Ten hydrogen atoms arranged in a row would be the length of a nanometer (nm), which is one billionth of a meter. In nanotechnology, structures, devices, and materials with the least length dimension of 1–100 nm is constructed, characterized, manufactured, and manipulated. When particle size is lessened below this threshold, the resulting material's physicochemical properties differ significantly from those of macroscale materials made of the same ingredient (Biswas et al., 2022; Kumar et al., 2025). Hence, materials smaller than 100 nm are known as nanomaterials, and they have unique chemical, biological, and physical characteristics which include surged surface area, improved reactivity, and unique optical or electrical properties (Ji et al., 2025; Sasikumar et al., 2026). Ultimately, the emergence of nanotechnology has ushered in a transformative era for food packaging, enabling unprecedented advancements in preservation, safety, and functionality. By engineering materials at the 1–100 nm, packaging systems are revealed to acquire enhanced barrier properties, mechanical strength, and active functionalities such as antimicrobial and antioxidant effects (Correia et al., 2024). These innovations hold significant promise to improving food quality, extending shelf stability, and decreasing waste. However, the unification of nanotechnology into food packaging also introduces complex and multifaceted considerations regarding consumer safety, environmental impact, and sustainability (Khalafi & Zandi, 2025). The toxicological profiles, migration potential, and health implications of nanoparticles used in packaging materials concerns recommend a rigorous risk assessment frameworks necessary to ensure consumer safety. In addition, concerns of nanotechnology in advance sustainable packaging solutions considers, the evaluation of nanotechnology contributions to package material efficiency, biodegradability, and circular economy goals. These actions tackle the ecological questions surrounding nanoparticle life cycles. By balancing the scientific innovations with a precautionary perspective, nanotechnology can be responsibly harnessed to suit the evolving quest of modern food systems. Thus, the synergistic relationship between nanotechnology and the advent of state-of-the-art food packaging systems (improved, active, smart, and intelligent packaging) remains groundbreaking in scientific research.

5.1 Insights into toxicities, risk, and safety of food package nanoparticles on consumer health

A rich ecosystem of technologies capable of enhancing food packaging functionality that fills interdisciplinary research gaps has been produced by the confluence of discoveries in materials science, food science, and engineering. Therefore, the merger of nanotechnology into food packaging act as a frontier of innovation, yet it raises critical questions regarding consumer safety. Nanoparticles, typically inorganic nanoparticles or organic nanomaterials (metal oxide, nano clay (bentonite, kaolin, montmorillonite), metals, and carbon nanotube) with no less than one dimension between 1 and 100 nm, manifest unique physicochemical characteristics due to their high surface area-to-volume ratio, which can enhance mechanical strength, barrier properties, and antimicrobial activity in packaging (Dejene, 2025b). Nanoparticles include ZnO, TiO_2, silicon dioxide (SiO_2), Al_2O_3, copper (Cu), CaO, MgO, cerium (IV) oxide (CeO_2), Ag_2O, graphene oxide (GO), gold (Au), Fe, ZrO_2, hybrid nanomaterials, nano protein, nano starch, nano cellulose, and nano chitosan (Dejene, 2025a; Juliet et al., 2025; Sasikumar et al., 2026; Wang et al., 2025; Zhang et al., 2026). Although, approaches in material science have demonstrated a substantial influence on the food packaging area yet traditional methods of creating nanoparticles based on their physical-chemical characteristics are hazardous and economical (Karuppan Perumal et al., 2025). The complex architecture employed to achieve sophisticated food package properties may pose potential health risks if nanoparticles migrate into food and are ingested. While some nanomaterials experience ion dissolution or aggregation, which greatly affects their bioavailability and safety profiles, the migration of nanomaterials into food matrices is a dynamic process entailing interlinkage with organic components (Wang et al., 2025). Additionally, migration happens by Fickian diffusion, just like with traditional polymer additives, or other material stress-based mechanisms such mechanical abrasion, material fatigue, ultra-violet (UV) exposure, hydrolysis, or swelling interplays that degrade the polymer matrix (Störmer et al., 2017). Further, the dissolution of nanomaterials into ionic species in the polymer and subsequent discharge of ions into the food can potentially result in a purely chemical release (Störmer et al., 2017). According to Wang et al. (2025) diffusion through polymer matrices, release mediated by matrix disintegration, ionic dissolution, and interfacial detachment are the four primary mechanisms that control the emission of nanoparticles in food systems. Key concerns include the ability of certain nanoparticles, such as Ag, ZnO, and TiO_2, to cross biological barriers, accumulate in bodily fluids (plasma, interstitial fluids, lymph), organs (kidney, spleen, liver), and induce oxidative stress, inflammation, or genotoxicity or neural disorders (Dejene, 2025a; Missaoui et al., 2018; Sanati et al., 2025). Toxicological research has highlighted that the migration of nanoparticles was influenced by food composition (e.g., pH, fat content), storage conditions, and the polymer matrix in which they were embedded (Missaoui et al., 2018). The study of the negative consequences of nanoparticles is known as nanotoxicology, and it is essential to understanding how nanoparticles interact with cellular systems. Nanoparticles have distinct physicochemical characteristics, including size, charge, content, and shape impacting their toxicity profile and cellular interactions (Kumar et al., 2025). The duration and routes of exposure, the amount consumed, and the size of the nanoparticles are all factors that affect the toxicological evaluation of nanomaterials. Nanomaterials can easily penetrate cell membranes and other biological barriers because of their smaller size, which enables them to enter living systems and perhaps interfere with regular cellular processes. Regulatory agencies are advancing risk assessment frameworks that require thorough characterization of nanoparticles, migration testing under realistic conditions, and long-term toxicological data to ensure safety. Future directions emphasize the development of "safe-by-design" nanomaterials and robust detection methods to monitor exposure and mitigate potential health impacts. In summary, advances in material science have had a substantial effect on the food packaging industry. Traditional methods of producing nanoparticles based on their physical-chemical characteristics are hazardous and economical. Accordingly, the food sector is concentrating progressively on less hazardous and environmentally friendly substitute packing (Fig. 4).

5.2 Relationship and related concerns of nanotechnology on sustainable packaged foods

A tool for assessing sustainability and life cycle assessment quantitatively assesses the cumulative ecological effects of every phase of a product's life cycle, such as raw material extraction and procurement, production, packing and transportation, application, and end-of-life scenarios (landfill, composting, incineration) (Banerjee et al., 2025). Nanotechnology offers promising pathways to enhance the sustainability of food packaging,

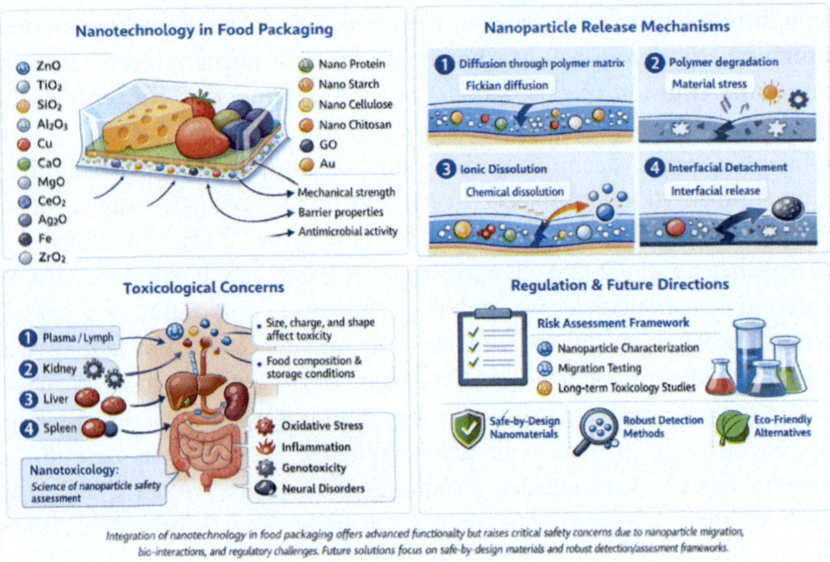

Fig. 4 Food package nanoparticles impact on consumer health.

though it also introduces complex environmental and safety trade-offs. On one hand, nano-enabled packaging can significantly reduce material usage, for instance, nanocomposites can provide superior barrier properties with thinner layers, lowering the overall plastic consumption and carbon footprint (Mishra et al., 2024; Srivastava et al., 2025). In addition, bio-based nanomaterials including nanocellulose and nano chitin are emerging as renewable and biodegradable alternatives to conventional polymers, aligning with circular economy principles (El-Sayed & Youssef, 2023). Active nano-packaging, incorporating nano-encapsulated antioxidants or antimicrobials, can extend the shelf stability of perishable foods, thereby minimizing food waste, a major contributor to environmental degradation (Boateng et al., 2025). Sustainability has an impact on the environment, society, and economy. However, concerns persist regarding the end-of-life behavior of nanomaterials. Melt mixing, solution casting, in-situ polymerization, and melt extrusion are some of the synthetic methods that have been developed to produce polymer matrix nanocomposites. These methods allow nanofillers, including carbon nanotubes, cellulose nanocrystals, or clay nanoparticles, to be dispersed throughout the polymer matrix to enhance sustainability, mechanical strength, and barrier attributes (Vafaeva et al., 2024). The potential for nano particles to seep into soil and water during composting or recycling, and their impacts on ecosystems and human wellbeing, remain areas of research exploration. Life cycle assessments are critical to evaluating the net ecological benefit of nano-packaging, considering not only resource efficiency but also the toxicity and persistence of nanomaterials on the part of regulatory bodies. Hence, a balanced approach, integrating green chemistry principles with comprehensive risk assessment, is essential to harnessing nanotechnology for sustainable packaging without unintended consequences.

6. Innovative packaging technologies on food products

The definitive assessment of every packaging innovation is its tangible effect on the preservation, safety, and quality of the food product it encases. This emphasizes the practical use of advanced packaging across many food categories, based on the packaging materials, mechanisms, and systems. Every food category, ranging from extremely perishable meats to shelf-stable grains, from respiring fresh fruit to delicate fats and oils, has distinct biochemical, microbiological, and physiological difficulties that necessitate customized packaging solutions. Therefore, a concise, product-centric analysis of how innovative packaging technologies that directly influence shelf-life, nutritional integrity, and sensory quality is presented. The section analyzes the unique interactions between packaging and food product, investigating how changed atmospheres, active components, and intelligent features are designed to mitigate the principal spoilage causes of various food categories. The discussion transitions from theoretical concepts to practical applications, emphasizing the implementation of packaging mechanisms such as oxygen scavenging, ethylene regulation, moisture control, and microbial inhibition to address the distinct preservation requirements of meats, seafood, dairy, fruits, vegetables, beverages, and other vital food categories. Moreover, customers' efforts to reduce waste across the supply chain are supported.

6.1 Shelf-life and nutritional impact of packaging on meat, fish and seafoods, fermented foods, and poultry products

Innovative packaging technologies have revolutionized the preservation of highly perishable protein-rich products (Fu et al., 2017a). For meat and poultry, advanced modified atmosphere packaging (MAP) with gas levels

(60–65 % O_2, 15–40 % CO_2) maintains myoglobin in its oxygenated form, preserving the desirable red color while inhibiting anaerobic spoilage bacteria (Habiba et al., 2025b; Rasheed et al., 2025). To preserve the appealing fresh color of beef in retail markets, MAP gas mixture comprising 80 % CO_2 and 20 % O_2 was utilized (Fu et al., 2017b; Liu et al., 2014). The integration of carbon monoxide-MAP (0.4 % CO) for improved color stability with nano-active packaging through the inclusion of ZnO nanoparticles for specific antimicrobial protection represents a key headway in packaging technology (Carpenter et al., 2024; Karuppan Perumal et al., 2025). For fish and seafood, low-O_2 MAP varying gas mixture (CO_2, N_2, O_2) combined with oxygen scavengers effectively reduce trimethylamine oxide conversion to trimethylamine, delaying off-odor development (Sivertsvik et al., 2002; Skura, 1991). Intelligent packaging with pH-sensitive indicators monitors real-time freshness by detecting volatile amines, while edible coatings containing essential oils provide additional protection against surface microbial augmentation (Fan et al., 2025; Krishnan et al., 2025; Panwar et al., 2024). Fermented foods require specialized packaging that maintains specific gas compositions. For example, kimchi packaging incorporates carbon dioxide emitters to maintain carbonation while including one-way valves to prevent package swelling, and probiotic dairy products utilize light-blocking materials with oxygen barriers to protect viable cultures (Lee et al., 2026). These technologies collectively preserve nutritional quality by minimizing protein oxidation, lipid peroxidation (Fu et al., 2015; Xiao et al., 2011), and vitamin degradation while extending shelf-life by 30–50 % compared to conventional packaging.

6.2 A Perspective of packaging on vegetables, mushrooms, legumes, and fruit

The packaging of fresh product focuses on managing respiration rates and ethylene sensitivity through tailored gas permeabilities and active components. Vegetables benefit from micro-perforated films that create optimal O_2 (2–5 %) and CO_2 (3–8 %) levels, preventing anaerobic respiration while reducing metabolic activity (Kargwal et al., 2020). Due to their high perishability, mushrooms must be packaged with high moisture permeability to avoid condensation, along with carbon dioxide emitters and ethylene absorbers to postpone cap opening and browning (Priya et al., 2025; Shi et al., 2020). Legumes in both fresh and dried forms utilize insect-resistant packaging incorporating pyrethrin-treated films or hermetic sealing with oxygen absorbers to prevent insect infestation and lipid oxidation. Fruits are categorized by climacteric behavior: non-climacteric fruits (citrus, grapes) primarily require humidity control and mechanical protection, while climacteric fruits (avocados, bananas) benefit from ethylene-absorbing technologies containing potassium permanganate or titanium dioxide-based photocatalytic systems. Emerging technologies include smart labels that change color based on ripeness indicators and biodegradable films incorporating plant extracts with antimicrobial properties, reducing post-harvest losses while maintaining phytonutrient content.

6.3 Insight into the quality assessment of packaged beverages, grains, nuts, seeds, eggs and dairy products

Diverse physicochemical characteristics across these product categories demand specialized packaging approaches. Beverages face challenges of flavor scalping and light-induced oxidation (Andersen & Skibsted, 2010), addressed through multilayer PET with UV blockers, oxygen scavenging closure liners, and nanocomposite barriers incorporating montmorillonite clay (Ahmed et al., 2021). Grains require insect-resistant packaging with modified atmospheres or hermetic sealed bags that create O_2 depleted environments, preventing insect development and rancidity (Labrot–Rhodes et al., 2025; Navarro et al., 2012). Nuts and seeds, with high lipid content, utilize packaging with high oxygen barriers (EVOH layers), integrated oxygen absorbers, and light-blocking materials to prevent lipid oxidation and preserve tocopherol content (da Silva et al., 2025; Nascimento et al., 2024). Eggs benefit from antimicrobial coatings containing chitosan or silver nanoparticles on shells, while specialized cartons with cushioning designs and temperature indicators protect against physical damage and thermal abuse (Huang et al., 2025; Yuceer & Caner, 2014). Dairy products present varied requirements: ultra-high temperature (UHT) milk employs aseptic brick packaging with Al foil layers; cheeses use MAP with specific CO_2/N_2 mixtures; and probiotic products utilize high-barrier containers (encapsulated material) with oxygen absorbers to maintain culture viability (Deeth, 2010; Mgomi et al., 2025; Singh & Singh, 2005). Quality assessment technologies include non-destructive techniques like hyperspectral imaging and electronic noses integrated with packaging systems for real-time quality monitoring.

6.4 Highlights of packaging functional mechanisms on cooking oils and fats, herbs and spices, and condiments

The preservation of flavor, color, and bioactive compounds in these sensitive products demands precise packaging mechanisms. Cooking oils and fats are highly susceptible to photo-oxidation and auto-oxidation (Gharby et al., 2025), necessitating packaging with exceptional light barriers (amber glass, metallized films), oxygen barriers (EVOH, PVdC coatings), and often integrated oxygen scavengers in bottle crowns or pouch layers. Herbs and spices contain volatile essential oils and pigments that degrade upon light, O_2, and moisture exposure, requiring packaging with high barriers to all these factors, often incorporating nitrogen flushing to remove O_2 and UV-filtering materials to preserve color and potency (Alizadeh Sani et al., 2024; Zachariah & Leela, 2006). Active packaging with antioxidant emitters (such as rosemary extract) provides additional protection against oxidative deterioration (He et al., 2026). Condiments with varied pH and water activity present unique challenges: high-acid products (vinegars, tomato-based sauces) require corrosion-resistant packaging, while emulsion-based products (mayonnaise, dressings) need light and oxygen barriers

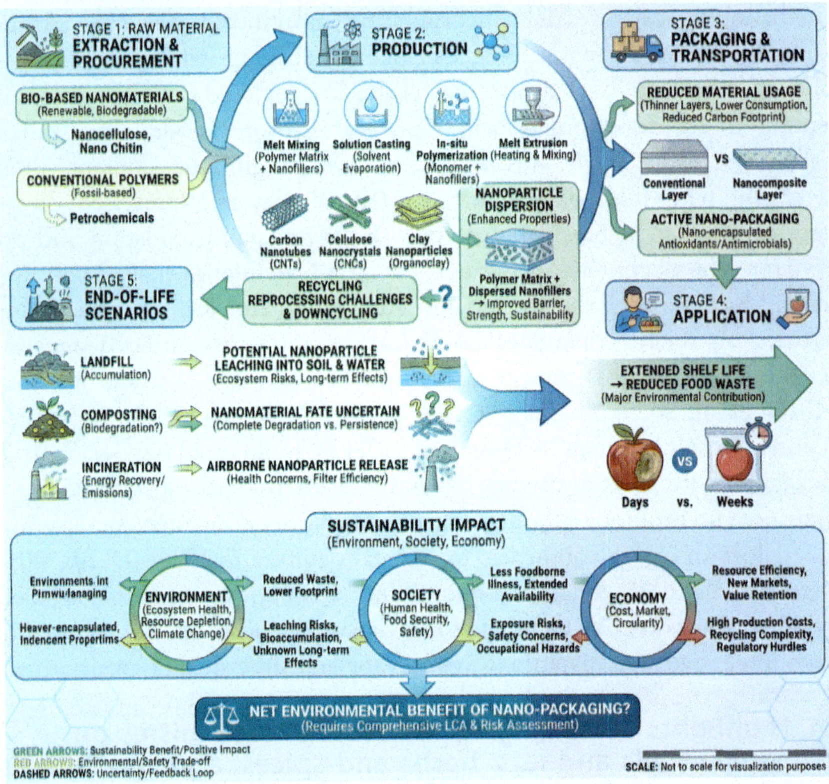

Fig. 5 Life cycle assessment of nano-enabled food packaging: sustainability benefits and environmental trade-offs.

to prevent lipid oxidation and phase separation. Innovative approaches include self-heating packaging for certain condiments, smart caps that indicate product freshness, and portion-controlled packaging that minimizes headspace O_2 (Fig. 5).

7. Future outlook

7.1 Prospects of packaging on food, consumer health concerns, environment, and sustainable food systems

The food packaging future is evolving toward multirole network that simultaneously address food preservation, consumer health, and environmental sustainability. Emerging trends indicate a shift toward bio-circular packaging systems utilizing algae-, mycelium-, and food waste-derived materials that offer compostability while maintaining functional performance (Dakhili et al., 2025; Pattoo, 2023). The integration of advanced active and intelligent technologies will become more sophisticated, with DNA-based freshness indicators, self-adjusting barrier materials, and nanotechnology-enabled preservation systems becoming commercially viable (Shao et al., 2021). Consumer health protection will be enhanced through migration-free packaging designs utilizing functional barriers and covalent bonding of active compounds, alongside improved detection methods for identifying NIAS at parts-per-trillion levels. Environmental considerations will drive the development of water-soluble polymers and edible electronics for monitoring food quality, significantly reducing packaging waste (Kataria et al., 2025). The notion of reusable packaging in the food ecosystems will gain traction, supported by digital tracking technologies and standardized cleaning protocols, potentially reducing single-use packaging by 40–60% in certain food product categories. Sustainable food systems will increasingly leverage blockchain-enabled traceability and dynamic shelf-life prediction using AI to optimize distribution and minimize food loss.

7.2 Economic regulation, efficiency, and industrial standardization of packaging foods

The economic and regulatory landscape for food packaging is undergoing significant transformation propelled by extended producer responsibility schemes and circular economy mandates. Global regulatory harmonization efforts are accelerating, with the development of international standards for biodegradable materials, recycling protocols, and safety assessment of nano-enabled packaging. Advanced recycling technologies, particularly chemical recycling and enzymatic depolymerization, will become economically viable, enabling true closed-loop recycling of complex multilayer materials. The implementation of digital product passports will become mandatory in many European Union jurisdictions, providing comprehensive lifecycle information and facilitating improved recycling rates for sustainable products regulations (Anonymous, 2026; Walden et al., 2021). Industrial efficiency will be enhanced through lightweighting 2.0 approaches utilizing AI driven design optimization and advanced materials to reduce material usage by 15–30% without compromising functionality. Smart manufacturing systems incorporating 4D printing and adaptive production lines will enable mass customization of packaging while maintaining economic efficiency. The economic model of packaging will shift toward packaging-as-a-service, where companies lease packaging systems and provide full lifecycle management, creating new business opportunities while ensuring proper end-of-life treatment. Standardization efforts will focus on universal sorting markers and cross-platform compatibility of smart packaging components to ensure interoperability across global supply chains.

8. Conclusion

The exploration of food packaging reveals a field that has experienced significant change, moving from simple containment to a complex and multipurpose technological system. This transition from passive protection to active interaction with food items shows how packaging has evolved into a crucial component of food safety, sustainability, and quality preservation in the global food value chain. The development of advanced packaging materials, which include edible films, bio-based composites, and better polymers, has significantly increased the protective functionalities of packaging. These materials now offer mechanical strength, environmental compatibility, and customized barrier qualities that greatly outperform conventional choices. Simultaneously, the advent of active and intelligent packaging structures has completely changed the way we monitor, store, and handle food items. These technologies, which range from freshness sensors and time-temperature indicators to oxygen scavengers and antimicrobial packaging, provide for dynamic protection and real-time quality evaluation across the supply chain. Opportunities in the nexus of nanotechnology with food packaging are very exciting but challenging. Although materials offered by nanotechnology provide new functions and improved barrier qualities, potential migration patterns and health effects must be carefully considered. This underscores the vital necessity for continuing research and vigorous safety evaluations to protect consumers and advance technological capabilities. In the future, food allocation and utilization are expected to be further replaced by the incorporation of digital technologies with packaging methods. The next phase in intelligent food packaging is represented by connected packaging that interacts throughout the supply chain, blockchain-enabled traceability, and smart labeling with dynamic expiration dating. Through improved inventory control and customer communication, these developments promote not just increased food safety and quality but also notable decreases in food waste. Innovation in circular packaging solutions is still being driven by the sustainability imperative. The industry's dedication to lessening its environmental impact is demonstrated by the formation of biodegradable materials, state-of-the-art recycling technology, and reusable packaging solutions. However, to achieve genuine sustainability, several elements must be balanced within thorough life cycle assessment frameworks, including material selection, manufacturing procedures, transportation efficiency, and end-of-life management. The ongoing convergence of materials science, digital technology, and sustainability principles will influence food packaging in the future. Packaging will become even more important in tackling the global issues of food security, safety, and waste reduction as it becomes more intelligent, linked, and ecologically conscious. To ensure that packaging innovations provide significant benefits while upholding the greatest standards of safety and environmental responsibility, success in this endeavor will depend on cooperation between research institutions, industry players, regulatory authorities, and consumers. Food packaging is more than just a protective container in this quickly changing environment. Instead, it is a vital integrated part of contemporary food systems that actively help to deliver safe and high-quality food while reducing environmental impact and improving consumer experience. In the years ahead, the development of packaging technologies is expected to provide even more advanced solutions that will revolutionize the production, distribution, and consumption of food.

Acknowledgements

The authors thank the Department of Food Science and Technology, University of Energy and Natural Resources, Sunyani, Ghana for their technical assistance and provision of resources throughout the preparation and completion of this project. During the preparation of this work, authors used the FigureLabs for preparing the figures. After using the tool, authors reviewed and edited the content as needed and took full responsibility for the content of the publication.

Ethics approval and consent to participate

Not applicable.

Funding statement

No external funding was received for this study.

Data availability

The data supporting this paper have been included in the paper.

Consent for publication

Not applicable.

CRediT authorship contribution statement

EFB: Conceptualization, Methodology, Software, Resources, Data Curation, Writing—Original Draft Preparation, Validation, Writing—Review, Editing, Investigation. **MMN:** Methodology, Data Curation, Validation, Investigation, Writing—Review, Editing. **RD:** Methodology, Editing, Proofread, Data Curation, Validation, Investigation. **APB:** Methodology, Software, Data Curation, Validation, Investigation, Writing—Review, Editing. **JO-K:** Conceptualization, Project Administration, Validation, Writing—Review, Proofread, Editing, Resources. **PKB:** Methodology, Software, Data Curation, Validation, Investigation, Writing—Review, Editing. **FA:** Validation, Writing—Review, Editing, Investigation, Software, Visualization.

Conflict of interest

On behalf of all the authors, the corresponding author states that there is no conflict of interest.

References

Adibi, A., Trinh, B. M., & Mekonnen, T. H. (2023). Recent progress in sustainable barrier paper coating for food packaging applications. *Progress in Organic Coatings, 181*, 107566. https://doi.org/10.1016/j.porgcoat.2023.107566.

Ahmed, A. M., Kainulainen, T. P., & Heiskanen, J. P. (2021). Furfural-based modification of pet for UV-blocking copolymers with decreased oxygen permeability. *Industrial & Engineering Chemistry Research, 60*(19), 7495–7504. https://doi.org/10.1021/acs.iecr.1c00629.

Akram, N., Saeed, M., Mansha, A., Bokhari, T. H., & Ali, A. (2023). Metal packaging for food items advantages, disadvantages and applications. *Green sustainable process for chemical and environmental engineering and science.* Elsevier 129–141. https://doi.org/10.1016/B978-0-323-95644-4.00019-X.

Alamri, M. S., Qasem, A. A. A., Mohamed, A. A., Hussain, S., Ibraheem, M. A., Shamlan, G., ... Qasha, A. S. (2021). Food packaging's materials: A food safety perspective. *Saudi Journal of Biological Sciences, 28*(8), 4490–4499. https://doi.org/10.1016/j.sjbs.2021.04.047.

Alizadeh Sani, M., Khezerlou, A., Tavassoli, M., Abedini, A. H., & McClements, D. J. (2024). Development of sustainable UV-screening food packaging materials: A review of recent advances. *Trends in Food Science & Technology, 145*, 104366. https://doi.org/10.1016/j.tifs.2024.104366.

Alnadari, F., Al-Dalali, S., Nasiru, M. M., Frimpong, E. B., Hu, Y., Abdalmegeed, D., ... Zeng, X. (2023a). A new natural drying method for food packaging and preservation using biopolymer-based dehydration film. *Food Chemistry, 404*(PB), 134689. https://doi.org/10.1016/j.foodchem.2022.134689.

Alnadari, F., Al-Dalali, S., Pan, F., Abdin, M., Frimpong, E. B., Dai, Z., ... Zeng, X. (2023b). Physicochemical characterization, molecular modeling, and applications of carboxymethyl chitosan-based multifunctional films combined with gum arabic and anthocyanins. *Food and Bioprocess Technology, 16*(12), 2984–3002. https://doi.org/10.1007/s11947-023-03122-0.

Alojaly, H., & Benyounis, K. Y. (2022). Packaging with plastics and polymeric materials. *Encyclopedia of materials: Plastics and polymers*. Elsevier 485–501. https://doi.org/10.1016/B978-0-12-820352-1.00025-0.

Andersen, M. L., & Skibsted, L. H. (2010). Light-induced quality changes in food and beverages. *Chemical deterioration and physical instability of food and beverages*. Elsevier 113–139. https://doi.org/10.1533/9781845699260.1.113.

Anonymous (2026). Available online at: https://www.fiegenbaum.solutions/en/blog/digital-product-passport-from-european-regulation-to-global-standard. Assessed on 16 January 2026.

Ardestani, F., McClements, D. J., & Hadidi, M. (2026). Pectin-based electrospun nanofiber materials for multifunctional food packaging: Bridging sustainability and safety. *Trends in Food Science & Technology, 167*, 105451. https://doi.org/10.1016/j.tifs.2025.105451.

Arshad, M. T., Naveed, F., Rasheed, A., Ikram, A., Maqsood, S., Hossain, M. S., & Gnedeka, K. T. (2025). Biopolymer-based edible films and coatings: Toward eco-friendly and safe food packaging. *International Journal of Food Science and Technology, 60*(2), 1–21. https://doi.org/10.1093/ijfood/vvaf213.

Ayvar-Gama, Y., Lara-Mejía, V., Munguía, A., Cruz-Casarrubias, C., Sagaceta-Mejía, J., Tolentino-Mayo, L., & Barquera, S. (2025). Sensory quality of six reformulated whole-grain cereal products consumed by children benefiting from a food assistance program after the implementation of front-of-package warning labeling in Mexico. *Food and Humanity, 5*, 100676. https://doi.org/10.1016/j.foohum.2025.100676.

Baku, A., Amadi, E.C., Eze, U.F., Chukwudebe, G.A., & Etus, C. (2024). Blockchain-based framework for agri-food supply chain traceability. 2024 IEEE 5th International Conference on Electro-Computing Technologies for Humanity (NIGERCON), 1–5. https://doi.org/10.1109/NIGERCON62786.2024.10927005.

Banerjee, R., Bandyopadhyay, J., & Ray, S. S. (2025). Functional Polymers and Their Nanocomposites for Sustainable Packaging Applications. *Macromolecular Materials and Engineering, 310*(9), https://doi.org/10.1002/mame.202500130.

Barilli, F., Fragni, R., Gelati, S., & Montanari, A. (2003). Study on the adhesion of different types of lacquers used in food packaging. *Progress in Organic Coatings, 46*(2), 91–96. https://doi.org/10.1016/S0300-9440(02)00215-1.

Beikzadeh, S., Assadpour, E., & Jafari, S. M. (2025). Active/smart packaging of bread and other bakery products; fundamentals, mechanisms, applications. *Future Foods, 12*, 100858. https://doi.org/10.1016/j.fufo.2025.100858.

Belasli, A., Aguerri, L., Ait Ouahioune, L., Becerril, R., Quintero, M., Canellas, E., ... Silva, F. (2025). Microbial safety improvement of date fruits using a Thymus numidicus essential oil-based active packaging. *Food Packaging and Shelf Life, 52*, 101673. https://doi.org/10.1016/j.fpsl.2025.101673.

Bher, A., & Auras, R. (2024). Life cycle assessment of packaging systems: A meta-analysis to evaluate the root of consistencies and discrepancies. *Journal of Cleaner Production, 476*, 143785. https://doi.org/10.1016/j.jclepro.2024.143785.

Bignardi, C., Cavazza, A., LaganÀ, C., Salvadeo, P., & Corradini, C. (2017). Release of non-intentionally added substances (NIAS) from food contact polycarbonate: Effect of ageing. *Food Control, 71*, 329–335. https://doi.org/10.1016/j.foodcont.2016.07.013.

Biswas, R., Alam, M., Sarkar, A., Haque, M. I., Hasan, M. M., & Hoque, M. (2022). Application of nanotechnology in food: processing, preservation, packaging and safety assessment. *Heliyon, 8*(11), e11795. https://doi.org/10.1016/j.heliyon.2022.e11795.

Boateng, E. F., Yang, Z., Zhang, J., Zhou, L., Xing, L., & Zhang, W. (2025). Incorporation of kiwifruit peel improved the property of carboxymethyl cellulose-gum Arabic active film and its effects on the quality of beef sausages. *Food Production, Processing and Nutrition,* 7(1), 10. https://doi.org/10.1186/s43014-024-00277-x.

Boz, Z. (2024). Packaging. *Encyclopedia of food safety*. Elsevier 631–647. https://doi.org/10.1016/B978-0-12-822521-9.00251-3.

Bradley, E., & Coulier, L. (2007). Report FD 07/01: An investigation into the reaction and breakdown products from starting substances used to produce food contact plastics. *Food standards agency*. London: Central Science Laboratory SAC Commercial Ltd. http://www.foodpackagingforum.org/wp-content/uploads/2014/06/Bradley-and-Coulier-2007.pdf.

Bumbudsanpharoke, N., & Boonsiriwit, A. (2025). *Moisture-regulating systems for active food packaging. Reference module in food science*. Elsevier, 1–9. https://doi.org/10.1016/B978-0-443-34158-8.00049-9.

Carpenter, B. J., Dobbins, T. W., Hernandez, M. S., Barker, S. N., Loomas, K. R., Osburn, W. N., & Legako, J. F. (2024). Comparison of gas treatments of high oxygen, carbon monoxide, and nitric oxide on ground beef color in modified atmosphere packaging. *Foods, 13*(6), 902. https://doi.org/10.3390/foods13060902.

Chen, W., Wang, X., Tao, Q., Wang, J., Zheng, Z., & Wang, X. (2013). Lotus-like paper/paperboard packaging prepared with nano-modified overprint varnish. *Applied Surface Science, 266*, 319–325. https://doi.org/10.1016/j.apsusc.2012.12.018.

Ciawi, Y., Tonyes, S. G., & Utami Dwipayanti, N. M. (2025). Multilayer packaging recycling: Challenges, current practices, and future prospects. *Academia Environmental Sciences and Sustainability, 2*(1), 1–12. https://doi.org/10.20935/AcadEnvSci7521.

Coja, T., & Steinwider, J. (2022). The new European transparency regulation: A panacea for EU risk assessment? *Journal of Consumer Protection and Food Safety,* 17(1), 1–3. https://doi.org/10.1007/s00003-022-01364-2.

Correia, L. F., da Silva Pinho, G., da Cruz Neves, T. J., de Oliveira Vieira, K. C., Maddela, N. R., Prasad, R., & Winkelstroter, L. K. (2024). Nanotechnology innovation combined with bacteriocins as emerging strategy for the development of active and intelligent food packaging. *Sustainable Chemistry and Pharmacy, 39*, 101551. https://doi.org/10.1016/j.scp.2024.101551.

Cozzini, P., Cavaliere, F., Spaggiari, G., Morelli, G., & Riani, M. (2022). Computational methods on food contact chemicals: Big data and in silico screening on nuclear receptors family. *Chemosphere, 292*, 133422. https://doi.org/10.1016/j.chemosphere.2021.133422.

D'Amore, T., Smaoui, S., & Varzakas, T. (2025). Chemical food safety in Europe under the spotlight: Principles, regulatory framework and roadmap for future directions. *Foods, 14*(9), 1628. https://doi.org/10.3390/foods14091628.

da Silva, J. P., de Cabassa, I. C. C., de Sousa, T. L., da Costa, D. S., Dall'Oglio, E. L., de Vasconcelos, L. G., ... Takeuchi, K. P. (2025). Active edible coating for lipid oxidation control in Brazil nuts. *Journal of Food Process Engineering, 48*(11), https://doi.org/10.1111/jfpe.70235.

Dakhili, S., Hashami, Z., Rostami, N., Moradi, S., Moslehishad, M., & Shojaee-Aliabadi, S. (2025). Innovative applications of microalgae in biodegradable food packaging: A review. *Journal of Agriculture and Food Research, 19*, 101723. https://doi.org/10.1016/j.jafr.2025.101723.

de Fátima Poças, M., & Hogg, T. (2007). Exposure assessment of chemicals from packaging materials in foods: A review. *Trends in Food Science & Technology, 18*(4), 219–230. https://doi.org/10.1016/j.tifs.2006.12.008.

Debeaufort, F., Galić, K., Kurek, M., Benbettaieb, N., & Ščetar, M. (2021). Packaging materials and processing for food, pharmaceuticals and cosmetics. *Packaging materials and processing for food, pharmaceuticals and cosmetics*. Wiley. https://doi.org/10.1002/9781119825081.

Deeth, H. (2010). Improving UHT processing and UHT milk products. *Improving the safety and quality of milk*. Elsevier 302–329. https://doi.org/10.1533/9781845699420.4.302.

Dejene, B. K. (2025a). Leveraging synergistic effects of metallic nanoparticles and essential oils in biopolymers: Emerging nanocomposites for food packaging applications–A review. *Journal of Agriculture and Food Research, 21*, 101885. https://doi.org/10.1016/j.jafr.2025.101885.

Dejene, B. K. (2025b). Eco-friendly synthesis of metallic nanoparticles from agri-food waste extracts: Applications in food packaging and healthcare–A critical review. *Materials Today Chemistry, 45*, 102619. https://doi.org/10.1016/j.mtchem.2025.102619.

Deshwal, G. K., & Panjagari, N. R. (2020). Review on metal packaging: Materials, forms, food applications, safety and recyclability. *Journal of Food Science and Technology, 57*(7), 2377–2392. https://doi.org/10.1007/s13197-019-04172-z.

Dhuey, E., Peterson, D., Barringer, S., Gao, C., Ruffley, K., Lammers, S., & Pascall, M. A. (2025). Volatile and non-volatile compounds in retort-processed green beans packaged in metallic cans coated with corrosion resistant polymers. *Packaging Technology and Science, 38*(5), 425–441. https://doi.org/10.1002/pts.2894.

Dobrucka, R. (2013). LogForum. *Scientific Journal of Logistics, 9*(2), 103–110.

Du, H., Jiang, Y., Simayi, A., Li, C., Liu, Y., Wen, Y., & Tang, Y. (2026). Material innovation and AI-driven smart packaging for fruit and vegetable precision preservation: Applications and future trends. *Trends in Food Science & Technology, 167*, 105441. https://doi.org/10.1016/j.tifs.2025.105441.

El-Sayed, S. M., & Youssef, A. M. (2023). Eco-friendly biodegradable nanocomposite materials and their recent use in food packaging applications: A review. *Sustainable Food Technology, 1*(2), 215–227. https://doi.org/10.1039/D2FB00021K.

Escobar, H. J., Garavito, J., & Castellanos, D. A. (2023). Development of an active packaging with an oxygen scavenger and moisture adsorbent for fresh lulo (*Solanum quitoense*). *Journal of Food Engineering, 349*, 111484. https://doi.org/10.1016/j.jfoodeng.2023.111484.

Falguera, V., Quintero, J. P., Jiménez, A., Muñoz, J. A., & Ibarz, A. (2011). Edible films and coatings: Structures, active functions and trends in their use. *Trends in Food Science & Technology, 22*(6), 292–303. https://doi.org/10.1016/j.tifs.2011.02.004.

Fan, B., Zhou, L., Xing, L., & Zhang, W. (2025). Fabricating cellulose nanocrystals from passion fruit peel to enhance the properties of electrospun zein/poly(ethylene oxide) nanofibrous films. *Food Hydrocolloids, 166*(1), 111355. https://doi.org/10.1016/j.foodhyd.2025.111355.

Fu, Q., Ge, Q., Liu, R., Wang, H., Zhou, G., & Zhang, W. (2017a). Influence of modified atmosphere packaging on protein oxidation, calpain activation and desmin degradation of beef muscles. *Journal of the Science of Food and Agriculture, 97*(13), 4508–4514. https://doi.org/10.1002/jsfa.8316.

Fu, Q., Liu, R., Zhou, G.-H., & Zhang, W. (2017b). Effects of packaging methods on the color of beef muscles through influencing myoglobin status, metmyoglobin reductase activity and lipid oxidation. *Journal of Food Processing and Preservation, 41*(1), e12740. https://doi.org/10.1111/jfpp.12740.

Fu, Q., Liu, R., Zhang, W., Li, Y., Wang, J., & Zhou, G. (2015). Effects of different packaging systems on beef tenderness through protein modifications. *Food and Bioprocess Technology, 8*(3), 580–588. https://doi.org/10.1007/s11947-014-1426-3.

Ganeson, K., Mouriya, G. K., Bhubalan, K., Razifah, M. R., Jasmine, R., Sowmiya, S., ... Ramakrishna, S. (2023). Smart packaging – A pragmatic solution to approach sustainable food waste management. *Food Packaging and Shelf Life, 36*, 101044. https://doi.org/10.1016/j.fpsl.2023.101044.

Geueke, B. (2016). Dossier – Can coatings. *Food Packaging Forum,* 1–10. https://doi.org/10.5281/zenodo.200633 December.

Geueke, B. (2018). Dossier - Non-intentionally added substances (NIAS). *Food Packaging Forum,* 1–10. https://doi.org/10.5281/zenodo.1265331.

Geueke, B., Groh, K., & Muncke, J. (2018). Food packaging in the circular economy: Overview of chemical safety aspects for commonly used materials. *Journal of Cleaner Production, 193*, 491–505. https://doi.org/10.1016/j.jclepro.2018.05.005.

Geueke, B., Parkinson, L. V., Groh, K. J., Kassotis, C. D., Maffini, M. V., Martin, O. V., ... Muncke, J. (2025). Evidence for widespread human exposure to food contact chemicals. *Journal of Exposure Science & Environmental Epidemiology, 35*(3), 330–341. https://doi.org/10.1038/s41370-024-00718-2.

Ghaffar, A., Iqbal, R., Niaz, M. T., Shafiq, K., Zulfiqar, S., Cochran, E. W., & Naqvi, S. M. I. H. (2025). Sustainable barrier coatings for paperboard packaging: Effect of VeoVa-10 on 2-EHA and MMA containing water-based emulsions. *Progress in Organic Coatings, 198*, 108908. https://doi.org/10.1016/j.porgcoat.2024.108908.

Gharby, S., Asbbane, A., Nid Ahmed, M., Gagour, J., Hallouch, O., Oubannin, S., ... Ibourki, M. (2025). Vegetable oil oxidation: Mechanisms, impacts on quality, and approaches to enhance shelf life. *Food Chemistry: X, 28*, 102541. https://doi.org/10.1016/j.fochx.2025.102541.

Gomes, V. E. de S., Baptista, D. P., Bócoli, P. F. J., Teixeira, F. G., Dantas, F. B. H., & Marangoni Júnior, L. (2025). Packaging materials for heat processed foods. *Reference module in food science*. Elsevier 1–11. https://doi.org/10.1016/B978-0-443-34158-8.00032-3.

Grayhurst, P. (2012). Glass packaging. *Packaging technology*. Elsevier 109–121. https://doi.org/10.1533/9780857095701.2.109.

Groh, K. J., Geueke, B., Martin, O., Maffini, M., & Muncke, J. (2021). Overview of intentionally used food contact chemicals and their hazards. *Environment International, 150*, 106225. https://doi.org/10.1016/j.envint.2020.106225.

Gunwant, D., Mahima, Parvez, Y., Bora, A., & Manuraj (2025). Edible films for food-grade packaging. In A. Verma, H. S. Gupta, & S. K. Sethi (Eds.). Polymers and composite materials for packaging: smart food packaging and solutions. *Materials horizons: From nature to nanomaterials* (pp. 21–63). https://doi.org/10.1007/978-981-95-0677-4_2.

Guo, M., Zhang, X., & Jin, T. Z. (2024). Active food packaging. *Encyclopedia of food safety* (2nd ed.). Elsevier 673–688. https://doi.org/10.1016/B978-0-12-822521-9.00078-2.

Gupta, D., Lall, A., Kumar, S., Patil, T. D., & Gaikwad, K. K. (2024). Plant-based edible films and coatings for food-packaging applications: Recent advances, applications, and trends. *Sustainable Food Technology, 2*(5), 1428–1455. https://doi.org/10.1039/D4FB00110A.

Gupta, K. (2023). Food safety guidelines for food packaging. *Green sustainable process for chemical and environmental engineering and science*. Elsevier 59–69. https://doi.org/10.1016/B978-0-323-95644-4.00014-0.

Gupta, R. K., Pipliya, S., Karunanithi, S., Eswaran U, G. M., Kumar, S., Mandliya, S., ... Kovács, B. (2024). Migration of chemical compounds from packaging materials into packaged foods: Interaction, mechanism, assessment, and regulations. *Foods, 13*(19), 3125. https://doi.org/10.3390/foods13193125.

Gurunathan, K. (2024). Time–temperature indicators (TTIs) for monitoring food quality. *Smart food packaging systems*. Wiley. 281–304. https://doi.org/10.1002/9781394189595.ch11.

Guzman-Puyol, S., Benítez, J. J., & Heredia-Guerrero, J. A. (2022). Transparency of polymeric food packaging materials. *Food Research International, 161*, 111792. https://doi.org/10.1016/j.foodres.2022.111792.

Habiba, U., Singh, R., Jeena, B., Kumar, V., Dmitriev, A. A., & Bisht, B. (2025a). Smart shields: A conceptual approach to bridging surface functionality and self-healing for next-generation food packaging 4.0. *Trends in Food Science & Technology, 166*, 105413. https://doi.org/10.1016/j.tifs.2025.105413.

Habiba, U., Bajpai, A., Shafi, Z., Pandey, V. K., & Singh, R. (2025b). Advancing sustainability through modified atmosphere packaging (MAP) for fresh food preservation: A critical review. *Journal of Stored Products Research, 112*, 102657. https://doi.org/10.1016/j.jspr.2025.102657.

Hasan, H. R., Musamih, A., Salah, K., Jayaraman, R., Omar, M., Arshad, J., & Boscovic, D. (2024). Smart agriculture assurance: IoT and blockchain for trusted sustainable produce. *Computers and Electronics in Agriculture, 224*, 109184. https://doi.org/10.1016/j.compag.2024.109184.

He, X., Tang, L., & Huang, R. (2026). Active packaging films from PBAT/PLA with rosmarinus officinalis l. extract: Antioxidant, UV-shielding, and compostable properties. *Molecules (Basel, Switzerland), 31*(2), 217. https://doi.org/10.3390/molecules31020217.

Hong, S. J., Riahi, Z., Khan, A., Lee, J. G., Kim, B. Y., Min, S. C., ... Kim, J. T. (2025). Active packaging film utilizing cerium metal-organic frameworks doped with titanium dioxide as ethylene scavengers for postharvest ripening control of avocados. *Food Chemistry, 489*, 145025. https://doi.org/10.1016/j.foodchem.2025.145025.

Hsu, C. L., & Chang, K. S. (2007). Development of a novel method for detecting the integrity of aseptic paperboard laminate packages containing aluminium foil. *Food Control, 18*(2), 102–107. https://doi.org/10.1016/j.foodcont.2005.08.012.

Huang, R., Liu, Z., Yuan, S., Yin, H., Dang, Z., & Wu, P. (2017). Worldwide human daily intakes of bisphenol A (BPA) estimated from global urinary concentration data (2000–2016) and its risk analysis. *Environmental Pollution, 230*, 143–152. https://doi.org/10.1016/j.envpol.2017.06.026.

Huang, S., Dong, Q., Che, S., Li, R., & Tang, K. H. D. (2025). Bioplastics and biodegradable plastics: A review of recent advances, feasibility and cleaner production. *Science of The Total Environment, 969*, 178911. https://doi.org/10.1016/j.scitotenv.2025.178911.

Huang, Y., Zhang, Q., Xu, Q., Li, H., Chen, S., & Feng, Z. (2025). Nanosilver–titanium dioxide-enhanced chitosan composite film for prolonged egg preservation: Antibacterial and antioxidant mechanisms. *Journal of Food Quality, 2025*(1), https://doi.org/10.1155/jfq/6651630.

Jat, S., Bains, A., Goksen, G., Dhull, S. B., Nagraik, R., Fareed, M., & Chawla, P. (2025). A review of plant-derived arabinogalactan-rich carbohydrate polymer (mucilage) complexed with other biopolymers to formulate sustainable antimicrobial packaging and coating material. *International Journal of Biological Macromolecules, 318*, 145027. https://doi.org/10.1016/j.ijbiomac.2025.145027.

Ji, Y., Liang, J., Hei, X., Chen, L., Yu, S., Fang, D., ... Zhang, S. (2025). Nanomaterial-based drug delivery systems for ocular diseases: A review. *Chinese Chemical Letters*, 112062. https://doi.org/10.1016/j.cclet.2025.112062.

Jia, X., Li, X., Alharbi, M. O., Ma, P., Ismail-Fitry, M. R., Xia, S., ... Wang, Q. (2026). AI awakens food packaging: Integrative advances, challenges, and future perspectives. *Trends in Food Science & Technology, 168*, 105466. https://doi.org/10.1016/j.tifs.2025.105466.

Jiang, H., Guo, P., Ju, H., Li, Z., Chen, L., Mo, H., & Liu, X. (2025). Lotus-leaf-bioinspired biomass-based films for intelligent/active packaging. *Food Chemistry, 496*, 146843. https://doi.org/10.1016/j.foodchem.2025.146843.

Juliet, I. B., Saheed, M., Oladejo, T. J., Saka, A. A., Adesola, O. O., & Yanda, S. E. (2025). Exploring ZnO and CeO_2 nanoparticle films for sustainable food packaging applications. *Food and Humanity, 5*, 100712. https://doi.org/10.1016/j.foohum.2025.100712.

Kaiser, K., Schmid, M., & Schlummer, M. (2017). Recycling of polymer-based multilayer packaging: A review. *Recycling, 3*(1), 1. https://doi.org/10.3390/recycling3010001.

Kargwal, R., Garg, M., Singh, V., Garg, R., & Kumar, N. (2020). Principles of modified atmosphere packaging for shelf life extension of fruits and vegetables: An overview of storage conditions. *International Journal of Chemical Studies, 8*(3), 2245–2252. https://doi.org/10.22271/chemi.2020.v8.i3af.9545.

Karuppan Perumal, M. K., Rajasekaran, M. B. S., Rajan Renuka, R., Samrot, A. V., & Nagarajan, M. (2025). Zinc oxide nanoparticles and their nanocomposites as an imperative coating for smart food packaging. *Applied Food Research, 5*(1), 100849. https://doi.org/10.1016/j.afres.2025.100849.

Kataria, P., Gupta, R. K., Gupta, A. K., Kaur, R., Sharma, M., Jha, A. K., ... Kaur, S. (2025). A state-of-the-art review on edible electronics: Next-generation technologies for biocompatible and ingestible devices. *Trends in Food Science & Technology, 156*, 104880. https://doi.org/10.1016/j.tifs.2025.104880.

Kato, L. S., & Conte-Junior, C. A. (2021). Safety of plastic food packaging: The challenges about non-intentionally added substances (NIAS) discovery, identification and risk assessment. *Polymers, 13*(13), 2077. https://doi.org/10.3390/polym13132077.

Kerry, J. (2012). Aluminium foil packaging. *Packaging technology*. Elsevier 163–177. https://doi.org/10.1533/9780857095701.2.163.

Khalafi, N., & Zandi, M. (2025). Utilizing natural pigments in electrospun nanofibers: Sustainable and smart food packaging solutions driven by nanotechnology. *Applied Food Research, 5*(2), 101220. https://doi.org/10.1016/j.afres.2025.101220.

Khule, P., Lu, G., Yang, Q., Beaudry, R., & Almenar, E. (2025). Enhanced barrier properties of bagasse paperboard for food packaging applications using shellac coating – Effect of concentration and number of layers. *Progress in Organic Coatings, 207*, 109395. https://doi.org/10.1016/j.porgcoat.2025.109395.

Kowser, M. A., Mahmud, H., Chowdhury, M. A., Hossain, N., Mim, J. J., & Islam, S. (2025). Fabrication and characterization of corn starch based bioplastic for packaging applications. *Results in Materials, 25*, 100662. https://doi.org/10.1016/j.rinma.2025.100662.

Krishnan, R. R., Olasupo, A. D., Patel, K. H., Bolarinwa, I. F., Oke, M. O., Okafor, C., ... Ifie, I. (2025). Edible biopolymers with functional additives coatings for postharvest quality preservation in horticultural crops: A review. *Frontiers in Sustainable Food Systems, 9*. https://doi.org/10.3389/fsufs.2025.1569458.

Kumar, A., Kumar, A., Wei, S., Chopra, S., Rudra, S. G., & Rabbani, A. (2025). Biodegradable and smart packaging films for food quality and safety: A review. *Applied Food Research, 5*(2), 101491. https://doi.org/10.1016/j.afres.2025.101491.

Kumar, A., Kumar, A., Chauhan, V., Singh, D., Vats, C., & Gehlot, R. (2025). Nanotechnology in sustainable food packaging. *Advancements in nanotechnology for food and packaging*. Elsevier 377–410. https://doi.org/10.1016/B978-0-443-21428-8.00013-8.

Kumar Chandra, D., Kumar, A., & Mahapatra, C. (2024). Ecofriendly bioplastics from biowaste: Antimicrobial and functional enhancements for sustainable packaging. *European Polymer Journal, 221*, 113557. https://doi.org/10.1016/j.eurpolymj.2024.113557.

Kumar, L., & Gaikwad, K. K. (2023). Advanced food packaging systems for space exploration missions. *Life Sciences in Space Research, 37*, 7–14. https://doi.org/10.1016/j.lssr.2023.01.005.

Kumar, R., Kumar, A., Bhardwaj, S., Sikarwar, M., Sriwastaw, S., Sharma, G., & Gupta, M. (2025). Nanotoxicity unveiled: Evaluating exposure risks and assessing the impact of nanoparticles on human health. *Journal of Trace Elements and Minerals, 13*, 100252. https://doi.org/10.1016/j.jtemin.2025.100252.

Kumari, S., Debbarma, R., Nasrin, N., Khan, T., Taj, S., & Bhuyan, T. (2024). Recent advances in packaging materials for food products. *Food Bioengineering, 3*(2), 236–249. https://doi.org/10.1002/fbe2.12096.

Labrot–Rhodes, L., Campo, E., & Poujaud, P. (2025). Review of monitoring systems for stored grains in a modified atmosphere. *Heliyon, 11*(3), e42347. https://doi.org/10.1016/j.heliyon.2025.e42347.

Lackner, M., Mukherjee, A., & Koller, M. (2023). What are "Bioplastics"? Defining renewability, biosynthesis, biodegradability, and biocompatibility. *Polymers, 15*(24), 4695. https://doi.org/10.3390/polym15244695.

Lee, H.-G., Jeong, S., Min, S. C., & Yoo, S. (2026). Expanded packaging material for removal of carbon dioxide and kimchi odor compounds using sodium-bicarbonate-powder-impregnated activated carbon. *Food Control, 183*, 111964. https://doi.org/10.1016/j.foodcont.2026.111964.

Lei, Y., Ter, Z. Y., Wan Mustapha, W. A., Mohd Razali, N. S., Md Nor, A. R., Lu, X., & Lim, S. J. (2025). Seaweed polysaccharide-based materials for advanced food packaging: A review. *Food Bioscience, 74*, 107832. https://doi.org/10.1016/j.fbio.2025.107832.

Li, J., Li, T., Wu, C., Fan, G., Zhou, D., Li, X., & Zhu, Y. (2026). Multipurpose liposome-based food active packaging: Carrier characteristics, preparation methods, and applications in freshness preservation. *Trends in Food Science & Technology, 168*, 105470. https://doi.org/10.1016/j.tifs.2025.105470.

Li, S., Łopusiewicz, Ł., Merino, D., Sani, M. A., Kalita, P., Roy, S., ... Zhang, W. (2025). High-performance ethyl cellulose composite films: Innovations in functional enhancement for sustainable food packaging. *Carbohydrate Polymers, 368*, 124154. https://doi.org/10.1016/j.carbpol.2025.124154.

Liu, C., Zhang, Y., Yang, X., Liang, R., Mao, Y., Hou, X., ... Luo, X. (2014). Potential mechanisms of carbon monoxide and high oxygen packaging in maintaining color stability of different bovine muscles. *Meat Science, 97*(2), 189–196. https://doi.org/10.1016/j.meatsci.2014.01.027.

Liu, S., & Naidu, R. (2025). Microplastics and nanoplastics released by peeling off aluminium foil cap sealing liner. *Journal of Hazardous Materials, 499*, 140265. https://doi.org/10.1016/j.jhazmat.2025.140265.

Madhu, B. (2025). AI-driven food packaging systems: A new frontier in intelligent food safety and shelf-life management. *Journal of Food Science, 90*(12), 1–23. https://doi.org/10.1111/1750-3841.70716.

Mahajan, P. V., Sonawane, A. D., Chaiwong, S., & Artés-Hernández, F. (2025). Active packaging for ethylene control in fresh produce: Materials, applications, and practical realities. *Reference module in food science*. Elsevier 1–10. https://doi.org/10.1016/B978-0-443-34158-8.00054-2.

Mahmud, R. U., Darda, M. A., Hossain, M. T., Habib, M. A., Nag, R. K., & Sarker, M. P. (2025). Fabrication of waste cotton reinforced corn and potato starch sustainable bioplastics for eco-friendly packaging material. *Sustainable Chemistry for the Environment, 11*, 100286. https://doi.org/10.1016/j.scenv.2025.100286.

Maxim, L., Mazzocchi, M., Van den Broucke, S., Zollo, F., Robinson, T., Rogers, C., ... Smith, A. (2021). Technical assistance in the field of risk communication. *EFSA Journal, 19*(4), https://doi.org/10.2903/j.efsa.2021.6574.

Meliana, C., Liu, J., Show, P. L., & Low, S. S. (2024). Biosensor in smart food traceability system for food safety and security. *Bioengineered, 15*(1), https://doi.org/10.1080/21655979.2024.2310908.

Mengozzi, A., Bot, F., Uysal-Unalan, I., & Chiavaro, E. (2025). Cured meat packaging: A whole system approach to the food product centric packaging solutions. *Food Research International, 221*, 117112. https://doi.org/10.1016/j.foodres.2025.117112.

Mgomi, F. C., Zhang, B., Lu, C., Yang, Z., & Yuan, L. (2025). Novel biofilm-inspired encapsulation technology enhances the viability of probiotics during processing, storage, and delivery. *Trends in Food Science & Technology, 160*, 105032. https://doi.org/10.1016/j.tifs.2025.105032.

Mishra, B., Panda, J., Mishra, A. K., Nath, P. C., Nayak, P. K., Mahapatra, U., ... Sridhar, K. (2024). Recent advances in sustainable biopolymer-based nanocomposites for smart food packaging: A review. *International Journal of Biological Macromolecules, 279*, 135583. https://doi.org/10.1016/j.ijbiomac.2024.135583.

Missaoui, W. N., Arnold, R. D., & Cummings, B. S. (2018). Toxicological status of nanoparticles: What we know and what we don't know. *Chemico-Biological Interactions, 295*, 1–12. https://doi.org/10.1016/j.cbi.2018.07.015.

Mittal, A., Kumar, M., & Garg, A. (2025). Artificial intelligence in agriculture: A review of applications, challenges and future directions. *International Journal of Statistics and Applied Mathematics, 10*(6), 220–224.

Mkhari, T., Adeyemi, J. O., & Fawole, O. A. (2025). Recent advances in the fabrication of intelligent packaging for food preservation: A review. *Processes, 13*(2), 539. https://doi.org/10.3390/pr13020539.

Mobahi, N., Razavi, M. A., Ekrami, M., Emam-Djomeh, Z., & Razavi, S. H. (2025). AI-integrated bio-nanocomposite food packaging: Sustainable, functional, and intelligent solutions. *Trends in Food Science & Technology, 165*, 105336. https://doi.org/10.1016/j.tifs.2025.105336.

Molina-Besch, K., & Olsson, A. (2022). Innovations in food packaging—Sustainability challenges and future scenarios. *Future foods*. Elsevier 375–392. https://doi.org/10.1016/B978-0-323-91001-9.00039-6.

Mordor Intelligence (2026). Packaged food market size and share analysis-growth trends and forecast (2026–2031). Available online at: https://www.mordorintelligence.com/industry-reports/packaged-food-market. Accessed on 02 April 2026.

Mousania, Z., & Atkinson, J. D. (2025). A cradle-to-grave life cycle assessment of multilayer plastic film food packaging materials, comparing to a paper-based alternative. *Waste Management, 200*, 114747. https://doi.org/10.1016/j.wasman.2025.114747.

Müller, P., & Schmid, M. (2019). Intelligent packaging in the food sector: A brief overview. *Foods, 8*(1), 16. https://doi.org/10.3390/foods8010016.

Muncke, J., Andersson, A. M., Backhaus, T., Belcher, S. M., Boucher, J. M., Carney Almroth, B., ... Scheringer, M. (2023). A vision for safer food contact materials: Public health concerns as drivers for improved testing. *Environment International, 180*, 108161. https://doi.org/10.1016/j.envint.2023.108161.

Muncke, J., Andersson, A. M., Backhaus, T., Boucher, J. M., Carney Almroth, B., Castillo Castillo, A., ... Scheringer, M. (2020). Impacts of food contact chemicals on human health: A consensus statement. *Environmental Health, 19*(1), 25. https://doi.org/10.1186/s12940-020-0572-5.

Mwenya, J., Sivaji, S., Saengrayap, R., Rattanakaran, J., Prahsarn, C., Mahajan, P., & Chaiwong, S. (2025). Food package design for thermal insulation. *Reference module in food science*. Elsevier 1–12. https://doi.org/10.1016/B978-0-443-34158-8.00013-X.

Nagalapur, A. N., & Byadagi, S. (2024). Migration of chemical compounds from packaging materials into food substances. *International Journal of Creativve Research Thoughts (IJCRT), 12*(12), f955–f972. www.ijcrt.org.

Nascimento, D. R., Mesquita, J., da Silva, T., Hernandes, T., Lengowski, E. C., & Takeuchi, K. (2024). A novel approach to protect brazil nuts from lipid oxidation: Efficacy of nanocellulose–tocopherol edible coatings. *Coatings, 14*(9), 1182. https://doi.org/10.3390/coatings14091182.

Navarro, S., Timlick, B., Demianyk, C. J., & White, N. D. G. (2012). Controlled or modified atmospheres: Part III management: monitoring-based methods. *Stored product protection.* K.State Research and Extension 1–11. www.ksre.ksu.edu.

Ndwandwe, B. K., Malinga, S. P., Kayitesi, E., & Dlamini, B. C. (2025). Avocado seed-derived bioactive compounds for functional food packaging solutions: current developments and future prospects – A review. *Food Bioscience, 72*, 107446. https://doi.org/10.1016/j.fbio.2025.107446.

Nerlekar, N., Satvekar, R., Nandre, V., & Dandge, P. (2026). Food packaging materials: Types, properties, and applications. *Safety and sustainability aspects of food packaging.* Elsevier 29–52. https://doi.org/10.1016/B978-0-443-24768-2.00019-3.

Ninčević, A., Pezzani, A., & Squitieri, G. (2007). Characterisation of different types of lacquers used in food packaging: Lacquer adhesion tests. *Acta Alimentaria, 36*(1), 27–37. https://doi.org/10.1556/AAlim.36.2007.1.5.

Noviello, M., Gattullo, C. E., Allegretta, I., Terzano, R., Gambacorta, G., & Paradiso, V. M. (2020). Synthetic zeolite materials from recycled glass and aluminium food packaging as potential oenological adjuvant. *Food Packaging and Shelf Life, 26*, 100572. https://doi.org/10.1016/j.fpsl.2020.100572.

Nuamduang, P., & Leelaphiwat, P. (2025). Active packaging: Emitters. *Reference module in food science.* Elsevier 1–12. https://doi.org/10.1016/B978-0-443-34158-8.00045-1.

Olawade, D. B., Wada, O. Z., & Ige, A. O. (2024). Advances and recent trends in plant-based materials and edible films: A mini-review. *Frontiers in Chemistry, 12*, 1–13. https://doi.org/10.3389/fchem.2024.1441650.

Omer, E., Bakiri, A., Hammel, Y. A., Sanders, M. J., Koster, S., & Ciclet, O. (2025). Deciphering the complexity of the chemicals in food packaging materials using molecular networks. *Food Chemistry, 462*, 140853. https://doi.org/10.1016/j.foodchem.2024.140853.

Oyegbade, S. A., Mameh, E. O., Oni, J. G., Gbate, M. A., Oluwaseyi, T. A., Wurie, A. W., ... Adetunji, C. O. (2025). Innovations in food technology: Intelligent packaging and advanced traceability solutions. *AI applications in food processing and packaging.* Royal Society of Chemistry 1–22. https://doi.org/10.1039/9781837679119-00001.

Özoğul, G. (2025). Applications of artificial intelligence technologies in agriculture: Advantages, challenges, risks, prospects, and recommendations. *Cogent Food & Agriculture, 11*(1), 1–25. https://doi.org/10.1080/23311932.2025.2568199.

Pan, J., Sun, J., Shen, Q., Zheng, X., & Zhou, W. (2025). Fabrication, properties, and improvement strategies of edible films for fruits and vegetables preservation: A comprehensive review. *Food Innovation and Advances, 4*(1), 43–52. https://doi.org/10.48130/fia-0025-0003.

Panwar, A., Kumar, S., Dhiman, A., Kumar, V., Gupta, D., & Sharma, A. (2024). pH sensitive indicator film based intelligent packaging systems of perishables: Developments and challenges of last decade. *Microchemical Journal, 207*, 111732. https://doi.org/10.1016/j.microc.2024.111732.

Pateiro, M., Domínguez, R., Bermúdez, R., Munekata, P. E. S., Zhang, W., Gagaoua, M., & Lorenzo, J. M. (2019). Antioxidant active packaging systems to extend the shelf life of sliced cooked ham. *Current Research in Food Science, 1*, 24–30. https://doi.org/10.1016/j.crfs.2019.10.002.

Pathiraja, A. S., & Munaweera, I. (2024). Innovative nanotechnology-based sustainable food packaging: A brief review. *JSFA Reports, 4*(1), 19–32. https://doi.org/10.1002/jsf2.179.

Pattoo, T. A. (2023). Bio-based materials: The next frontier in green chemistry for sustainable packaging. *Acta Biology Forum, 2*(3), 8–14. https://doi.org/10.51470/ABF.2023.2.3.8.

Pavelková, A. (2013). Time temperature indicators as devices intelligent packaging. *Acta Universitatis Agriculturae et Silviculturae Mendelianae Brunensis, 61*(1), 245–251. https://doi.org/10.11118/actaun201361010245.

Podd, B. (2022). Paper and paperboard. *Global legislation for food contact materials* (2nd ed.). Elsevier 99–121. https://doi.org/10.1016/B978-0-12-821181-6.00004-1.

Poddar, D., Pandey, K., Singh, A., & Yoo, H. (2025). Impact of particle structure on the storage stability and release kinetics of essential oil encapsulated in shellac particle for high-barrier paper coatings for food packaging applications. *International Journal of Biological Macromolecules, 323*, 146221. https://doi.org/10.1016/j.ijbiomac.2025.146221.

Priya, S., Nandhakumar, S. K, Aditya, Koshariya, A. K., Nehul, J. N., & Sharma, Y. (2025). Postharvest innovations in mushroom shelf life enhancement. *Plant Science Review, 6*(1), 18–23. https://doi.org/10.51470/PSR.2025.06.01.18.

Putra, B. T. W., & Kuswandi, B. (2025). Smart food packaging. *Sensor technologies for food quality and safety*. Royal Society of Chemistry 336–354. https://doi.org/10.1039/9781837674787-00336.

Qian, S., Ji, H., Wu, X., Li, N., Yang, Y., Bu, J., ... Zhang, C. (2018). Detection and quantification analysis of chemical migrants in plastic food contact products. *PLoS One, 13*(12), e0208467. https://doi.org/10.1371/journal.pone.0208467.

Radan, N., Ghobadi Nejad, Z., Ghasemi, S., & Yaghmaei, S. (2025). Boosting antibacterial efficiency of gelatin/chitosan composite films through synergistic interaction of ag nanoparticles and ZIF-8 metal-organic frameworks for food packaging. *International Journal of Biological Macromolecules, 305*(P2), 141175. https://doi.org/10.1016/j.ijbiomac.2025.141175.

Rasheed, J. L., Evangeline, C. A., Azeez, M. N., & Samuel, J. (2025). Modified atmospheric packaging in various food products. *American Journal of Food Science and Technology, 4*(1), 1–10. https://doi.org/10.54536/ajfst.v4i1.2942.

Riley, A. (2012). Paper and paperboard packaging. *Packaging technology*. Elsevier 178–239. https://doi.org/10.1533/9780857095701.2.178.

Rodrigues, C., Souza, V. G. L., Coelhoso, I., & Fernando, A. L. (2021). Bio-based sensors for smart food packaging—Current applications and future trends. *Sensors, 21*(6), 2148. https://doi.org/10.3390/s21062148.

Rozo, D. F., Alvarado, J. F., Chaparro, L. M., Medina, J. A., & Salcedo, F. (2024). Modeling oxidation kinetics of linseed oil in oxygen scavenger nanocapsules to be potentially used in active food packaging. *Food Packaging and Shelf Life, 42*, 101256. https://doi.org/10.1016/j.fpsl.2024.101256.

Sanati, S., Bakhti, A., & Mohammadipanah, F. (2025). Long-term toxic effects of nanoparticles on human microbiota. *Journal of Trace Elements in Medicine and Biology, 91*, 127723. https://doi.org/10.1016/j.jtemb.2025.127723.

Santana, R. F., Muniz, I. de C. B., Lima, C. M. G., Veloso, C. M., Santos, F. G., & Bonomo, R. C. F. (2025). Starch-based bioplastics and natural antimicrobials: A sustainable alternative for active packaging. *Food Bioscience, 74*, 107786. https://doi.org/10.1016/j.fbio.2025.107786.

Santos, I. J. B., Gonçalves, M. M., Schogl, A. E., de Paiva, G. M., & Coimbra, J. S. dos R. (2024). Nanomaterial migration into the food matrix. *Nanostructured materials for food packaging applications*. Elsevier 553–573. https://doi.org/10.1016/B978-0-323-99525-2.00020-7.

Sasikumar, R., Selva Kumar, T., Hamid, Jaiswal, A. K., & Jaiswal, S. (2026). Nanotechnology-enabled smart and intelligent food packaging systems. *Smart and intelligent food packaging*. Elsevier 81–107. https://doi.org/10.1016/B978-0-443-24724-8.00010-7.

Satheesh, N., Geremew, M., & Parmar, A. (2026). Edible packaging: Sustainable solutions for food packaging. *Dietary, sensory and gastronomic applications*. Elsevier 659–682. https://doi.org/10.1016/B978-0-443-33345-3.00040-7.

Saud AlMulaifi, R. (2024). Exploring the chemical composition, production, and uses of glass across industries. *International Journal of Life Sciences Research, 12*(4), 11–18. https://doi.org/10.5281/zenodo.13986382.

Schaefer, D., & Cheung, W. M. (2018). Smart packaging: Opportunities and challenges. *Procedia CIRP, 72*, 1022–1027. https://doi.org/10.1016/j.procir.2018.03.240.

Schyns, Z. O. G., & Shaver, M. P. (2021). Mechanical recycling of packaging plastics: A review. *Macromolecular Rapid Communications, 42*(3), 1–27. https://doi.org/10.1002/marc.202000415.

Seref, N., & Cufaoglu, G. (2025). Food packaging and chemical migration: A food safety perspective. *Journal of Food Science, 90*(5), 1–23. https://doi.org/10.1111/1750-3841.70265.

Shao, P., Liu, L., Yu, J., Lin, Y., Gao, H., Chen, H., & Sun, P. (2021). An overview of intelligent freshness indicator packaging for food quality and safety monitoring. *Trends in Food Science & Technology, 118*, 285–296. https://doi.org/10.1016/j.tifs.2021.10.012.

Shao, X., Zhang, X., & Wen, H. (2025). From waste to taste: Understanding consumer perceptions and marketing strategies for edible food packaging. *British Food Journal, 127*(13), 820–837. https://doi.org/10.1108/BFJ-06-2025-0791.

Sharif, M. K., Zahid, A., & Shah, F. H. (2018). Role of food product development in increased food consumption and value addition. *Food processing for increased quality and consumption*. Elsevier 455–479. https://doi.org/10.1016/B978-0-12-811447-6.00015-1.

Sharma, N., Lamichhane, P., Love, P., Poudel, P. K., Russell, C., & Bell, C. (2025). Nutrients of concern for older infants and young children in packaged food products available in Nepali retail stores. *The Lancet Regional Health - Southeast Asia, 39*, 100637. https://doi.org/10.1016/j.lansea.2025.100637.

Sharma, S., Kumar, G., Bhardwaj, S. K., Chauhan, A., Rathi, A., Kumar, J., ... Vishvamitera, S. (2025). Bioplastics as next-generation eco-friendly materials: a comprehensive review. *Clean Technologies and Environmental Policy, 27*(10), 5009–5027. https://doi.org/10.1007/s10098-025-03214-9.

Shekarchizadeh, H., & Nazeri, F. S. (2020). Active nanoenabled packaging for the beverage industry. *Nanotechnology in the beverage industry*. Elsevier 587–607. https://doi.org/10.1016/B978-0-12-819941-1.00020-1.

Shi, Y., Zhang, W., & Zhou, G. (2020). Effects of different moisture-permeable packaging on the quality of aging beef compared with wet aging and dry aging. *Foods, 9*(5), 649. https://doi.org/10.3390/foods9050649.

Shirzad, K., & Joodaky, A. (2024). Enhancing sustainable food packaging design: A machine learning approach to predict ventilated corrugated paperboard strength. *Biosystems Engineering, 247*, 26–41. https://doi.org/10.1016/j.biosystemseng.2024.08.012.

Siciliano, S., Lopresto, C. G., CarnÌ, D. L., & Lamonaca, F. (2025). Optical gas sensors in smart food bio-packaging: Innovation for monitoring the product freshness and safety. *Measurement: Food, 19*, 100245. https://doi.org/10.1016/j.meafoo.2025.100245.

Siddiqui, S. A., Yang, X., Deshmukh, R. K., Gaikwad, K. K., Bahmid, N. A., & Castro-Muñoz, R. (2024). Recent advances in reinforced bioplastics for food packaging – A critical review. *International Journal of Biological Macromolecules, 263*, 130399. https://doi.org/10.1016/j.ijbiomac.2024.130399.

Singh, M. B., Singh, P., & Jain, P. (2023). Market demand for smart packaging versus consumer perceptions. *Green sustainable process for chemical and environmental engineering and science*. Elsevier 115–128. https://doi.org/10.1016/B978-0-323-95644-4.00011-5.

Singh, N., Mann, B., Sharma, R., Verma, A., Panjagari, N. R., & Gandhi, K. (2022). Identification of polymer additives from multilayer milk packaging materials by liquid-solid extraction coupled with GC-MS. *Food Packaging and Shelf Life, 34*, 100975. https://doi.org/10.1016/j.fpsl.2022.100975.

Singh, R. K., & Singh, N. (2005). Quality of packaged foods. *Innovations in food packaging*. Elsevier 24–44. https://doi.org/10.1016/B978-012311632-1/50035-8.

Sivertsvik, M., Jeksrud, W. K., & Rosnes, J. T. (2002). A review of modified atmosphere packaging of fish and fishery products – significance of microbial growth, activities and safety. *International Journal of Food Science and Technology, 37*(2), 107–127. https://doi.org/10.1046/j.1365-2621.2002.00548.x.

Skura, B. J. (1991). Modified atmosphere packaging of fish and fish products. *Modified atmosphere packaging of food*. Springer US148–168. https://doi.org/10.1007/978-1-4615-2117-4_6.

Sobhan, A., Hossain, A., Wei, L., Muthukumarappan, K., & Ahmed, M. (2025). IoT-enabled biosensors in food packaging: A breakthrough in food safety for monitoring risks in real time. *Foods, 14*(8), 1403. https://doi.org/10.3390/foods14081403.

Soundara Rajan, S., & Wani, K. M. (2025). A review of smart food and packaging technologies: Revolutionizing nutrition and sustainability. *Food and Humanity, 4*, 100593. https://doi.org/10.1016/j.foohum.2025.100593.

Spinei, M., & Oroian, M. (2025). Identification and quantification of intentionally and non-intentionally added substances occurring from biodegradable food packaging: A review. *Food Research International, 218*, 116880. https://doi.org/10.1016/j.foodres.2025.116880.

Srivastava, V., Garg, S., & Saran, A. D. (2025). Advanced polymer nanocomposites in packaging applications. *Journal of Polymer Engineering, 45*(6), 480–505. https://doi.org/10.1515/polyeng-2024-0248.

Störmer, A., Bott, J., Kemmer, D., & Franz, R. (2017). Critical review of the migration potential of nanoparticles in food contact plastics. *Trends in Food Science & Technology, 63*, 39–50. https://doi.org/10.1016/j.tifs.2017.01.011.

Sun, W., Zhou, S., Zhou, Y., Ramezan, Y., Tabrizi, A. N., Kamkari, A., ... Zhang, Y. (2025). Bridging functionality and sustainability: Ferulic acid-driven advances in bioactive food packaging films. *Trends in Food Science & Technology, 163*, 105205. https://doi.org/10.1016/j.tifs.2025.105205.

Syed, T. R., Sharaby, M. R., Salmieri, S., & Lacroix, M. (2025). Innovations in active food packaging: Harnessing waste valorization and irradiation-assisted functionalization for sustainable food preservation. *Trends in Food Science & Technology, 163*, 105183. https://doi.org/10.1016/j.tifs.2025.105183.

Tamizhdurai, P., Mangesh, V. L., Santhosh, S., Vedavalli, R., Kavitha, C., Bhutto, J. K., ... Kumaran, R. (2024). A state-of-the-art review of multilayer packaging recycling: Challenges, alternatives, and outlook. *Journal of Cleaner Production, 447*, 141403. https://doi.org/10.1016/j.jclepro.2024.141403.

Taylor, S., Colonna, A., Jung, J., Gutierrez, J., & Zhao, Y. (2024). Consumer perception and acceptance of edible packaging for various food products. *Journal of Food Science, 89*(4), 2423–2437. https://doi.org/10.1111/1750-3841.16992.

Tumu, K., Vorst, K., & Curtzwiler, G. (2024). Understanding intentionally and non-intentionally added substances and associated threshold of toxicological concern in post-consumer polyolefin for use as food packaging materials. *Heliyon, 10*(1), e23620. https://doi.org/10.1016/j.heliyon.2023.e23620.

Usman, I., Sana, S., Jaffar, H. M., Munir, M., Afzal, A., Sukhera, S., ... Urugo, M. M. (2025). Recent progress in edible films and coatings: Toward green and sustainable food packaging technologies. *Applied Food Research, 5*(2), 101070. https://doi.org/10.1016/j.afres.2025.101070.

Vafaeva, K. M., Chhetri, A., Sudan, P., Mishra, M., Babu, B. S., & Mongal, B. N. (2024). Polymer matrix nanocomposites for sustainable packaging: A green approach. *E3S Web of Conferences, 537*, 08001. https://doi.org/10.1051/e3sconf/202453708001.

Vaferi, B., Dehbashi, M., Yousefzadeh, R., & Alibak, A. H. (2025). Prediction of the packaging chemical migration into food and water by cutting-edge machine learning techniques. *Scientific Reports, 15*(1), 7806. https://doi.org/10.1038/s41598-025-92459-x.

Vardhan, H., Singhal, N., Vashistha, P., Jain, R., Bist, Y., Gaur, A., & Wagri, N. K. (2025). Starch–biomacromolecule complexes: A comprehensive review of interactions, functional materials, and applications in food, pharma, and packaging. *Carbohydrate Polymer Technologies and Applications, 11*, 101001. https://doi.org/10.1016/j.carpta.2025.101001.

Walayat, N., Campagnol, P. C., Zhang, W., Su, Z., & Lorenzo, J. M. (2024). Films with antimicrobial activity for meat products. *Current Opinion in Food Science, 56*, 101127. https://doi.org/10.1016/j.cofs.2024.101127.

Wang, J., Liang, X., Hou, H., Deng, J., Lin, Q., Li, W., & Bai, J. (2025). A review on migration behavior and safety regulation of inorganic nanoparticles in food packaging: Multi-scale mechanisms, risk assessment, and innovative strategies. *Food Control, 178*, 111490. https://doi.org/10.1016/j.foodcont.2025.111490.

Walden, J., Steinbrecher, A., & Marinkovic, M. (2021). Digital product passports as enabler of the circular economy. *Chemie Ingenieur Technik, 93*(11), 1717–1727. https://doi.org/10.1002/cite.202100121.

Wei, L., Zhu, S., Xiong, G., Li, J., & Zhang, W. (2025). Citric acid vapor-assisted crosslinking of zein/PEG composite nanofiber membrane embedded with nisin by electrospinning for the cooled goose meat preservation. *Current Research in Food Science, 10*, 100983. https://doi.org/10.1016/j.crfs.2025.100983.

Wrona, M., & Nerin, C. (2019). Risk assessment of plastic packaging for food applications. In M. Suman (Vol. Ed.), *Food contact materials analysis: Mass spectrometry techniques. 8. Food contact materials analysis: Mass spectrometry techniques* (pp. 163–191). The Royal Society of Chemistry. https://doi.org/10.1039/9781788012973-00163.

Wu, Y., Ruffley, K., & Pascall, M. A. (2024). Investigation and mechanism of headspace corrosion in metal cans filled chicken noodle soup. *Packaging Technology and Science, 37*(3), 187–197. https://doi.org/10.1002/pts.2786.

Xiao, S., Zhang, W. G., Lee, E. J., Ma, C. W., & Ahn, D. U. (2011). Lipid and protein oxidation of chicken breast rolls as affected by dietary oxidation levels and packaging. *Journal of Food Science, 76*(4), 612–617. https://doi.org/10.1111/j.1750-3841.2011.02137.x.

Xuan, L., Ju, Z., Skonieczna, M., Zhou, P., & Huang, R. (2023). Nanoparticles-induced potential toxicity on human health: Applications, toxicity mechanisms, and evaluation models. *MedComm, 4*(4), 1–39. https://doi.org/10.1002/mco2.327.

Yar, M. S., Ibeogu, I. H., Regmi, A., Zhang, N., & Li, C. (2025). Advances in intelligent time-temperature indicators for cold chain monitoring: Mechanisms, challenges, and applications. *Trends in Food Science & Technology, 163*, 105128. https://doi.org/10.1016/j.tifs.2025.105128.

Yenidoğan, S., Aydemir, C., & Doğan, C. E. (2023). Packaging–food interaction and chemical migration. *Cellulose Chemistry and Technology, 57*(9–10), 1029–1040. https://doi.org/10.35812/CelluloseChemTechnol.2023.57.91.

Yu, X., Wang, Z., Gao, Z., Lang, H., Shao, X., & Liu, Z. (2025). Applications, challenges, and prospects of AI-driven intelligent packaging technologies in the food supply chain under Industry 4.0. *Trends in Food Science & Technology, 165*, 105367. https://doi.org/10.1016/j.tifs.2025.105367.

Yuceer, M., & Caner, C. (2014). Antimicrobial lysozyme–chitosan coatings affect functional properties and shelf life of chicken eggs during storage. *Journal of the Science of Food and Agriculture, 94*(1), 153–162. https://doi.org/10.1002/jsfa.6322.

Zachariah, T. J., & Leela, N. K. (2006). Volatiles from herbs and spices. *Handbook of herbs and spices*. Elsevier 177–218. https://doi.org/10.1533/9781845691717.2.177.

Zaheer, M., Awais, M., Rautkari, L., & Sorvari, J. (2018). Finite element analysis of paperboard package under compressional load. *Procedia Manufacturing, 17*, 1162–1170. https://doi.org/10.1016/j.promfg.2018.10.008.

Zhang, Q., Huo, Y., Yang, Q., Zhao, F., Li, M., & Ju, J. (2025). Migration of chemical substances from packaging materials to food. *Food Chemistry, 485*, 144544. https://doi.org/10.1016/j.foodchem.2025.144544.

Zhang, W., Yang, J., Ghasemlou, M., Riahi, Z., Khan, A., Goksen, G., ... Rhim, J. W. (2025). Recent progress on biopolymer-based food packaging films/edible coatings functionalized with catechol derivatives based on mussel biomimetics. *Materials Science and Engineering: R: Reports, 166*, 101068. https://doi.org/10.1016/j.mser.2025.101068.

Zhang, X., Xia, L., Li, Y., Li, X., Ren, C., Yin, J., & Zhang, Z. (2026). Dispersion mechanism of sodium dodecylbenzene sulfonate on nanoparticles in TiO2 water based lubricant. *Journal of Molecular Liquids, 441*, 129032. https://doi.org/10.1016/j.molliq.2025.129032.

Zhou, L., Zhang, W., & Wang, J. (2022). Recent advances in the study of modified cellulose in meat products: Modification method of cellulose, meat quality improvement and safety concern. *Trends in Food Science & Technology, 122*(1), 140–156. https://doi.org/10.1016/j.tifs.2022.02.024.

Zimmermann, L., Scheringer, M., Geueke, B., Boucher, J. M., Parkinson, L. V., Groh, K. J., & Muncke, J. (2022). Implementing the EU chemicals strategy for sustainability: The case of food contact chemicals of concern. *Journal of Hazardous Materials, 437*, 129167. https://doi.org/10.1016/j.jhazmat.2022.129167.

Printed and bound by CPI Group (UK) Ltd, Croydon, CR0 4YY
16/09/2026
02217657-0005